Student Survival and Solutions

The Nature of Mathematics

TWELFTH EDITION

Karl J. Smith
Santa Rosa Junior College

Prepared by

Karl J. Smith
Santa Rosa Junior College

BROOKS/COLE
CENGAGE Learning™

Australia • Brazil • Japan • Korea • Mexico • Singapore • Spain • United Kingdom • United States

BROOKS/COLE
CENGAGE Learning

For product information and technology assistance, contact us at **Cengage Learning Customer & Sales Support, 1-800-354-9706**

For permission to use material from this text or product, submit all requests online at **www.cengage.com/permissions** Further permissions questions can be emailed to **permissionrequest@cengage.com**

ISBN-13: 978-0-538-49528-8
ISBN-10: 0-538-49528-6

Brooks/Cole
20 Davis Drive
Belmont, CA 94002-3098
USA

Cengage Learning is a leading provider of customized learning solutions with office locations around the globe, including Singapore, the United Kingdom, Australia, Mexico, Brazil, and Japan. Locate your local office at: **www.cengage.com/global**

Cengage Learning products are represented in Canada by Nelson Education, Ltd.

To learn more about Brooks/Cole, visit **www.cengage.com/brookscole**

Purchase any of our products at your local college store or at our preferred online store **www.cengagebrain.com**

Printed in the United States of America
1 2 3 4 5 6 7 15 14 13 12 11

CONTENTS

INTRODUCTION

It has been said that "Mathematics is not a Spectator Sport" and this means you cannot learn by simply attending class, but instead you must build a body of information that will enable you to do problem solving in the real world. There are several things you must do if you wish to be successful with this course.

- Attend every class.
- Read the book. Regardless of how clear and lucid your professor's lecture on a particular topic may be, do not attempt to do the problems without first reading the text and studying the examples. It will serve to reinforce and clarify the concepts and procedures.

- ## Problems, Problems, Problems, problems...

  You must work the assigned problems. Even if you are sick or absent from class, you must work each assignment before moving on. Mathematics is sequential and one day's lesson is built on the previous lessons. Look over the entire problem set (even those problems which are not assigned).
- Ask questions when you are stuck (and you will get stuck — that is part of the process).
- Keep asking questions until you receive answers that are understandable to you.
- Today's calculators and computers are good at obtaining answers, and if all that is desired is an answer, then you have relegated yourself to the level of a machine. Do not work problems to obtain answers. It is the *concepts* that are important. Even though a solutions manual is basically a "how to" document, always ask *why* a particular approach was used, and understand the concept the problem is illustrating.
- Check for homework hints on the website **www.mathnature.com**.

When I am teaching this course, I ask all of my students to make a commitment to this course. I believe that you should also consider making a commitment. There should be two aspects to your commitment:

1. You must make a commitment to attend each class. Obviously, unforeseen circumstances can come up, but you must plan on being here for every class. There is no such thing as a class that is not important. If you cannot be in class, you must make up the material you missed before continuing with the class. Generally, the make up work for missing a class meeting is to outline the section you missed. Include in the outline each example worked in the text, and in addition make up an additional example of your own for each example in the book. Then work your own example. Check with your instructor by phone or e-mail for any additional information about the class that you missed.

2. Also you must make a commitment to daily work. It takes one or two hours for each classroom hour; you cannot save up and do 5 or 10 hours on the weekend or just before an exam. This works best if you set aside a fixed time for doing your math homework each day. If you do miss some assignment (for whatever reason) it should be made up as soon as possible.

I would be happy to hear from you. Let me know both the positive and negative comments about my book.

Dr. Karl J. Smith
email:
SMITHKJS@mathnature.com

Prologue

Overview
This prologue asks the question "Why Math" whereas the epilogue (at the end of the book) asks the question "Why Not Math?" Using a historical framework, the prologue gives perspective to the vast topic called *mathematics*.

Prologue Problem Set

1. The seven chronological periods are:

Egyptian, Babylonian and Native American Periods	3000 BC to 601 BC
Greek, Chinese, and Roman Periods	600 BC to AD 499
Hindu and Persion Period	500 to 1199
Transition Period	1200 to 1599
Age of Reason	1600 to 1699
Early Modern Period	1700 to 1799
Modern Period	1800 to present

   You need to select one of these periods and write about a page about why you think this is the most interesting period. Give reasons for your opinion.

3-7. Throughout the book you will find problems labeled **IN YOUR OWN WORDS**, and as the title implies, has no right or wrong answer. As long you present coherent well-though out answers you should be given credit for a correct answer.

9. A fence of 8 ft requires $\frac{8}{8} + 1 = 2$ poles; 16 ft needs $\frac{16}{8} + 1 = 3$ poles; 1,440 ft needs

$$\frac{1,440}{8} + 1 = 181 \text{ poles}$$

11. The worst-case scenario is for the first $3 \cdot 366$ people to have birthdays so that 3 have birthdays on each different day of the year (including leap years); thus would need to have $3 \cdot 366 + 1$ to be certain that there are four people with the same birthday. This number is 1,099 people.

13. February and March; February is the only month with exactly 28 days (exactly 4 weeks).

15. Begin by finding the prime factors of each number: $210 = 2 \cdot 3 \cdot 5 \cdot 7$ and $330 = 2 \cdot 3 \cdot 5 \cdot 11$. The largest number that divides both 210 and 330 is the product of the common factors, namely $2 \cdot 3 \cdot 5 = 30$.

17. Guesses vary. To calculate the date, first find the growth rate using the given information and the formula $P = P_0 e^{rt}$.

$$6.248 = 6e^{3r} \qquad \text{\textit{The time, t, from Oct. 12, 1999 to Oct. 12, 2002 is 3.}}$$

$$\frac{6.248}{6} = e^{3r} \qquad \text{\textit{Divide both sides by 6.}}$$

$$3r = \ln\left(\frac{6.248}{6}\right) \qquad \text{\textit{Definition of logarithm.}}$$

$$r = \frac{1}{3}\ln\left(\frac{6.248}{6}\right) \qquad \text{\textit{Divide both sides by 3.}}$$

$$\approx 0.0135 \qquad \text{\textit{By calculator.}}$$

Now use this value of r to calculate the date.

$$7 = 6.248e^{rt} \qquad \text{\textit{Use the population formula where P = 7 and P_0 = 6.248.}}$$

$$rt = \ln\left(\frac{7}{6.248}\right) \qquad \text{\textit{Definition of logarithm.}}$$

$$t = \frac{1}{r}\ln\left(\frac{7}{6.248}\right) \qquad \text{\textit{Divide both sides by r.}}$$

$$\approx 8.418 \qquad \text{\textit{By calculator.}}$$

Subtract 8 (for 8 years), and then multiply by 12 (to obtain the nearest month), 5.016. Therefore, eight years five months from October 12, 2002 puts the predicted date at March, 2011.

19. Since $(a, b) = a \times b + a + b$ we find:

$(1, 2) = 1 \times 2 + 1 + 2 = 5$ and $(3, 4) = 3 \times 4 + 3 + 4 = 19$

Thus, $((1, 2), (3, 4)) = (5, 19) = 5 \times 19 + 5 + 19 = 119$.

21. If $2n - 1$ represents all odd numbers, then for $n = 1$ we have $2(1) - 1 = 1$; for $n = 2$ we have $2(2) - 1 = 3$; $\cdots$ for $n = 473$ we have $2(473) - 1 = 945$

23. After cutting, there are $4^3 = 64$ one-inch cubes; the corners are painted on 3 sides and the edges are painted on 2 sides; there are four other cubes on each face for a total of $6 \cdot 4 = 24$ cubes that are painted on one side only. The fraction is

$$\frac{24}{64} = \frac{3}{8}$$

25. For $m(1, 2) = 1$ because we select the smaller number; similarly, $m(2, 3) = 2$. Thus, $M(m(1, 2), m(2, 3)) = M(1, 2)$. To find $M(1, 2)$ select the larger number, so $M(1, 2) = 2$.

27. **a.** tetrahedron **b.** cube (hexahedron) **c.** octahedron **d.** dodecahedron **e.** icosahedron

29. Consider a pattern where we cross out multiples of 3:

$$1, 2, \cancel{3}, 4, 5, \cancel{6}, 7, 8, \cancel{9}, 10, 11, \cancel{12}, 13, 14, \cancel{15}, 16, 17, \cancel{18}, \cdots$$

Let n be the nth positive integer that is not divisible by 3.

When n is even, the nth number that is not divisible by 3 is $\frac{3}{2}n - 1$;
when n is odd, the nth number that is not divisible by 3 is $\frac{3}{2}n - \frac{1}{2}$.
If $n = 1$, then $\frac{3}{2}(1) - \frac{1}{2} = 1$ is the 1st number not divisible by 3.
If $n = 2$, then $\frac{3}{2}(2) - 1 = 2$ is the 2nd number not divisible by 3.
If $n = 3$, then $\frac{3}{2}(3) - \frac{1}{2} = 4$ is the 3rd number not divisible by 3.
If $n = 4$, then $\frac{3}{2}(4) - 1 = 5$ is the 4th number not divisible by 3.
If $n = 1,000$, then $\frac{3}{2}(1,000) - 1 = 1,499$ is the 1000th number not divisible by 3.

31. If you are checking the calculation for 450 dollar bills per pound, you find 1 lb $\approx$ 454 g, but there will be some waste in the process, so 450 is correct. Note, \$1 trillion would require $\frac{1 \times 10^{12}}{450}$ lb of wood or about 2.22×10^9 lb. To find the volume of a tree, we assume that it is a cylinder so that $V = Bh$. B is the area of the base (12 in. diameter is 1 ft diameter or 0.5 ft radius) and h is the height of the tree so that

$$V = 0.5^2\pi(50) \approx 39.26990817 \text{ ft}^3 \text{ or about } 1,963.5 \text{ lb.}$$

Thus, the number of trees necessary to print the money is

$$\frac{2.22 \times 10^9}{V} \approx 1,131,768$$

That is, more than a million trees must be cut down to print a trillion one-dollar bills.

33. Instead of calculating the area of the shaded region directly, notice that the desired area is the area of the square less the area of the circle.

$$8^2 - \pi(4)^2 = 64 - 16\pi$$

This is an area of approximately 13.7 in.^{2}.

35. Areas of the three circles:
$A = 2^2\pi$ (smallest area); $B = 3^2\pi$ (middle area); $C = 5^2\pi$ (largest area)
Area of shaded region: $C - (A + B) = 25\pi - (4\pi + 9\pi) = 12\pi$
Ratio of the smallest area to the shaded region is $\frac{4\pi}{12\pi} = \frac{1}{3}$.

37. Let x be the amount she started with. Then at the end of the first day she had $2x - 30$. After the second day she had $3(2x - 30) - 20 = 6x - 110$. Thus,

$$6x - 110 = x$$
$$5x = 110$$
$$x = 22$$

She had \$22 at the start of the first day.

Page 3

39. Assume the cost of the charter is the same regardless of the number of travelers. Let x be the cost per traveler if 100 persons travel. Then,

$$100x = 125(x - 78)$$
$$100x = 125x - 9{,}750$$
$$25x = 9{,}750$$
$$x = 390$$

The cost is $390 if 100 make the trip.

41. Let x be the number of weeks enrolled, and y be the total cost. Then $y = 45(10 - x)$. The graph is shown.

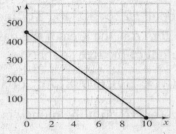

43. The rate at which alcohol is changing with respect to time is

$$\frac{dC}{dt} = 0.3(-\frac{1}{2})e^{-t/2} = -0.15e^{-t/2}$$

45. Label the number of possible paths to each vertex.

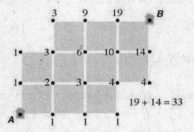

There are 33 paths.

47. $\dfrac{13 \times 10^{12}}{3.081 \times 10^{8}} \approx 42{,}194.092827$ days; divide by 365 to find 115.600254321 years. This will be the year $2010 + 115 = 2125$ and $0.600254321 \times 365 \approx 219$ days. This is February 28 (in the following year) it would take until March 2125 to pay off this debt (without paying off accrued interest in the meantime). This assumes the population does not change over the 115 years.

Page 4

49. $\log_2 x + \log_4 x = \log_b x$ *Given equation.*

$\log_2 x + \dfrac{\log_2 x}{\log_2 4} = \log_b x$ *Change base 4 to base 2.*

$\log_2 x + \dfrac{\log_2 x}{2} = \log_b x$ *$\log_2 4 = 2$ since $2^2 = 4$.*

$\dfrac{3}{2}\log_2 x = \log_b x$ *Add fractions.*

$\dfrac{3}{2}\log_2 x = \dfrac{\log_2 x}{\log_2 b}$ *Change base b to base 2.*

$\log_2 b = \dfrac{2\log_2 x}{3\log_2 x}$ *Multiply both sides by $2\log_2 b$ and divide both sides by $3\log_2 x$.*

$\log_2 b = \dfrac{2}{3}$ *Reduce fraction.*

$b = 2^{2/3}$ *Definition of logarithm.*

51. There are at least three cubes to be seen:
  (1) A little cube nestled in the corner of a larger cube.
  (2) A big cube with a cubical chunk removed from one corner.
  (3) Two cubes meeting externally at a corner.
If you see any other configurations, please send them the author (email: **SMITHKJS@mathnature.com**).

53. $_5P_5 = 5! = 120$, so the probability is $\frac{1}{120}$. The number of cards in the original deck does not matter.

55. This is what is called a magic square; the sum of the numbers in all of the rows, columns, and diagonals is 15. Wow!

57. When $t = 0$ (corresponding to the year 2000), we find $P = 153,000e^0 = 153,000$. The graph is shown.

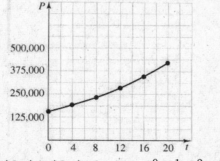

59. The pattern is $1, $2, $4, $8, $16, $\cdots$ or $2^0, 2^1, 2^2, 2^3, \cdots$ so:

1 day:	$1	This is $2^1 - 1$
2 days:	$1 + $2 = $3	This is $2^2 - 1$
3 days:	$1 + $2 + $4 = $7	This is $2^3 - 1$
4 days:	$1 + $2 + $4 + $8 = $15	This is $2^4 - 1 = 15$

For 30 days, this must be $2^{30} - 1$ or $1,073,741,823

CHAPTER 1 The Nature of Problem Solving

Chapter Overview

This chapter introduces the flavor of this text. You are not expected to see an example and then apply it to another similar problem. Instead, the goal of this text is to get you to think critically, to view mathematics as a worthwhile endeavor, and to create a mind set that mathematics is not just for classrooms, but for real life. Throughout this book there are problems labeled IN YOUR OWN WORDS. These are included to stimulate your writing skills and to help ensure understanding of important concepts. You many be taking this course and using this book because it is a requirement, and you may be thinking that you need to do only what is necessary to "get through" with a passing grade. Whatever your motivation, if you begin with the correct attitude, ask questions, seek help, and focus on learning, you should find the material not only to be worthwhile, but enjoyable.

1.1 Problem Solving, page 3
New Terms Introduced in this Section

Pascal's triangle   Problem-solving procedure   Street problem

Level 1, page 14

1. There is, of course, no right or wrong answer for this question. By asking this question, however, my intent is to let you know that this may not be the math course you thought it would be. The book really *was* written for people who think they do not like mathematics. It is also about ideas and not about "processes." In your previous mathematics courses you were probably learning "how to" do things in or with math, like solving equations for example, whereas in this course you will see why you might want to learn how to solve equations or when it is appropriate to set up or solve an equation. For this answer, write a half page or so about your feelings of agreement or disagreement with the questions.

3. The reason for this question is to make sure you read "around the edges." Many students come to my office after the first examination and tell me that they did all the homework but did not do as well as they would have hoped. My goal in writing this book was to ensure your success, but this goal cannot be reached if you don't read the book. Many students "read" math books by first reading the assigned problems until you reach a problem that you can't work. Next, they look for a similar example. If that does not work, the material of the section is scanned seeking the answer. This is backwards. This book was written as a result of thirty years of teaching this course, and the questions you have are probably answered somewhere in the overviews, summaries, or author's notes included to help you

with the presented material. For this answer, write a half page or so about whether those descriptions apply to you or not.

5. Have you written down Pascal's triangle "from scratch"? If not, you should do so now. Did you see the slogan on the first page of this section? "Mathematics is not a Spectator Sport" means no passive spectators. To work this problem you need to "see" the pattern of generating new rows for Pascal's triangle. Take a look at Figure 1.4, and the illustration just above it. Add the two adjacent numbers in a row to find the number in the following row. The next two rows after row 15 are found by addition:

    1 16 120 560 1,820 4,368 8,008 11,440 12,870 11,440 8,008 4,368 1,820 560 120 16 1
    1 17 136 680 2,380 6,188 12,376 19,448 24,310 24,310 19,448 12,376 6,188 2,380 680 136 17 1

    Now, that you have a triangle of numbers on your paper, look for the numbers 1, 2, 3, 4, 5, ... Look at the diagonals to see that they are found in the first diagonal.

7. Answers vary; yes, it is the same, even though you would be in a different location in the triangle. The reason for this is because the triangle is symmetric; this means that the triangle reads the same if you read from the left or if you read from the right.

9. **a.** Look at Pascal's triangle, row 7, diagonal 3 to find the entry 35. Don't forget that the first diagonal is called the 0th diagonal. Look at the headings at the top of Figure 1.4.

    b. Look at Pascal's triangle, row 8, diagonal 3 to find the entry 56.

11. From the given figure in the problem, count up 2 and over 3 to get from *A* to *B*; then look at Pascal's triangle and count down 2, over 3 to find the entry 10. Thus, there are 10 paths from *A* to *B*.

13. From the given figure in the problem, count up 4 and over 6 to get from *A* to *B*; then look at Pascal's triangle and count down 4, over 6 to find the entry 210.

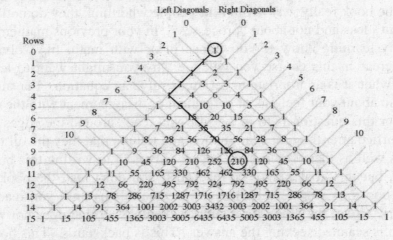

Thus, there are 210 paths from *A* to *B*.

15. Using the given map, start at point A and count down 4, over 3; now look at Pascal's triangle and count down 4, over 3 to find 35 paths.

17. Looking at the map, note that to move from A to H you need to go 2 up, 4 over; now look at Pascal's triangle and count down 2, and 4 over to find 15 paths.

Level 2, page 15

19. Some will choose A for superficial reasons (*i.e.*, both A and Problem 18 deal with penguins and bears). Better problem solvers will select C, which resembles Problem 18 in several regards; namely there are two flower types (which is comparable to two body parts) and halving/doubling plays a role in both problems. Finally, others who are focused on the underlying structure of Problem 18 will choose the best possible answer, namely B. Any of these answers would be acceptable, as long as you can supply reasons for the answer you have chosen.

21. You might begin with a diagram;

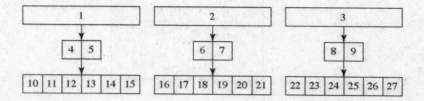

We see there are 27 boxes.

23. This is an "I gotcha" problem. The blind person says, "I need a pair of scissors." Try this one on a friend or family member!

25. a. $1 + 1 = 2$
 b. $1 + 2 + 1 = 4$
 c. $1 + 3 + 3 + 1 = 8$
 d. $1 + 4 + 6 + 4 + 1 = 16$

27. Draw a little "map" from A to J and label the vertices:

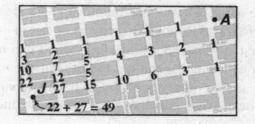

There are 49 paths.

29. Draw a little "map" from A to L and label the vertices:

There are 284 paths.

31. The are twelve stamps (of any denomination) in a dozen.

33. You have 7 cards.

35. Let's begin with another question. If you want to put up a 6-ft fence, how many posts do you need? Answer: 2. If you want to put up a 12-ft fence with a post every 6 ft, how many posts do you need? Answer: 3. If you want to put up a 24-ft fence with a post every 6 ft, how many posts do you need? Answers: $24 \div 6 = 4$ plus one post at the end; total 5 posts. Thus 100 ft of fence requires $100 \div 10 = 10$ plus one post for a total of 11 posts.

37. The person arrived home at 12:30; then one bong for 1 o'clock; and then one bong for 1:30.

39. $6 \times 9 = 54$ outs; if the game is a full nine innnings, there are 6 outs for each inning.

41. The bookworm needs to travel through two covers or 2 mm $+ 2$ mm $= 4$ mm.

43. Answers vary; The first cup has one lump; the second cup has four lumps; and the third cup has five lumps. This accounts for the ten lumps, but the second cup does not have an odd number of lumps — not to worry; put the first cup (containing the one lump) inside the second cup. Now, the second cup has five lumps, so the conditions of the problem are satisfied.

Page 10

Level 3, page 16

45. Answers vary; you can add the columns up or down or you can select out pairs of numbers whose sum is ten.

47. Answers vary; how many items are in the cart in the other checkout stand; how experienced are the grocery clerks; is one of them answering the phone?

49. This is a fun problem.... if you begin by drawing a picture, you may be confused by the complexity of the problem. Remember Pólya's problem solving principles and try to think simply about the problem. It takes the cyclists one hour to reach each other ($6 + 4 = 10$). If the fly flies at 20 mph, then the fly must travel 20 miles.

51. Draw a picture; there are two possible answers, 8 miles or 12 miles.

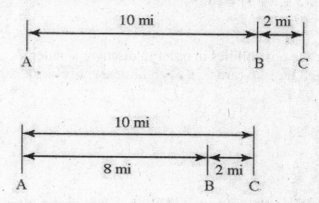

and

As you can see, there two possible answers, 8 miles and 12 miles since we do not know if Cal lives between Alex and Beverly, or if Beverly lives between Alex and Cal.

53. *puff*

55. Try several examples; 1.62

Problem Solving, page 17

57. Let's begin by numbering the vertices.

    a. There are 26 paths.   **b.** There are 23 paths.

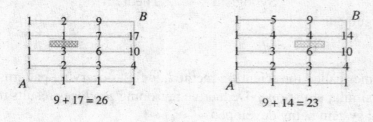

Page 11

c. There are 27 paths.

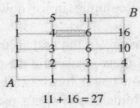

$$11 + 16 = 27$$

If we look for a pattern, we find the number of routes from the start to the destination minus the number of routes passing through the barricade.

a. $35 - (3 + 3 + 3) = 26$

b. $35 - (6 + 6) = 23$

c. $35 - (4 + 4) = 27$

59. Consider the possibilities in order to discover a pattern.

Sum of the 3 face-up cards	# of additional cards	# of cards remaining	# of carded needed
30	0	23	$30 = 23 + 7$
29	1	22	$29 = 22 + 7$
28	2	21	$28 = 21 + 7$
$\vdots$	$\vdots$	$\vdots$	$\vdots$
8	22	1	$8 = 1 + 7$
7	23	0	$7 = 0 + 7$

Compare the columns on the ends to see the pattern. If the three face-up cards total x, then we need to count $23 - (30 - x) + 7 = x$ cards down.

1.2 Inductive and Deductive Reasoning, page 18

New Terms Introduced in this Section

Axiom	Conclusion	Conjecture	Counting number
Cube	Deductive reasoning	Euler Circles	Exponentiation
Inductive reasoning	Invalid argument	Natural number	Order of operations
Premise	Square	Syllogism	Theorem
Undefined term	Valid argument		

Level 1, page 24

1. Inductive reasoning, sometimes called the scientific method, is first observing patterns and then predicting answers for similar situations. Deductive reasoning produces results that are certain within the logical system being developed.

3. First, perform any operations enclosed in parentheses. Next, perform multiplication and divisions as they occur by working from left to right. Finally, perform additions and subtractions as they occur by working from left to right.

5. Inductive reasoning involves reasoning from particular facts or individual cases to a general conjecture — a statement you think may be true. That is, inductive reasoning is a generalization made on the basis of some observed occurrences.

7. **a.** $5 + 2 \times 6 = 5 + 12$
    $= 17$
    b. $7 + 3 \times 2 = 7 + 6$
    $= 13$

9. **a.** $3 \times 8 + 3 \times 7 = 24 + 21$
    $= 45$
    b. $3(8 + 7) = 3(15)$
    $= 45$

11. **a.** $12 + \dfrac{6}{3} = 12 + 2$
    $= 14$
    b. $\dfrac{12 + 6}{3} = \dfrac{18}{3}$
    $= 6$

13. **a.** $\dfrac{20}{2} \times 5 = 10 \times 5$
    $= 50$
    b. $\dfrac{20}{(2 \times 5)} = \dfrac{20}{10}$
    $= 2$

15. **a.** $10 + 5 \times 2 + 6 \times 3 = 10 + 10 + 18$
    $= 38$

    b. $4 + 3 \times 8 + 6 + 4 \times 5 = 4 + 24 + 6 + 20$
    $= 28 + 26$
    $= 54$

17. **a.** $3 + 9 \div 3 \times 2 + 2 \times 6 \div 3 = 3 + 3 \times 2 + 12 \div 3$
    $= 3 + 6 + 4$
    $= 13$

    b. $[(3 + 9) \div 3] \times 2 + [(2 \times 6) \div 3] = [12 \div 3] \times 2 + [12 \div 3]$
    $= [4] \times 2 + [4]$
    $= 8 + 4$
    $= 12$

19. The cartoon character is using inductive reasoning to conclude that no two snowflakes are the same, since all those he looked at were different.

21. Consider successive multiplications by 3:
    $3 \times 1 = 3$
    $3 \times 2 = 6$
    $3 \times 3 = 9$
    $3 \times 4 = 12$ and $1 + 2 = 3$
    $3 \times 5 = 15$ and $1 + 5 = 6$

Page 13

$3 \times 6 = 18$ and $1 + 8 = 9$
$3 \times 7 = 21$ and $2 + 1 = 3$
$3 \times 8 = 24$ and $2 + 4 = 6$
$3 \times 9 = 27$ and $2 + 7 = 9$
$3 \times 10 = 30$ and $3 + 0 = 3$
$3 \times 11 = 33$ and $3 + 3 = 6$
$3 \times 12 = 36$ and $3 + 6 = 9$

The pattern continues giving the *three pattern*: $3, 6, 9, 3, 6, 9, \ldots$

23. Consider successive multiplications by 5:

$5 \times 1 = 5$
$5 \times 2 = 10$ and $1 + 0 = 1$
$5 \times 3 = 15$ and $1 + 5 = 6$
$5 \times 4 = 20$ and $2 + 0 = 2$
$5 \times 5 = 25$ and $2 + 5 = 7$
$5 \times 6 = 30$ and $3 + 0 = 3$
$5 \times 7 = 35$ and $3 + 5 = 8$
$5 \times 8 = 40$ and $4 + 0 = 4$
$5 \times 9 = 45$ and $4 + 5 = 9$
$5 \times 10 = 50$ and $5 + 0 = 5$
$5 \times 11 = 55$ and $5 + 5 = 10$ and $1 + 0 = 1$
$5 \times 12 = 60$ and $6 + 0 = 6$

The pattern continues giving the *five pattern*: $5, 1, 6, 2, 7, 3, 8, 4, 9, 5, 1, 6, 2, 7, \ldots$

25. Begin by looking for a pattern (see Example 2).

    a. $25^2 = 625$ **b.** $250^2 = 62{,}500$

27.

2	7	6
9	5	1
4	3	8

Level 2, page 25

29. Circle "M" represents mathematicians; circle "E" represents eccentrics, and "R" represents rich people.

The conclusion is valid.

31. Circle "C" represents cats, circle "A" represents animals. Write "x" somewhere outside of the circle labeled A.

The conclusion is valid.

33. Circle "S" represents students, circle "E" represents enthusiastic people.

The conclusion is valid.

35. Circle "C" represents candy, circle "F" represents fattening foods, and "D" represents delicious foods.

Write "x" inside of circle "F" but outside other circles to shown the reasoning is not valid.

37. Circle "P" represents professors, "I" represents ignorant people, and "V" represents vain people.

Write "x" inside the circles labeled P and V to show that the reasoning is not valid.

39. Circle "L" represent lions, "F" represents fierce creatures, and "C" represents creatures that drink coffee. Write "x" somewhere outside circle C.

The placement of our "x" shows the reasoning is not valid.

Page 15

Level 3, page 26

41. You may need to consult a map to see the travel is from Phoenix to Albuquerque which is traveling east. This is deductive reasoning.

43. If the mode of travel is automobile (see the previous problem), if we, once again, look at a map and project a line from Phoenix to Albuquerque *backwards* we see that the most probable starting point is Las Vegas. This is inductive reasoning.

45. The speaker, speaker's brother, mama, Billy Joe McAllister, papa, Tom, preacher Brother Taylor, and Becky Thompson. There are eight persons mentioned in the song. This is deductive reasoning.

47. We know that it is on Choctaw Ridge, and from the name of the bridge and the language used, we can guess it is in the south (inductive reasoning). If we look at an atlas, we see several Choctaw Ridges, and the one that seems most likely is in Alabama.

49. Look for the door.... which can't be seen. Thus, the U. S. the bus is traveling to the left, or west (by deductive reasoning).

51. a. Continue the given pattern:
$$9 \times 54321 - 1 = 488,888$$

b.

$9 \times 54321 - 1 = 488888$	$(\mathbf{5} - 1 = 4)$ *followed by five eights*
$9 \times 654321 - 1 = 5888888$	$(\mathbf{6} - 1 = 5)$ *followed by six eights*
$9 \times 7654321 - 1 = 68888888$	$(\mathbf{7} - 1 = 6)$ *followed by seven eights*
$9 \times 87654321 - 1 = 788888888$	$(\mathbf{8} - 1 = 7)$ *followed by eight eights*
$9 \times 10987654321 - 1 = 8888888888$	$(\mathbf{9} - 1 = 8)$ *followed by nine eights*

The answer is 8,888,888,888

c. Continue the pattern:
$$9 \times 10,987,654,321 - 1 = 98,888,888,888$$
This is $(10 - 1 = 9)$ followed by ten eights.

53. Look for a pattern:

$$4^2 = 16; \text{ sum of the digits is } 7$$
$$34^2 = 1,156; \text{ sum of the digits is } 13$$
$$334^2 = 111,556; \text{ sum of the digits is } 19$$
$$3334^2 = 11,115,556; \text{ sum of the digits is } (4 \times 1) + (3 \times 5) + 6 = 25$$

Now, we see a pattern; let's check it for the first three cases (working backwards)

$(4 \times 1) + (3 \times 5) + 6 = 25$; note the number of ones is the number of digits in the number being squared

$(3 \times 1) + (2 \times 5) + 6 = 19$; note the number of fives is one less than the number of ones.

$(2 \times 1) + (1 \times 5) + 6 = 13$

$(1 \times 1) + (0 \times 5) + 6 = 7$

The pattern seems to check, so we now calculate the requested sum:

Page 16

$$(9 \times 1) + (8 \times 5) + 6 = 55$$

We are using inductive reasoning.

Level 3, Problem Solving, page 27

55. Look at the hint to formula a pattern:

36	1×1 squares
25	2×2 squares
16	3×3 squares
9	4×4 squares
4	5×5 squares
1	6×6 squares

The total is 91 squares.

57. Balance 3 with 3; if it balances, the heavier one is in the three not weighed; if it does not balance, then the pan balance will show which set of 3 is the heavier one. In any case, the first weighing narrows it down to a set of 3 coins, one of which contains the heavy coin. Take 2 from this set of heavy 3 and balance them. If they balance, the heavy coin is the one not weighed; if they do not balance, then the pan balance will show the heavier one. In any case, it takes two weighings.

59. a. One diagonal is 38,307; other sums are 21,609. **b.** One diagonal is 13,546,875; other sums are 20,966,014.

1.3 Scientific Notation and Estimation, page 28

New Terms Introduced in this Section

Addition law of exponents	Base	Billion
Decimal notation	Distributive law of exponents	Estimate
Exponent	Exponential	Exponential notation
Extended order of operations	Fixed-point form	Floating-point form
Googol	Laws of exponents	Million
Multiplication law of exponents	Power	Scientific notation
Trillion		

Level 1, page 40

1. Answers vary; for any number b and any counting number n,

$$b^n = \underbrace{b \cdot b \cdot b \cdot \;\cdots\; \cdot b}_{n \text{ factors}}, \qquad b^0 = 1, \qquad b^{-n} = \frac{1}{b^n}$$

The number b is called the base, the number n in b^n is called the exponent, and the number b^n is called a power or exponential.

3. The reason for this question is to force you to have an appropriate calculator for the class. For example, press $\boxed{2}\ \boxed{+}\ \boxed{3}\ \boxed{\times}\ \boxed{4}\ \boxed{=}$ and look at the display. If the answer is 20, then you have a calculator with *arithmetic logic* and this calculator is not appropriate for this text. If the answer is 14, then your calculator is programmed for the order of operations agreement and you have an calculator with *algebraic logic*. Finally, if your calculator is an HP calculator, you have one which uses *RPN logic*. For the answer to this question, select one of the three of these possibilities.

5. You should select the largest number that you *know*. For example, if you say one million, then you should be able to explain *in your own words* the meaning of this word (like we did by asking how large a room it would take to hold a million inflated balloons). If you say one trillion, but cannot explain what you mean by a trillion, then you cannot use this as the largest number you know.

7. **a.** $3,200 = 3.2 \times 10^3$; floating-point notation: 3.2 03
   b. $0.0004 = 4 \times 10^{-4}$; floating-point notation: 4 $-$04
   c. $64,000,000,000 = 6.4 \times 10^{10}$; floating-point notation: 6.4 10

9. **a.** $5,629 = 5.629 \times 10^3$; floating-point notation: 5.629 03
   b. $630,000 = 6.3 \times 10^5$; floating-point notation: 6.3 05
   c. $0.00000\ 0034 = 3.4 \times 10^{-8}$; floating-point notation: 3.4 $-$08

11. **a.** $7^2 = 49$
   b. $7.2 \times 10^{10} = 72,000,000,000$; move the decimal point 10 places to the right
   c. $4.56 + 3 = 4,560$; move the decimal point 3 places to the right

13. **a.** $6^3 = 216$
   b. $4.1 \times 10^{-7} = 0.00000\ 041$; move the decimal point 7 places to the left
   c. $4.8\ -7 = 0.00000\ 048$; move the decimal point 7 places to the left

15. $93,000,000 = 9.3 \times 10^7$; $365 = 3.65 \times 10^2$

17. $220,000,000 = 2.2 \times 10^8$; $0.0000025 = 2.5 \times 10^{-6}$

19. $3.6 \times 10^6 = 3,600,000$

21. $3 \times 10^{-8} = 0.00000\ 003$

Estimates in Problems 23-30 *may vary.*

23. Estimate 35 miles

25. Estimate $2,000 \times 500 = 1,000,000$; actual answer: $1,850 \times 487 = 900,950$

27. Estimate $20 \times 15 = 300$; actual answer: $23 \times 15 = 345$ mi

29. Estimate $600,000 \div 60 = 10,000$ hours; $10,000 \div 25 = 400$ days; estimate one year; actual answer: $525,600 \div 60 \div 24 = 365$ days

Level 2, page 41

31. a. $(6 \times 10^5)(2 \times 10^3) = (6 \times 2)(10^5 \times 10^3)$
$$= 12 \times 10^{5+3}$$
$$= 1.2 \times 10^9$$

b. $\dfrac{6 \times 10^5}{2 \times 10^3} = \dfrac{6}{2} \times 10^{5-3}$
$$= 3 \times 10^2$$

33. a. $\dfrac{(6 \times 10^7)(4.8 \times 10^{-6})}{2.4 \times 10^5} = \dfrac{6 \times 4.8}{2.4} \times 10^{7+(-6)-5}$
$$= 12 \times 10^{-4}$$
$$= (1.2 \times 10) \times 10^{-4}$$
$$= 1.2 \times 10^{-3}$$

b. $\dfrac{(2.5 \times 10^3)(6.6 \times 10^8)}{8.25 \times 10^4} = \dfrac{2.5 \times 6.6}{8.25} \times 10^{3+8-4}$
$$= 2 \times 10^7$$

35. a. $\dfrac{0.00016 \times 500}{2,000,000} = \dfrac{1.6 \times 10^{-4} \times 5 \times 10^2}{2 \times 10^6}$ *Write the numbers in scientific notation.*
$$= \dfrac{1.6 \times 5}{2} \times 10^{-4+2-6} \quad \textit{Properties of exponents}$$
$$= 4 \times 10^{-8} \quad \textit{Simplify.}$$

b. $\dfrac{15,000 \times 0.0000004}{0.005} = \dfrac{1.5 \times 10^4 \times 4 \times 10^{-7}}{5 \times 10^{-3}}$ *Write in scientific notation.*
$$= \dfrac{1.5 \times 4}{5} \times 10^{4+(-7)-(-3)} \quad \textit{Properties of exponents.}$$
$$= 1.2 \times 10^0$$
$$= 1.2$$

37. Mark off a square centimeter and count the number of oranges in this square. Estimate $10/\text{cm}^2$, so about 120 oranges.

39. Mark off one square inch (say the bottom left corner); we count approximately 60 chairs in this part of the photograph (which includes chairs and part of aisle). Since the photo is 3 in. by 1.5 in. we calculate an area of $3 \times 1.5 = 4.5$. Thus, we estimate about $60 \times 4.5 \approx 270$ chairs.

41. Assume one balloon is about 1 ft^3 and also assume a typical classroom is $30 \text{ ft} \times 50 \text{ ft} \times 10 \text{ ft} = 15,000 \text{ ft}^3$, so $10^6 \div (1.5 \times 10^4) \approx 67$; we estimate that it would take 60-70 classrooms.

43. Assume that 233 $1 bills forms a stack one inch high. $1,000,000/233 \approx 4,292$ in. or 358 ft or 119 yd. Thus, a million dollars would form a stack about 119 yd high.

45. About 135,600,000,000,000,000 grains:

$$(\text{GRAINS PER LB}) \cdot (\text{POUNDS PER TON}) \cdot (\text{TONS OF SUGAR}) = (2.26 \times 10^6) \cdot (2.00 \times 10^3) \cdot (3 \times 10^7)$$
$$= (2.26)(2)(3)(10^{16})$$
$$= 1.356 \times 10^{17}$$

47. Assume that a box of chalk contains 12 four inch pieces, so one box is about 4 ft of chalk. Also assume that a box of chalk costs about $1. $400,000$ ft ≈ 75 miles.

Level 3, page 42

49. a. Answers vary (see part c below).

    b. 9^{9^9}

    c. This number is too large for your calculator; $9^{9^9} = 9^{387,420,489}$; the largest power of 9 most calculators will handle is $9^{104} \approx 1.7 \times 10^{99}$. We can make an estimate by noting that a larger base 10 will have a smaller exponent so that $9^{387,420,489} \approx 10^{387,000,000}$. In Chapter 8, we will learn how to find a better approximation: $9^{9^9} \approx 10^{369,692,902}$

51. $23,451,789$ yd $\times 5,642,732$ yd $\times 54,465$ yd $= 7.207471108 \times 10^{18}$ yd^3; we can also calculate that 1 yd^3 = 36 in. $\times$ 36 in. $\times$ 36 in. = 46,656 in.^{3}. Thus,

$$7.207471108 \times 10^{18} \times 46,656 = 3.36271772 \times 10^{23}$$

(By the way, the answer Jedidiah gave was 336,271,772,009,375,336,113,920.)

53. If we assume a brick is 8 in. by $3\frac{1}{2}$ in. by 2 in. we can "build" a square using $4\frac{1}{2}$ bricks which is 2 in. tall.

Now, this means that the 100 ft length (2 in. tall) would take about $100 \times 4\frac{1}{2} = 450$ bricks. Since the wall is 10 ft tall, we see that there needs to be 60 layers (6 per ft times 10 ft), Thus, our estimate is

$$450 \times 60 = 27,000 \text{ bricks}$$

If you use strictly bricks and volumes (no mortar), then the volume of the wall is

$$100 \text{ ft} \times 10 \text{ ft} \times 1 \text{ ft} = 1{,}000 \text{ ft}^3$$

Since one brick is

$$\frac{8}{12} \text{ ft} \times \frac{3.5}{12} \text{ ft} \times \frac{2}{12} \text{ ft} \approx 0.0324 \text{ ft}^3$$

This means that the wall will hold

$$\frac{1{,}000 \text{ ft}^3}{0.0324 \text{ ft}^3} \approx 30{,}857.14$$

A good estimate would be a number between 27 and 31 thousand bricks.

Level 3 Problem Solving, page 43

55. The units column will contain   1,000 digits
The tens column will contain   1,000 − 9 digits
The hundreds column will contain   1,000 − 99 digits
The thousands column will contain   1,000 − 999 digits
Sum: 4,000 − 1,107 = 2,893 digits

Thus, the time required is 2,893 seconds or 2,893 ÷ 60 minutes = 48 minutes, 13 seconds.

57. Look for a pattern; consider each column separately.
The units column will contain   100 zeros
The tens column will contain   100 − 9 zeros
The hundreds column will contain   100 − 99 zeros
Sum: 300 − 108 = 192 zeros

59. Answers vary;
California $\approx 1.586 \times 10^5$ sq mi
$\approx 4.422 \times 10^{12}$ sq ft (1 mi = 5,280 ft, so multiply by $5{,}280^2$)
Divide by world population 6.3×10^9 and obtain about 7.018×10^2 or about 700 sq ft per person; choose D.

Page 21

Chapter 1 Review Questions, page 44

STOP   Studying for a chapter examination is a personal process, one which nobody else can do for you. Simply take the time to review what you have done. Here are the new terms in Chapter 1.

Addition law of exponents [1.3]

Axiom [1.2]

Base [1.3]

Billion [1.3]

Conclusion [1.2]

Conjecture [1.2]

Counting number [1.2]

Cube [1.2]

Decimal notation [1.3]

Deductive reasoning [1.2]

Distributive laws of
exponents [1.3]

Estimate [1.3]

Exponent [1.3]

Exponential [1.3]

Exponential notation [1.3]

Exponentiation [1.2]

Extended order of
operations [1.3]

Euler circle [1.2]

Fixed-point form [1.3]

Floating-point form [1.3]

Googol [1.3]

Inductive reasoning [1.2]

Invalid argument [1.2]

Laws of exponents [1.3]

Million [1.3]

Multiplication law of
exponents [1.3]

Natural number [1.2]

Order of operations [1.2]

Pascal's triangle [1.1]

Power [1.3]

Premise [1.2]

Problem-solving
procedure [1.1]

Scientific notation [1.3]

Square [1.2]

Street problem [1.1]

Subtraction law of
exponents [1.3]

Square [1.2]

Street problem [1.1]

Subtraction law of
exponents [1.3]

Syllogism [1.2]

Theorem [1.2]

Trillion [1.3]

Undefined term [1.2]

Valid argument [1.2]

If you can describe the term, read on to the next one; if you cannot, then look it up in the text (the section number is shown in brackets). Next, review the types of problems in Chapter 1.

TYPES OF PROBLEMS

Use Pólya's method to solve a problem. [1.1]

Use Pascal's triangle as an aid to problem solving. [1.1]

Answer questions by using inductive reasoning. [1.2]

Simplify an expression using the order of operations. [1.2, 1.3]

Distinguish inductive from deductive reasoning. [1.2]

Use Euler circles to determine the validity of a syllogism [1.2]

Write out exponential numbers without using exponents. [1.3]

Write a large or small number in scientific notation. [1.3]

Use a calculator to answer numerical questions. [1.3]

Estimate answers to numerical questions. [1.3]

Simplify numerical problems by using the laws of exponents. [1.3]

Describe the relative sizes of large and small numbers. [1.3]

Once again, see if you can verbalize (to yourself) how to do each of the listed types of problems.

    Work all of Chapter 1 Review Questions (whether they are assigned or not). Work through all of the problems before looking at the answers, and *then* correct each of the problems. The entire solution is shown in the answer section at the back of the text. If you worked the problem correctly, move on to the next problem, but if you did not work it correctly (or you did not know what to do), look back in the chapter to study the procedure, or ask your instructor.

    Finally, go back over the homework problems you have been assigned. If you worked a problem correctly, move on to the next problem, but if you missed it on your homework, then you should look back in the book or talk to your instructor about how to work the problem.

    If you follow these steps, you should be successful with your review of this chapter.

We give all of the answers to the chapter review questions (not just the odd-numbered questions).

Chapter 1 Review Questions, page 44

1. Understand the problem, devise a plan, carry out the plan, and then look back.
2. Use Pascal's triangle; look at 5 blocks down and 4 blocks over to find 126.
3. Turn the chessboard so that it forms a triangle with the rook at the top. To get the appropriate square, look at 7 blocks down and 4 blocks over using Pascal's triangle to find 330 paths.
4. By patterns:

$$1 \times 1 = 1$$
$$11 \times 11 = 121$$
$$111 \times 111 = 12,321$$
$$\vdots$$
$$111,111,111^2 = 12,345,678,987,654,321$$

5. Order of operations: (1) First, perform any operations enclosed in parentheses. (2) Next, perform multiplications and divisions as they occur by working from left to right. (3) Finally, perform additions and subtractions as they occur by working from left to right.
6. The scientific notation for a number is that number written as a power of 10 times another number x, such that $1 \leq x < 10$.
7. Inductive reasoning; the answer was found by looking at the pattern of questions.
8. Answers vary. **a.** $\boxed{2}\ \boxed{y^x}\ \boxed{63}\ \boxed{=}$ or $\boxed{2}\ \boxed{\wedge}\ \boxed{63}\ \boxed{\text{ENTER}}$ 9.223372037E18
    b. $\boxed{9.22}\ \boxed{\text{EE}}\ \boxed{18}\ \boxed{\div}\ \boxed{6.34}\ \boxed{\text{EE}}\ \boxed{6}\ \boxed{=}$ 1.454258675E12

9. A. $50/person times 263 million is about $13 billion; not even close.

 B. $1 × 60 sec × 60 min × 24 hr × 365.25 days × 1,000 years is about $3.1 billion; not even close.

 C. 1 million people × $80,000 + 1 billion people × 200 ≈ 280 billion; choice C comes closest.

10. $\frac{3}{4}$ hr × 365 = 273.75 hr. This is approximately $2\frac{3}{4}$ hr/book. Could not possibly be a complete transcription of each book.

11. Answers vary. An ice cube is about 1 in.^{3} and a cubic foot has 12^3 in.^{3}. A classroom 30 ft × 50 ft × 10 ft would hold about $2.592 × 10^7$ ice cubes. $\frac{7×10^{16}}{2.592×10^7} ≈ 2.7 × 10^9$ classrooms; this is not meaningful, but is shown here because it is typical of first attempts. It is important to look for a meaningful comparison. Let's try again. Lake Mead is about 35,154,000 m^3 (**www.infoplease.com**). To covert this to "ice cubes," we look up a cubic meter (in a dictionary) to find it is 35.314667 ft^3. Thus, we have:

$$(3.5154 × 10^7)(35.314667)(12^3) ≈ 2.145 × 10^{12} \text{ in.}^3$$

Finally, we divide the reported size of the iceberg by the capacity of Lake Mead to find:

$$(7 × 10^{16}) ÷ (2.145 × 10^{12}) ≈ 32,630$$

It would take more than 32,600 dams the size of the Hoover Dam (which forms Lake Mead) to hold the capacity of the iceberg. This is, no doubt, larger than the capacity of all of lakes behind all the world's dams, even the world's largest one on the Yangze River in China which was completed in 2009.

12. $1.38 × 10^{12}$; $(1.38 × 10^{12}) ÷ (3.10 × 10^8) ≈ 4.4416 × 10^4$; each person's share is about $44,516.

13. Arrange the cards as $\begin{bmatrix} 2 & 7 & 6 \\ 9 & 5 & 1 \\ 4 & 3 & 8 \end{bmatrix}$.

14. Measure a dollar bill (2.5 in. by 6 in.). The floor is 240 in. by 360 in. This is 96 bills by 60 bills, or 5,760 bills on the floor. The ceiling is 10 ft = 120 in. with 233/in., so the total number of bills in the room is

$$5,760 × 120 × 233 = 161,049,600$$

The national debt is $13,300,000,000,000 so

$$\frac{1.38 × 10^{12}}{161,049,600} ≈ 85,688 \text{ classrooms}$$

Page 24

This number of classrooms is still hard to comprehend, so we will fill them with $100 bills instead. Now we can estimate it will take 857 classrooms. How many classrooms are in your city? Imagine them *filled* with $100 bills.

15. Step 1: draw a circle representing birds (labeled "B") inside a circle representing creatures with wings (labeled "W").

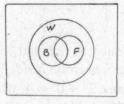

Step 2: draw a circle representing files (labeled "F") inside the circle labeled W. It may or may not contain parts of B.

No conclusion can be draw about the relationship of the circles B and F, so the reasoning expressed in this problem is not valid.

16. Step 1: draw two nonoverlapping circles representing apples (labeled "A") and bananas (labeled "B").

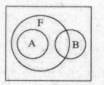

Step 2: draw a circle representing fruit (labeled "F") so that it contains circle A. It may or may not include any part of B.

No conclusion about B (bananas) and F (fruit) can be drawn, so the reasoning expressed is not valid.

17. Step 1: draw a circle representing artists (labeled "A") inside a circle representing creative people (labeled "C").

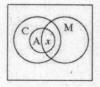

Step 2: draw a circle representing musicians (labeled "M") overlapping the previously drawn circles.

Step 3: place an "x" inside the circles labeled A and M.

The conclusion is valid because the x (which is in M) *must* be inside C.

18. Step 1: draw a circle representing rectangles (labeled "R") inside a circle representing polygons (labeled "P").

   Step 2: draw a circle representing squares (labeled "S") inside the circle R.
   The conclusion is valid because S *must* be inside P.

19. Answers vary; 20 is unhappy because $2^2 + 0^2 = 4$, which is unhappy. 100 is happy because $1^2 + 0^2 + 0^2 = 1$, which is happy.

20. By patterns:
   $7^1 = 7$;
   $7^2 = 49$ ends in 9;
   $7^3 = 343$ ends in 3;
   $7^4 = 2,401$ ends in 1;
   7^5 ends in 7;
   7^6 ends in 9;
   7^7 ends in 3;
   7^8 ends in a 1, $\cdots$.
   Looking ahead, we see that 7^{1000} must end in 1.

CHAPTER 2 The Nature of Sets

Chapter Overview

This chapter introduces some basic building blocks for mathematics, namely the notion of sets and set theory This chapter will be particularly important in Chapter 3 (logic) and Chapter 13 (probability), but the terminology of sets is found in every chapter of this book.   Remember, the focus of this book is everyday usage of mathematics, so it is important to relate the everyday (nonmathematical) usage of the words *and, or*, and *not* to their mathematical counterparts, *intersection, union*, and *complement*. These same words will be discussed in a logical context in the next chapter.

2.1 Sets, Subsets, and Venn Diagrams, page 49

New Terms Introduced in this Section

Belongs to	Cardinal number	Cardinality	Circular definition
Complement	Contained in	Counting number	Description method
Disjoint	Element	Empty set	Equal sets
Equivalent sets	Improper subset	Integer	Member
Natural number	Proper subset	Rational number	Roster method
Set	Set-builder notation	Set theory	Subset
Universal set	Venn diagram	Well-defined set	Whole number

Level 1, page 57

1. Answers vary. Not every word can be defined, so in order to build a mathematical system, there must be a starting place, with some words given as undefined.

3. Answers vary. The universal set is the set that includes all the elements under consideration for a particular discussion.

5. Answers vary.   Set of people over 10 feet tall; set of negative integers between 0 and 1; set of living cats with seven heads. You should try to come up with your own examples.

7. Answers vary. The number of people attending the University of California. You should try to come up with your own examples.

9. **a.** We assume that there is a list (roster) somewhere which lists all the students attending the University of California (all campuses), so this set is well defined.

   b. Not everyone would agree whether or not a particular small speck of dirt is a "grain" or not, so this set is not well defined.

11. **a.** Not everyone would agree on whether a particular bet at Hialeah is a "good bet" or not; if that were not the case everyone would bet on the same horses. It is not well

© 2012 Cengage Learning. All Rights Reserved. May not be scanned, copied or duplicated, or posted to a publicly accessible website, in whole or in part.

defined.

    b. A "bumper crop" may be different things to different people, so this set is not well defined.

13. a. $\{m, a, t, h, e, i, c, s\}$ Be sure not to list the same element twice.

    b. {Barack Obama}

15. a. To answer this question you need to know what we mean by a counting number (see page 50) and what it means to be greater than. The set is $\{7, 8, 9, 10, \cdots\}$.

    b. $\{1, 2, 3, 4, 5\}$; note that the counting number 6 is not included in your answer for either part of this problem.

17. a. $\{p, i, e\}$; be sure you do not list "p" twice.

    b. To answer this question you need to know what we mean by a counting number (see page 50) and what it means to be greater than. The set is $\{151, 152, 153, \cdots\}$.

Answers to Problems 19-24 may vary.

19. This is the set of counting numbers less than 10.

21. This is the set of multiples of 10 between 0 and 101.

23. This is the set of odd numbers between 100 and 170.

25. The set of all numbers, x, such that x is an odd counting number; $\{1, 3, 5, 7, \cdots\}$

27. The set of natural numbers, x, such that x is not equal to 8; $\{1, 2, 3, 4, 5, 6, 7, 9, 10, \cdots\}$

29. The set of all whole numbers, x, such that x is less than eight; $\{0, 1, 2, 3, 4, 5, 6, 7\}$

31. a. $\varnothing$ **b.** $\varnothing, \{1\}$ **c.** $\varnothing, \{1\}, \{2\}, \{1, 2\}$ **d.** $\varnothing, \{1\}, \{2\}, \{3\}, \{1, 2\}, \{1, 3\}, \{2, 3\},$
$\{1, 2, 3\}$ **e.** $\varnothing, \{1\}, \{2\}, \{3\}, \{4\}, \{1, 2\}, \{1, 3\}, \{1, 4\}, \{2, 3\}, \{2, 4\}, \{3, 4\}, \{1, 2, 3\},$
$\{1, 2, 4\}, \{1, 3, 4\}, \{2, 3, 4\}, \{1, 2, 3, 4\}$

33. 32 subsets; yes

Level 2, page 58

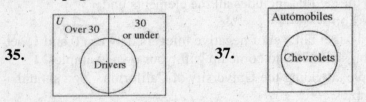

35.   **37.**

39. a. The cardinality is the number of elements in each set.

    $|A| = 3, |B| = 1, |C| = 3, |D| = 1, |E| = 1, |F| = 1$

    b. Two sets are equivalent if they have the same cardinality.

    $A \leftrightarrow C; B \leftrightarrow D \leftrightarrow E \leftrightarrow F$

    c. Two sets are equal if they are the same set. $A = C; D = E = F$

Page 28

41. a. Subset symbol; it is a true statement.   **b.** Element symbol; it is a false statement

43. a. The sets are equal, and every set is a subset of itself, so it is a true statement.

  b. Proper subset symbol does not include equality; it is a false statement.

45. This is the proper subset symbol, and since these are both high school subjects, but are not the only ones, the statement is true.

Contrast Problems 47-51 and make sure you can distinguish among the symbols $\in$, $\subset$, *and* $\subseteq$.

47. Since "1" is an element of the given set, the statement is true.

49. Since the elements of the given set are themselves sets, we see that "1" is not an element of the given set, so the statement is false.

51. Since the elements of the given set are themselves sets, any subset of the given set would contain sets, such as {{1},{2}} or {{1}}. so the given statement is false.

Contrast Problems 52-54.

53. The empty set is named on both the left and the right sides of the equal sign, so the statement is true.

Level 3, page 58

55. Answers vary; the set of real numbers, or the set of people alive in China today. Try to think of an original example.

57. There will be five different ways, since each group of four will have one missing person.

Problem Solving, page 58

59. The set of Chevrolets is a subset of the set of automobiles.

2.2 Operations with Sets, page 59
New Terms Introduced in this Section

  And   Intersection   Or   Union

Level 1, page 62

1. *Union*: $A \cup B$ is the set of all elements in either A or B (or possibly in both).

  Intersection: $A \cap B$ is the set of all elements in both A and B.

  Complementation: $\overline{A}$ is the set of all elements not in A.

3. a. The English word "or" is used to describe union.

  b. The English word "and" is used to describe intersection.

  c. The English word "not" is used to describe complementation.

5. Union is anything in either set: $\{2, 6, 8\} \cup \{6, 8, 10\} = \{2, 6, 8, 10\}$
   Do not list the element "8" twice.

7. Union is anything in either set. $\{2, 5, 8\} \cup \{3, 6, 9\} = \{2, 3, 5, 6, 8, 9\}$

9. Intersection is anything in both sets: $\{1, 2, 3, 4, 5\} \cap \{3, 4, 5, 6, 7\} = \{3, 4, 5\}$

11. Complement is everything not in the given set: $\overline{\{2, 8, 9\}} = \{1, 3, 4, 5, 6, 7, 10\}$

13. This is the set of all even numbers, subject to the given universe: $\{2, 4, 6, 8, 10\}$

15. $\{x \mid x \text{ is a multiple of } 2\} \cup \{x \mid x \text{ is a multiple of } 3\} = \{2, 4, 6, 8, 10\} \cup \{3, 6, 9\}$
$$= \{2, 3, 4, 6, 8, 9, 10\}$$

17. $\overline{\{x \mid x \text{ is even}\}} = \overline{\{2, 4, 6, 8, 10\}}$
$$= \{1, 3, 5, 7, 9\}$$

19. We know that the integers are made up of the positive integers, the negative integers, and the integer 0. Since this is a union we list everything except for zero: $\{x \mid x \text{ is a nonzero integer}\}$

21. Since the set of counting numbers ($\mathbb{N}$) is a subset of the set of whole numbers ($\mathbb{W}$), we see the answer is the set of whole numbers, namely $\mathbb{W}$.

23. $\overline{U} = \varnothing$

25. $X \cap \varnothing = \varnothing$

27. $U \cup \varnothing = U$

29. $A \cup B = \{1, 2, 3, 4\} \cup \{1, 2, 5, 6\}$
$$= \{1, 2, 3, 4, 5, 6\}$$

31. $A \cup C = \{1, 2, 3, 4\} \cup \{3, 5, 7\}$
$$= \{1, 2, 3, 4, 5, 7\}$$

33. $B \cap C = \{1, 2, 5, 6\} \cap \{3, 5, 7\}$
$$= \{5\}$$

35. $\overline{A} = \overline{\{1, 2, 3, 4\}}$
$$= \{5, 6, 7\}$$

37. $\overline{C} = \overline{\{3, 5, 7\}}$
$$= \{1, 2, 4, 6\}$$

39. $\{1, 2, 3, 4\}$ Don't forget that your answer is subject to the given universe.

41. $\{x \mid x \text{ is less than } 5\} \cup \{x \mid x \text{ is greater than } 5\} = \{1, 2, 3, 4\} \cup \{6, 7\}$
$$= \{1, 2, 3, 4, 6, 7\}$$

43. $\varnothing \cup A = A$

45. $A \cap B$

47. $\overline{A \cap B}$

49.   **51.**

Level 2, page 63

53. We work this problem without Venn diagrams to show an alternate method of solution. Let $B = \{$people in band$\}$ and $S = \{$people in orchestra$\}$.

$$|B \cup S| = |B| + |S| - |B \cap S|$$
$$= 50 + 36 - 14$$
$$= 72 \qquad \text{Yes, they can get by with two buses.}$$

55. 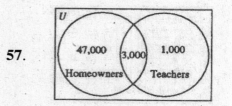   Three people are not taking either.

57.

From the Venn diagram, we see 51,000 booklets are needed.

Level 3 Problem Solving, page 63

59. We use patterns;

1 set is $2 = 2^1$ regions (inside and outside);

2 sets are $4 = 2^2$

3 sets are $8 = 2^3$

It looks like 4 sets should be $2^4 = 16$ and 5 sets should be $2^5 = 32$. Our hypotheses for n sets is 2^n regions.

2.3 Applications of Sets, page 64

New Terms Introduced in this Section

Associative property for union and intersection   De Morgan's laws

Level 1, page 70

1. *De Morgan's laws* relate the operations of complements, unions, and intersections. There are two statements: $\overline{X \cup Y} = \overline{X} \cap \overline{Y}$ and $\overline{X \cap Y} = \overline{X} \cup \overline{Y}$.

3. $(A \cup B) \cap C = (\{1, 2, 3, 4\} \cup \{1, 2, 5, 6\}) \cap \{3, 5, 7\}$
$= \{1, 2, 3, 4, 5, 6\} \cap \{3, 5, 7\}$
$= \{3, 5\}$

5. $\overline{A \cup B} \cap C = \overline{\{1, 2, 3, 4, 5, 6\}} \cap \{3, 5, 7\}$
$= \{7\} \cap \{3, 5, 7\}$
$= \{7\}$

7. $\overline{A} \cup (B \cap C) = \overline{\{1, 2, 3, 4\}} \cup \{5\}$
$= \{5, 6, 7\} \cup \{5\}$
$= \{5, 6, 7\}$

9. $\overline{(A \cup B) \cap C} = \overline{\{3, 5\}}$ *From Problem 3.*
$= \{1, 2, 4, 6, 7\}$

11. $\overline{X} \cup \overline{Y}$ **13.** $\overline{X} \cap Y$ **15.** $\overline{X \cup Y}$ **17.** $\overline{X} \cap \overline{Y}$

Level 2, page 70

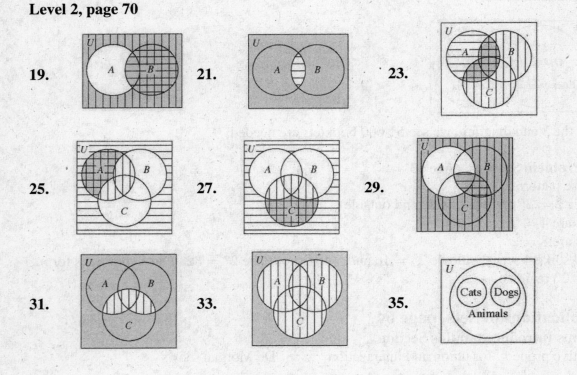

19.   **21.**   **23.**

25.   **27.**   **29.**

31.   **33.**   **35.**

37.

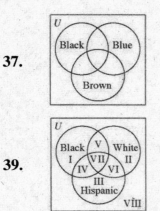

39.

In order to calculate the numbers in the appropriate categories, let $U = \{$U. S. population$\}$ where $|U| = 263$ (assume all numbers are in millions).

As we can see from the Venn diagram, there are 8 regions.
Region VII is empty since everyone has two parents and are accounted for elsewhere. Let V be the region representing persons with one white and one black parent:

$$0.005(263) = 1.315$$

IV the region representing persons with one black and one Hispanic parent:

$$0.02(263) = 5.26$$

VI the region representing persons with one white and one Hispanic parent:

$$0.01(263) = 2.63$$

Let I be the region representing persons with 2 black parents [0.12(263) = 31.56]:

$$\text{Region I:}\ \ 31.56 - 1.315 - 5.26 - 0 = 24.985$$

Let II be the region representing persons with 2 white parents [0.79(263) = 207.77]:

$$\text{Region II:}\ \ 207.77 - 1.315 - 2.63 - 0 = 203.825$$

Let III be the region representing persons with 2 Hispanic parents [0.09(263) = 23.67]:

$$\text{Region III:}\ \ 23.67 - 5.26 - 2.63 - 0 = 15.78$$

Region VIII includes everyone not otherwise classified:

$$263 - (24.985 + 203.825 + 15.78 + 5.26 + 1.315 + 2.63 + 0) = 9.205$$

Here are the answers:

Page 33

I (two black parents): 24,985,000;

II (two white parents): 203,825,000;

III (two Hispanic parents): 15,780,000;

IV (one black and one Hispanic parent): 5,260,000;

V (one black and one white parent): 1,315,000;

VI (one white and one Hispanic parent): 2,630,000;

VII (empty): 0

VIII (one parent is nonwhite, nonblack, or nonhispanic): 9,205,000

41. Look at Figure 2.18 to place each person in the proper region.

Steffi Graf, II;   Michael Stich, V;

Martina Navratilova, III;   Stefan Edberg, V;

Chris Evert Lloyd, VI;   Boris Becker, V;

Evonne Goolagong, II;   Pat Cash, V;

Virginia Wade, II;   John McEnroe, IV

43. We see that the shaded region is outside of A, but inside B: $\overline{A} \cap B$

45. We see that the shaded region is the common region of A and B which is inside of C along with the portion of C that is not inside either A and B. We rephrase this:

(Common region of A and B which is inside of C) $\cup$ (Region inside of C and outside of A and B)

$$[(A \cap B) \text{ and } C] \cup [C \cap \text{ outside } (A \cup \text{B})]$$

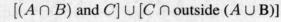

$$[(A \cap B) \cap C] \cup [C \cap \overline{(A \cup B)}]$$

47. 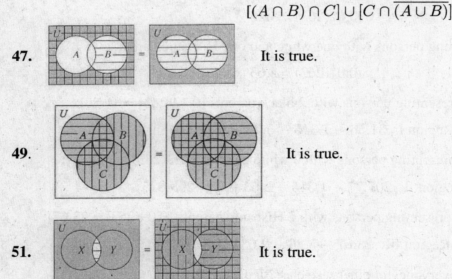 It is true.

49. It is true.

51. It is true.

53. Let $C = \{\text{like comedies}\}$; $V = \{\text{like variety}\}$; $D = \{\text{like drama}\}$

Page 34

Draw the Venn diagram.

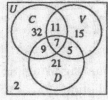

We now add all of the numbers in the Venn diagram:

$$32 + 11 + 15 + 9 + 7 + 5 + 21 + 2 = 102$$

No, there were 102 persons polled, not 100, or perhaps he made up data.

55. a. Let $E = \{$favor Prop. 8$\}$; $T = \{$favor Prop. 13$\}$; $F = \{$favor Prop. 5$\}$

Draw the Venn diagram.

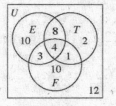

b. Look at the Venn diagram to see it is 12.

c. "And" means both E and F, so we find $3 + 4 = 7$

Level 3 Problem Solving, page 72

57. The key in drawing this Venn diagram is to "give up" the idea that the sets need to be circles.

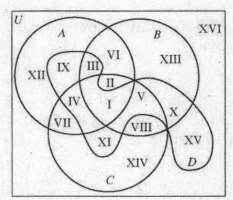

59. a. region 32   **b.** region 2

Page 35

2.4 Finite and Infinite Sets, page 72
New Terms Introduced in this Section

Cartesian product	Countable set	Countably infinite
Finite set	Fundamental counting principle	Infinite set
One-to-one correspondence	Proof by contradiction	Uncountable set
Uncountably infinite		

Level 1, page 77

1. It is a very powerful method for counting very large and complex sets.
3. The Cartesian product of set A and B, denoted by $A \times B$, is the set of all ordered pairs (x, y) where $x \in A$ and $y \in B$.
5. $A \times B = \{(c, w), (c, x), (d, w), (d, x), (f, w), (f, x)\}$
7. $F \times C = \{(1, a), (1, b), (1, c), (2, a), (2, b), (2, c), (3, a), (3, b), (3, c), (4, a), (4, b), (4, c), (5, a), (5, b), (5, c)\}$.
9. $X \times V = \{(2, a), (2, e), (2, i), (2, o), (2, u), (3, a), (3, e), (3, i), (3, o), (3, u), (4, a), (4, e), (4, i), (4, o), (4, u)\}$
11. $(190 - 48) + 1 = 143$
13. $42 + 42 + 1 = 85$
15. $26 - 2 - 3 = 21$
17. The empty set has zero (0) elements.
19. Since this set can be put into a 1-1 correspondence with the set of natural numbers, the cardinality is $\aleph_0$.
21. $20 \times 23 = 460$
23. Since the odd integers can be put into a 1-1 correspondence with the set of natural numbers, the cardinality is $\aleph_0$.

Level 2, page 78

25. $|A| = 26$ and $|B| = 10$, so $|A \times B| = 26 \times 10 = 260$
27. $|C| = 100$ and $|D| = 50$, so $|C \times D| = 100 \times 50 = 5,000$
29. $\{m, a, t\}$; $\{m, a, t\}$; there are others.
$$\updownarrow \; \updownarrow \; \updownarrow \qquad \updownarrow \; \updownarrow \; \updownarrow$$
$$\{1, 2, 3\}; \qquad \{3, 2, 1\}$$
31. $\{1, \quad 2, \quad 3, \quad \cdots, \quad n, \quad n+1, \quad \cdots, \quad 353, \quad 354, \quad 355, 356, \cdots 586, 587\}$
$$550, \; 551, \; 552, \cdots \quad n + 549, \; n + 550, \cdots \qquad 902 \quad 903\}$$
As you can see, the elements $355, 356, \cdots, 586, 587$ are left without a "partner." Thus, the sets do not have the same cardinality.
33. **a.** finite; a search on the Internet estimate this is about 400 billion stars.

b. infinite; the set of counting numbers is infinite, and as long as we remove only a finite number of elements, it is still infinite.

Level 3, page 78

35. A set will have cardinality $\aleph_0$ if it can be placed into a 1-1 correspondence with the set of counting numbers.

$$\{1, 2, 3, \cdots, n, \cdots\}$$
$$\updownarrow \quad \updownarrow \quad \updownarrow \qquad \updownarrow$$
$$\{-1, -2, -3, \cdots, -n, \cdots\}$$

Since these sets can be put into a 1-1 correspondence, they have the same cardinality; namely $\aleph_0$.

37. A set will have cardinality $\aleph_0$ if it can be placed into a 1-1 correspondence with the set of counting numbers.

$$\{1, 2, 3, \cdots, n, \cdots\}$$
$$\updownarrow \quad \updownarrow \quad \updownarrow \qquad \updownarrow$$
$$\{\tfrac{1}{1}, \tfrac{1}{2}, \tfrac{1}{3}, \cdots, \tfrac{1}{n}, \cdots\}$$

Since these sets can be put into a 1-1 correspondence, they have the same cardinality; namely $\aleph_0$.

39. A set will have cardinality $\aleph_0$ if it can be placed into a 1-1 correspondence with the set of counting numbers.

$$\{0, 1, 2, 3, \cdots, n, \cdots\}$$
$$\updownarrow \quad \updownarrow \quad \updownarrow \quad \updownarrow \qquad \updownarrow$$
$$\{1, 2, 3, 4, \cdots n+1, \cdots\}$$

Since these sets can be put into a 1-1 correspondence, they have the same cardinality; namely $\aleph_0$.

41. A set will have cardinality $\aleph_0$ if it can be placed into a 1-1 correspondence with the set of counting numbers.

$$\{1, 3, 5, 7, \cdots, 2n-1, \cdots\}$$
$$\updownarrow \quad \updownarrow \quad \updownarrow \quad \updownarrow \qquad \updownarrow$$
$$\{1, 2, 3, 4, \cdots n, \cdots\}$$

Since these sets can be put into a 1-1 correspondence, they have the same cardinality; namely $\aleph_0$.

43. A set will have cardinality $\aleph_0$ if it can be placed into a 1-1 correspondence with the set of counting numbers.

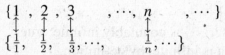

$$\{1, 2, 3, \cdots, n, \cdots\}$$
$$\updownarrow \quad \updownarrow \quad \updownarrow \qquad \updownarrow$$
$$\{1, 3, 9, \cdots 3^{n-1}, \cdots\}$$

Since these sets can be put into a 1-1 correspondence, they have the same cardinality; namely $\aleph_0$.

Page 37

45. $\mathbb{W} = \{\ 0\ ,\ 1\ ,\ 2\ ,\ 3\ \ \ ,\ \cdots,\ \ n\ \ \ \ \ ,\ \cdots\ \}$ Thus, this set is infinite.

$\{1,\ \ 2,\ \ 3,\ \ 4,\cdots \ \ \ n+1,\cdots\}$

47. $\{12\ ,\ 14\ ,\ 16\ \ \ \ ,\ \cdots,\ n\ \ \ \ \ ,\ \cdots\ \}$ Thus, this set is infinite.

$\{14,\ \ 16,\ \ 18,\cdots,\ \ \ n+2,\cdots\}$

49. $\mathbb{W} = \{\ \ 0\ ,\ 1\ ,\ 2\ ,\ \ 3\ \ \ ,\ \cdots,\ n\ \ \ \ \ \ \ ,\ \cdots\ \}$

$\{1,\ \ 2,\ \ 3,\ \ 4,\cdots,\ \ \ n+1,\cdots\}$

Thus, $\mathbb{W}$ is countably infinite.

51. $\mathbb{Q}$ is countably infinite because it can be put into a one-to-one correspondence with the counting numbers (see Example 3).

53. $\{2\ ,\ 4\ ,\ 8\ \ \ \ ,\ \cdots,\ 2^n\ \ \ \ ,\ \cdots\ \}$

$\{1,\ \ 2,\ \ 3,\cdots,\ \ \ \ \ \ n,\cdots\}$

Thus, the set $\{2, 4, 8, 16, 32, \cdots\}$ is countably infinite; true.

55. If a set is uncountable, then is infinite, by definition.

57. Some infinite sets have cardinality $\aleph_0$, while other infinite sets have larger cardinality; false.

Level 3 Problem Solving, page 79

59. The set, E, of even integers has cardinality $\aleph_0$; the set, O, of odd integers has cardinality $\aleph_0$. If we add the cardinality of the even integers to the cardinality of the odd integers, we have $\aleph_0 + \aleph_0$. However, if we put the even integers together with the odd integers, we have the set of counting numbers, which has cardinality $\aleph_0$. Thus, our example illustrates why $\aleph_0 + \aleph_0 = \aleph_0$.

Chapter 2 Review Questions, page 80

STOP   Studying for a chapter examination is a personal process, one which nobody else can do for you. Simply take the time to review what you have done. Here are the new terms in Chapter 2.

And [2.2]	Cartesian product [2.4]	Counting number [2.1]
Associative property for union and intersection [2.3]	Circular definition [2.1]	De Morgan's laws [2.3]
	Complement [2.1]	Description method [2.1]
Belongs to [2.1]	Contained in [2.1]	Disjoint [2.1]
Cardinal number [2.1]	Countable set [2.4]	Element [2.1]
Cardinality [2.1]	Countably infinite [2.4]	Empty set [2.1]

Equal sets [2.1]	Natural number [2.1]	Set theory [2.1]
Equivalent sets [2.1]	One-to-one	Subset [2.1]
Finite set [2.4]	correspondence [2.4]	Uncountable set [2.4]
Fundamental Counting	Or [2.2]	Uncountably infinite [2.4]
Principle [2. 4]	Proof by contradiction [2.4]	Universal set [2.1]
Improper subset [2.1]	Proper subset [2.1]	Union [2.2]
Infinite set [2.4]	Rational number [2.1]	Venn diagram [2.1]
Integer [2.1]	Roster method [2.1]	Well-defined set [2.1]
Intersection [2.2]	Set [2.1]	Whole number [2.1]
Member [2.1]	Set-builder notation [2.1]	

If you can describe the term, read on to the next one; if you cannot, then look it up in the text (the section number is shown in brackets). Next, review the types of problems in Chapter 2.

TYPES OF PROBLEMS
Tell whether a set is well defined. [2.1]
Specify sets by roster and by description. [2.1]
Understand and use set-builder notation. [2.1]
Draw Venn diagrams showing subsets, equal sets, or disjoint sets. [2.1]
Distinguish between equal and equivalent sets, and find the cardinality of a given set. [2.1]
Distinguish the symbols $\subset$, $\subseteq$, and $\in$. [2.1]

Find the complement of a set [2.1,2.2]
Find the union of two sets. [2.2]
Find the intersection of two sets. [2.2]
Recognize and draw the Venn diagrams for union, intersection, and complement. [2.2]
Solve survey problems involving two sets. [2.2]

Perform mixed operations using union, intersection, and complement [2.3]
Draw Venn diagrams for mixed operations using union, intersection, and complement. [2.3]
Draw Venn diagrams using three or more sets. [2.3]
Prove or disprove set statements using Venn diagrams. [2.3]
Solve survey problems involving three or more sets. [2.3]

Find the Cartesian product of two sets, and determine its cardinality. [2.4]
Find the cardinality of a given set. [2.4]
Determine whether sets have the same cardinality by placing them in a one-to-one
correspondence. [2.4]

Classify a given set at finite or infinite. [2.4]

Show that a given set has cardinality $\aleph_0$.[2.4]

Show that a given set is infinite. [2.4]

Once again, see if you can verbalize (to yourself) how to do each of the listed types of problems.
    Work all of Chapter 2 Review Questions (whether they are assigned or not). Work through all of the problems before looking at the answers, and *then* correct each of the problems. The entire solution is shown in the answer section at the back of the text. If you worked the problem correctly, move on to the next problem, but if you did not work it correctly (or you did not know what to do), look back in the chapter to study the procedure, or ask your instructor.
    Finally, go back over the homework problems you have been assigned. If you worked a problem correctly, move on the next problem, but if you missed it on your homework, then you should look back in the book or talk to your instructor about how to work the problem.
    If you follow these steps, you should be successful with your review of this chapter.

We give all of the answers to the chapter review questions (not just the odd-numbered questions).

Chapter 2 Review Questions, page 80

1. a. $A \cup B = \{1, 3, 5, 7, 9\} \cup \{2, 4, 6, 9, 10\}$
$$= \{1, 2, 3, 4, 5, 6, 7, 9, 10\}$$
  b. $A \cap B = \{1, 3, 5, 7, 9\} \cap \{2, 4, 6, 9, 10\}$
$$= \{9\}$$

2. a. $\overline{B} = \overline{\{2, 4, 6, 9, 10\}}$
$$= \{1, 3, 5, 7, 8\}$$
  b. There are no elements in the empty set, so the cardinality is 0.

3. a. The cardinality (number of elements) in the universal set is 10.
  b. The cardinality of the set A is 5.

4. a. $|A \times B| = |A| \times |B| = 5 \times 5 = 25$
  b. $|U \times B| = |U| \times |B| = 10 \times 5 = 50$

Page 40

5. a. $\overline{A \cap B} = \overline{\{1, 3, 5, 7, 9\} \cap \{2, 4, 6, 9, 10\}}$

$\qquad = \overline{\{9\}}$

$\qquad = \{1, 2, 3, 4, 5, 6, 7, 8, 10\}$

b. $\overline{A} \cup \overline{B} = \overline{\{1, 3, 5, 7, 9\}} \cup \overline{\{2, 4, 6, 9, 10\}}$

$\qquad = \{2, 4, 6, 8, 10\} \cup \{1, 3, 5, 7, 8\}$

$\qquad = \{1, 2, 3, 4, 5, 6, 7, 8, 10\}$

c. Yes, from De Morgan's law.

6. $\overline{(A \cup B) \cap A} = \overline{(\{1, 3, 5, 7, 9\} \cup \{2, 4, 6, 9, 10\}) \cap \{1, 3, 5, 7, 9\}}$

$\qquad = \overline{\{1, 2, 3, 4, 5, 6, 7, 9, 10\} \cap \{1, 3, 5, 7, 9\}}$

$\qquad = \overline{\{1, 3, 5, 7, 9\}}$

$\qquad = \{2, 4, 6, 8, 10\}$

7. $\overline{A} \cap (B \cup A) = \overline{\{1, 3, 5, 7, 9\}} \cap (\{2, 4, 6, 9, 10\} \cup \{1, 3, 5, 7, 9\})$

$\qquad = \overline{\{1, 3, 5, 7, 9\}} \cap \{1, 2, 3, 4, 5, 6, 7, 9, 10\}$

$\qquad = \{2, 4, 6, 8, 10\} \cap \{1, 2, 3, 4, 5, 6, 7, 9, 10\}$

$\qquad = \{2, 4, 6, 10\}$

8. a. $\{2\}$ is a subset of B, and it is also a proper subset of B, so you can use either $\subset$ or $\subseteq$. Note, the correct answer includes *both* of these symbols.

b. 2 is an element of B, so the correct symbol is $\in$. You should remember the distinction displayed in parts **a** and **b**.

c. Both represent the empty set, so you can use the symbol $=$. Also, since any set is a subset of itself, you can also use the symbol $\subseteq$. Note, the correct answer includes *both* of these symbols.

d. If we rearrange the elements of a set, the set does not change, so you can use the symbol $=$. Also, since any set is a subset of itself, you can also use the symbol $\subseteq$. Note, the correct answer includes *both* of these symbols.

e. Every set is a subset of the universal set, so you can use the symbol $\subseteq$. Also, since B is not the universal set itself, you can also use the symbol $\subset$. Note, the correct answer includes *both* of these symbols.

9. a. Since $|U| = 50$, and since the odd numbers include exactly half the elements in U, we conclude $|N| = 25$.

b. There are 9 one digit numbers so $|P| = 50 - 9 = 41$.

c. $|N \cap P| = 41 - 21 = 20$

d. $|N \cup P| = |N| + |P| - |N \cap P|$
$$= 25 + 41 - 20$$
$$= 46$$

10. a.
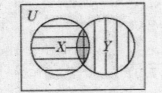

Draw two overlapping circles labeled X and Y. Shade X (horizontal, for example), and then shade Y (vertical). The intersection is everything with both horizontal and vertical shading.

b.
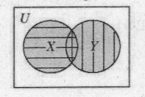

Draw two overlapping circles labeled X and Y. Shade X (horizontal, for example), and then shade Y (vertical). The union is everything with either horizontal or vertical shading.

11. a.

Draw a circle labeled X. Shade X (horizontal). The complement is everything not shaded.

b.
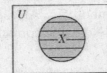

Draw a circle labeled X. Shade X (horizontal). The complement is everything not shaded. The complement of that is back to everything with horizontal lines.

12. a.

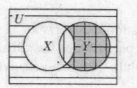

Draw two overlapping circles labeled X and Y. Shade $\overline{X}$ (horizontal, for example), and then shade Y (vertical). The intersection is everything with both horizontal and vertical shading

b.

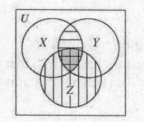

Draw two overlapping circles labeled X and Y. Shade $X \cup Y$ (horizontal, for example). The complement is everything without horizontal shading.

13. a.

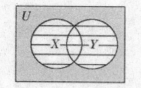

Draw three overlapping circles labeled X, Y, and Z. Do the parentheses first; that is, shade $X \cap Y$ (horizontal, for example). Next, shade Z (vertical). The intersection is everything with both horizontal and vertical shading.

b.

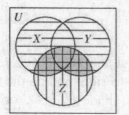

Draw three overlapping circles labeled X, Y, and Z. Do the parentheses first; that is, shade $X \cup Y$ (horizontal, for example). Next, shade Z (vertical). The intersection is everything with both horizontal and vertical shading.

Page 43

14. Begin by drawing a Venn diagram for the left-side of the equation: $A \cup (B \cap C)$.
Draw three overlapping circles labeled A, B, and C. Do the parentheses first; that is,
shade $B \cap C$ (vertical, for example). Next, shade A (horizontal). The union is everything
with any horizontal or vertical shading.

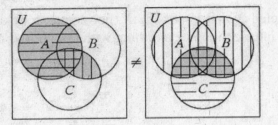

Next, draw a Venn diagram for the right-side of the equation: $(A \cup B) \cap C$.
Draw three overlapping circles labeled A, B, and C. Do the parentheses first; that is,
shade $A \cup B$ (vertical, for example). Next, shade C (horizontal). The intersection is
everything with both horizontal and vertical shading.
Since the Venn diagrams have the different *final* shading, the statement has been
disproved.

15. Begin by drawing a Venn diagram for the left-side of the equation: $(A \cup B) \cap C$.
Draw three overlapping circles labeled A, B, and C. Do the parentheses first; that is,
shade $A \cup B$ (vertical, for example). Next, shade C (horizontal). The intersection is
everything with both horizontal and vertical shading.

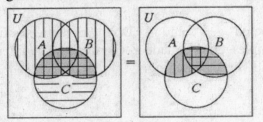

Next, draw a Venn diagram for the right-side of the equation: $(A \cap C) \cup (B \cap C)$. Draw
three overlapping circles labeled A, B, and C. Read from left-to-right and do the first
parentheses; that is, shade $A \cap C$ (vertical, for example). Next, shade $B \cap C$ (horizontal).
The union is everything with either horizontal and vertical shading.
Since both Venn diagrams have the same *final* shading, the statement has been proved.

16. a. The set of rational numbers is the set of all numbers of the form $\frac{a}{b}$ such that a is an integer and b is a counting number.

b. $\frac{2}{3}$; answers vary; try to think of an original one on your own.

17. $\{\; 5 \;,\; 10 \;,\; 15 \quad,\; \cdots,\; 5n \qquad,\; \cdots \;\}$
$\quad\;\; \updownarrow \quad\;\; \updownarrow \quad\;\; \updownarrow \qquad\qquad\quad \updownarrow$
$\;\;\{10,\;\; 20,\;\; 30, \cdots, \qquad 10n, \cdots\}$

Since the second set is a proper subset of the first set, we see that the set F is infinite.

18. a. $\{\; n \mid (-n) \in \mathbb{N} \}$; answers vary.

  b. $\{\; 1 \;\;,\;\; 2 \;,\;\; 3 \quad,\; \cdots,\; n \qquad,\; \cdots \;\}$
$\qquad \updownarrow \quad\;\; \updownarrow \quad\;\; \updownarrow \qquad\qquad \updownarrow$
$\quad\;\;\{-1,\;\; -2,\;\; -3, \cdots \qquad -n, \cdots\}$

Since the first set is the set of counting numbers, it has cardinality $\aleph_0$; so the given set also has cardinality $\aleph_0$ since it can be put into a one-to-one correspondence with the set of counting numbers.

19. Draw three overlapping circles, labeled C (those who have a car), TV (those who have TVs), and B (those who have a bicycle). Begin by filling in "15" in the region which is the intersection of all three circles.

  Next, fill in $25 - 15 = 10$ in the region of TV and B which is not in C. Fill in $35 - 15 = 20$ in the region of C and TV which is not B. Fill in $17 - 15 = 2$ in the region of C and B which is not in TV.

  Finally, fill in the region which is C, but not in any of the other circles: $42 - 20 - 15 - 2 = 5$.   Fill in the region which is TV, but not in any of the other circles: $50 - 20 - 15 - 10 = 5$. Fill in the region which is B, but not in any of the other circles: $30 - 2 - 15 - 10 = 3$.

  These numbers are shown in this Venn diagram:

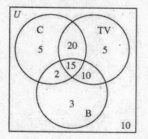

  In order to see how many students have none of the three items, subtract all of the numbers shown in the Venn diagram from the total number:

$$70 - 5 - 20 - 15 - 2 - 5 - 10 - 3 = 10$$

Ten students had none of the items.

20. a. The eight possible blood types are shown in Figure 2.20:

A^-, AB^-, B^-, A^+, AB^+, B^+, O^+, O^-

b. A^-, AB^-, B^-, A^+, AB^+, B^+, O^-

c. A, B, O^+

CHAPTER 3 The Nature of Logic

Chapter Overview
This chapter introduces the idea of building patterns of logical thought. You are not expected to see an example and then apply it to another similar problem. Instead, the goal of this text is to get you to think critically, to view mathematics as a worthwhile endeavor, and to create a mind set that mathematics is not just for classrooms, but for real life.

3.1 Deductive Reasoning, page 83
New Terms Introduced in this Section

And	Argument	Compound Statement	Conclusion
Conjunction	Connective	Deductive reasoning	Disjunction
Either...or	Exclusive Or	Fuzzy Logic	Hypotheses
Inclusive or	Invalid argument	Law of the excluded middle	Logic
Negation	Not	Operators	Or
Premise	Simple statement	Statement	Symbolic logic
Truth value	Valid argument		

Level 1, page 89

1. A *logical operator* is a way of connecting simple statements. The fundamental operators are *and, or,* and *not*. Other operators include *because, either ... or, neither... nor,* and *unless.*
3. *Disjunction* is the operator defined by Table 3.2. It is defined so it is false only when both of the simple propositions are false, and it is true otherwise.
5. a, b, and d are statements; c is not a statement because it is a question, and consequently cannot be classified as true or false.
7. a, b, and d are statements; c is not a statement because it is a command, and consequently cannot be classified as true or false.
9. Claire has both glasses and a hat.   11. Charles and Joe are both blond, so they qualify.
13. The males (Alfred, Sam, and Paul) together with the blue-eyed persons (Anita, Alfred) make up the entire set of people, so the statement is true.
15. All the faces have eyebrows, so the statement is false.
17. The only person with a mustache is Alfred and he does not have glasses, so the statement is true.
19. There is one person (Alfred) who does not have white hair, so the statement is false.

Level 2, page 90

Use Table 3.4 *for Problems 20-27.*

21. Some dogs do not have fleas.

23. Some triangles are squares.

25. Some counting numbers are not divisible by 1.

27. All integers are odd.

29. For this problem, we assume that p is true and q is also true.

    a. $p \vee \sim q$ **b.** $p \wedge \sim q$ **c.** $p \wedge q$ **d.** $\sim p \wedge q$

    $T \vee \sim T$ $T \wedge \sim T$ $T \wedge T$ $\sim T \wedge T$

    $T \vee F$ $T \wedge F$ T $F \wedge T$

    T F F

31. **a.** Prices or taxes will rise.

    b. Prices will not rise and taxes will rise.

    c. Prices will rise or taxes will not rise.

    d. Prices will not rise or taxes will not rise.

33. For this problem, we assume that p is true and q is also true.

    a. $p \vee q$ **b.** $\sim p \wedge q$ **c.** $p \vee \sim q$ **d.** $\sim p \vee \sim q$

    $T \vee T$ $\sim T \wedge T$ $T \vee \sim T$ $\sim T \vee \sim T$

    T $F \wedge T$ $T \vee F$ $F \vee F$

                F T F

35. **a.** $(r \vee s) \vee t$ **b.** $(r \wedge s) \wedge \sim t$

    $(F \vee T) \vee T$ $(F \wedge T) \wedge \sim T$

    $T \vee T$ $F \wedge \sim T$

    T $F \wedge F$

                       F

37. **a.** $(p \vee q) \wedge r$ **b.** $(p \wedge q) \wedge \sim r$

    $(T \vee T) \wedge F$ $(T \wedge T) \wedge \sim F$

    $T \wedge F$ $T \wedge T$

    F T

39. **a.** $(p \vee q) \vee (r \wedge \sim q)$ **b.** $\sim (\sim p) \vee (p \wedge q)$

    $(T \vee T) \vee (T \wedge \sim T)$ $\sim (\sim T) \vee (T \wedge T)$

    $T \vee (T \wedge F)$ $T \vee T$

    $T \vee F$ T

    T

Level 3, page 91

41. Let e: W. C. Fields is eating;
  d: W. C. Fields is drinking;
  t: W. C. Fields is having a good time.
Then, $e \wedge d \wedge t$.

43. Let t: Jack will go tonight;
  m: Rosamond will go tomorrow.
Then, $\sim t \wedge \sim m$.

45. Let j: The decision will depend on judgment;
  i: The decision will depend on intuition;
  m: The decision will depend on who paid the most.
Then, $(j \vee i) \wedge \sim p$.

47. Let d: The winner must have an A. A. degree in drafting;
  e: The winner has three years professional experience.
Then, $d \vee e$.

49. Let s: Marsha finished the sign;
  t: Marsha finished the table;
  c: Marsha finished a pair of chairs.
  a. $(s \wedge t) \vee c$ **b.** $s \wedge (t \vee c)$

51. $p \vee (q \wedge r)$ **53.** $(p \wedge \sim q) \vee (r \wedge \sim q)$ **55.** $\sim (r \wedge q) \wedge (q \vee \sim q)$
  $\text{T} \vee (\text{F} \wedge \text{F})$ $(\text{T} \wedge \sim \text{F}) \vee (\text{F} \wedge \sim \text{F})$ $\sim (\text{F} \wedge \text{F}) \wedge (\text{F} \vee \sim \text{F})$
  $\text{T} \vee \text{F}$ $(\text{T} \wedge \ \text{T}) \vee (\text{F} \wedge \ \text{T})$ $\sim (\text{F}) \ \wedge (\text{F} \vee \ \text{T})$
   T $\text{T} \quad \vee \quad \text{F}$ $\text{T} \quad \wedge \quad \text{T}$
        T T

57. $(q \vee \sim q) \wedge [(p \wedge \sim q) \vee (\sim r \vee r)]$
  $(\text{F} \vee \sim \text{F}) \wedge [(\text{T} \wedge \sim \text{F}) \vee (\sim \text{F} \vee \text{F})]$
  $(\text{F} \vee \ \text{T}) \ \wedge [(\text{T} \wedge \text{T}) \quad \vee (\ \text{T} \vee \text{F})]$
   $\text{T} \qquad \wedge [\quad \text{T} \qquad \vee \qquad \text{T}]$
   $\text{T} \qquad \wedge \qquad\qquad \text{T}$
        T

Level 3 Problem Solving, page 91

59. Each phrase is equivalent to a negation: Dr. Smith, I wish to explain that...

  I was   I didn't
  joking... mean... reconsider... not... change mind
   $\sim$ $\sim$ $\sim$ $\sim$

By the law of double negation (Problem 58 used twice) this is equivalent to changing her mind. Yes, Melissa did change her mind.

3.2 Truth Tables and the Conditional, page 91

New Terms Introduced in this Section

Antecedent	Conditional	Consequent	Contrapositive
Converse	Fundamental operators	Inverse	Law of contraposition
Law of double negation	Truth table		

Level 1, page 98

1. A truth table is a device used in logic which lists all the possible truth and false conditions for a given argument.

3. The *law of double negation* is: $\sim(\sim p)$ may be replaced by p in any logical expression.

5.

p	q	$\sim p$	$\sim p \vee q$
T	T	F	T
T	F	F	F
F	T	T	T
F	F	T	T

7.

p	q	$p \wedge q$	$\sim(p \wedge q)$
T	T	T	F
T	F	F	T
F	T	F	T
F	F	F	T

9.

r	$\sim r$	$\sim(\sim r)$
T	F	T
F	T	F

11.

p	q	$\sim q$	$p \wedge \sim q$
T	T	F	F
T	F	T	T
F	T	F	F
F	F	T	F

13.

p	q	$\sim p$	$\sim p \wedge q$	$\sim q$	$(\sim p \wedge q) \vee \sim q$
T	T	F	F	F	F
T	F	F	F	T	T
F	T	T	T	F	T
F	F	T	F	T	T

15.

p	q	$p \rightarrow q$	$p \vee (p \rightarrow q)$
T	T	T	T
T	F	F	T
F	T	T	T
F	F	T	T

17.

p	q	$p \vee q$	$p \wedge (p \vee q)$	$[p \wedge (p \vee q)] \rightarrow p$
T	T	T	T	T
T	F	T	T	T
F	T	T	F	T
F	F	F	F	T

19.

p	q	r	$p \vee q$	$(p \vee q) \vee r$
T	T	T	T	T
T	T	F	T	T
T	F	T	T	T
T	F	F	T	T
F	T	T	T	T
F	T	F	T	T
F	F	T	F	T
F	F	F	F	F

Page 50

21.

p	q	r	$p \vee q$	$\sim r$	$(p \vee q) \wedge \sim r$	$[(p \vee q) \wedge \sim r] \wedge r$
T	T	T	T	F	F	F
T	T	F	T	T	T	F
T	F	T	T	F	F	F
T	F	F	T	T	T	F
F	T	T	T	F	F	F
F	T	F	T	T	T	F
F	F	T	F	F	F	F
F	F	F	F	T	F	F

23. If the slash through "P" means "no parking," then a slash through "No Parking" must mean park here (law of double negation).

Level 2, page 99

25. Statement: $\sim p \rightarrow \sim q$ Converse: $\sim q \rightarrow \sim p$
   Inverse: $p \rightarrow q$ Contrapositive: $q \rightarrow p$

27. Statement: $\sim t \rightarrow \sim s$ Converse: $\sim s \rightarrow \sim t$
   Inverse: $t \rightarrow s$ Contrapositive: $s \rightarrow t$

29. Statement: If I get paid, then I will go Saturday.
   Converse: If I go Saturday, then I got paid.
   Inverse: If I do not get paid, then I will not go Saturday.
   Contrapositive: If I do not go Saturday, then I do not get paid.

31. If it is a triangle, then it is a polygon.

33. If you are a good person, then you will go to heaven.

35. If we make a proper use of these means which the God of Nature has placed in our power, then we are not weak.

37. If it is work, then it is noble.

39. F $\rightarrow$ F is true

41. F $\rightarrow$ F is true

43. p is T and q is T.

    a. $\sim p \vee q$ **b.** $\sim p \wedge \sim q$ **c.** $\sim (p \wedge q)$

        $\sim T \vee T$ $\sim T \wedge \sim T$ $\sim (T \wedge T)$

          $F \vee T$ $F \wedge F$ $\sim (T)$

            T F F

45. p is F and q is T.

    a. $(p \wedge \sim q) \wedge p$ **b.** $p \vee (p \to q)$ **c.** $(p \wedge q) \to p$

      $(F \wedge \sim T) \wedge F$ $F \vee (F \to T)$ $(F \wedge T) \to F$

        $(F \wedge F) \wedge F$ $F \vee \qquad T$ $F \to F$

             $F \wedge F$ T T

                F

47. Let q: The qualifying person is a child;

          d: This child is your dependent;

          n: You enter your child's name.

    Then, $(q \wedge \sim d) \to n$.

49. Let b: The amount on line 32 is \$86,025;

          c: The amount on line 32 is less than \$86,025;

          n: You multiply the number of exemptions on line 6e by \$2,500.

    Then, $(b \vee c) \to n$.

51. Let m: You are a student; d: You are a person with disabilities;

          s: You see line 6 of instructions.

    Then, $(m \vee d) \to s$.

Level 3, page 99

53. a. Rephrase the statement as "If you are an applicant for the position, then you have a two-year college degree or five years of experience in the field." Thus, $a \to (e \vee f)$.

    b. Rephrase the statement as "If you are an applicant for the position and you have a two-year college degree in drafting, then you are qualified for the position." Thus, $(a \wedge e) \to q$.

    c. Rephrase the statement as "If you are an applicant for the position and you have five years of experience in the field, then you are qualified for the position." Thus, $(a \wedge f) \to q$.

Page 52

55. The problem calls for an implication which yields a conclusion of q, so we begin by rewriting the compound statement using the if-then connective with the consequent being "you qualify for the special fare." There are at least two ways of dong this:

(1) "If you fly on Monday, Tuesday, Wednesday, or Thursday and you stay over a Saturday evening, then you qualify for the special rate."

(2) "If you stay over a Saturday evening and you fly on Monday, Tuesday, Wednesday, or Thursday, then you qualify for the special rate."

The symbolic representations of these statements are, respectively,

(1) $[(m \vee t \vee w \vee h) \wedge s] \rightarrow q$ (2) $[s \wedge (m \vee t \vee w \vee h)] \rightarrow q$

Either one of these statements is acceptable.

57. The problem calls for an implication with antecedent t, so we begin by rewriting the compound statement using the if-then connective with the antecedent "you are a tenant." The following statement makes sense: "If you are a tenant, then you lease the premises for 12 months beginning on September 1 and at a monthly rental charge of $800."

This statement suggests that a tenant must do *all* three of the following: (1) lease the premises for 12 months, (2) begin on September 1, *and* (3) pay $800 a month. Thus, the consequent is a three-part conjunction: $t \rightarrow (m \wedge s \wedge p)$.

Level 3 Problem Solving, page 100

59. This is not a statement, since "Apollo can do anything" gives rise to a paradox and is neither true nor false.

3.3 Operators and Laws of Logic, page 100

New Terms Introduced in this Section

Because	Biconditional	De Morgan's Law	Implication
Logical equivalence	Negation of a conditional	Neither...nor	No p is q
Tautology	Truth set	Unless	

Level 1, page 105

1. Answers vary; the conditional has the form $p \rightarrow q$, whereas the biconditional $p \leftrightarrow q$ not only means $p \rightarrow q$ but also $q \rightarrow p$.

3. The symbol $\leftrightarrow$ is used to symbolize the biconditional, whereas the symbol $\Leftrightarrow$ is used to symbolize a tautology (a statement that is always true).

5. Answers vary

"*Either* you will do what your mother says *or* I will not pay you an allowance."

"*Neither* your brother *nor* his wife will attend her wedding."

"I will be there *unless* my boss calls me into work."

"I will go *because* I love you."

"*No* person *is* ten foot tall."

7. a. Look at the top sign; you can stay there until 2 A.M., which is 9 hours.

   b. If you look at the middle sign, 10 hours; if you look at the lower sign, 15 minutes. If you stay more than 15 minutes, you will be in violation of at least one sign.

9.

p	q	$p \vee q$	$\sim p$	$q \to \sim p$	$(p \vee q) \wedge (q \to \sim p)$
T	T	T	F	F	F
T	F	T	F	T	T
F	T	T	T	T	T
F	F	F	T	T	F

No, it is not a tautology because not all the entries in the final column are true.

11.

p	q	$\sim q$	$\sim q \to p$	$(\sim q \to p) \to q$
T	T	F	T	T
T	F	T	T	F
F	T	F	T	T
F	F	T	F	T

No, it is not a tautology because not all the entries in the final column are true.

13.

p	q	$p \to q$	$\sim p$	$\sim p \vee q$	$(p \to q) \leftrightarrow (\sim p \vee q)$
T	T	T	F	T	T
T	F	F	F	F	T
F	T	T	T	T	T
F	F	T	T	T	T

Yes, it is a tautology because all the entries in the final column are true.

15.

p	q	$\sim q$	$p \vee \sim q$	$\sim p$	$\sim p \vee q$	$(p \vee \sim q) \leftrightarrow (\sim p \vee q)$
T	T	F	T	F	T	T
T	F	T	T	F	F	F
F	T	F	F	T	T	F
F	F	T	T	T	T	T

No, it is not a tautology because not all the entries in the final column are true.

17.

p	q	$p \vee q$	$\sim (p \vee q)$
T	T	T	F
T	F	T	F
F	T	T	F
F	F	F	T

19.

p	q	$\sim q$	$p \to \sim q$
T	T	F	F
T	F	T	T
F	T	F	T
F	F	T	T

21. Let h: I will buy a new house;

   p: All provisions of the sale are clearly understood.

Page 54

Not h unless p: $\sim p \rightarrow \sim h$.

23. Let r: I am obligated to pay the rent;
         s: I signed the contract.

r because s: $(r \wedge s) \wedge (s \rightarrow r)$

25. Let m: It is a man;
         i: It is an island.

No m is i: $m \rightarrow \sim i$

27. Let n: You are nice to people on your way up;
         m: You will meet people on your way down.

n because m: $(n \wedge m) \wedge (m \rightarrow n)$.

29. Let f: The majority, by mere force of numbers, deprives a minority of a clearly written
          constitutional right;
         r: Revolution is justified.

If f then r: $f \rightarrow r$.

Answers to Problems 30-35 may vary.

31. Let p: The cherries have turned red;
         q: The cherries are ready to be picked.

Given: $p \rightarrow q$; this is equivalent to $\sim p \vee q$:

The cherries have not turned red or they are ready to be picked.

33. Let p: Hannah watches Jon Stewart;
         q: Hannah watches the NBC late-night orchestra.

Given $\sim p \vee q$; this is equivalent to $p \rightarrow q$:

If Hannah watches Jon Stewart, then she watches the NBC late-night orchestra.

35. Let p: The money is available;
         q: I will take my vacation.

Given $p \vee \sim q$; this is equivalent to $\sim p \rightarrow \sim q$:

If the money is not available, then I will not take my vacation.

Level 3, page 106

37. See Problem 58, Section 3.1 for a proof without using a truth table.

p	$\sim p$	$\sim (\sim p)$	$p \leftrightarrow \sim (\sim p)$
T	F	T	T
F	T	F	T

Thus, $p \Leftrightarrow \sim (\sim p)$.

Page 55

39.

p	q	$p \vee q$	$\sim (p \vee q)$	$\sim p$	$\sim q$	$\sim p \wedge \sim q$	$\sim (p \vee q) \leftrightarrow (\sim p \wedge \sim q)$
T	T	T	F	F	F	F	T
T	F	T	F	F	T	F	T
F	T	T	F	T	F	F	T
F	F	F	T	T	T	T	T

Thus, $\sim (p \vee q) \Leftrightarrow \sim p \wedge \sim q$.

41.

p	q	$p \rightarrow q$	$\sim p$	$\sim p \vee q$	$(p \rightarrow q) \leftrightarrow (\sim p \vee q)$
T	T	T	F	T	T
T	F	F	F	F	T
F	T	T	T	T	T
F	F	T	T	T	T

Thus, $(p \rightarrow q) \Leftrightarrow (\sim p \vee q)$.

43. $\sim (p \rightarrow \sim q) \Leftrightarrow \sim (\sim p \vee \sim q) \Leftrightarrow \sim (\sim p) \wedge \sim (\sim q) \Leftrightarrow p \wedge q$

45. $\sim (\sim p \rightarrow \sim q) \Leftrightarrow \sim [\sim (\sim p) \vee \sim q] \Leftrightarrow \sim p \wedge \sim (\sim q) \Leftrightarrow \sim p \wedge q$

47. Let b: Jane went to the basketball game;
   s: Jane went to the soccer game.
The given statement (symbolically) is: $j \vee s$.
   The negation is $\sim (j \vee s) \Leftrightarrow \sim j \wedge \sim s$, by De Morgan's law. Thus, the translated statement is: Jane did not go to the basketball game and she did not go to the soccer game.

49. Let t: Missy is on time;
   b: Missy missed the boat.
The given statement (symbolically) is: $\sim t \wedge b$.
   The negation is $\sim (\sim t \wedge b) \Leftrightarrow \sim (\sim t) \vee (\sim b) \Leftrightarrow t \vee \sim b$. Thus, the translated statement is: Missy is on time or she did not miss the boat.

51. Let s: You have Schlitz.
   b: You have beer.
The given statement (symbolically) is: $\sim s \rightarrow \sim b$.
   The negation is $\sim (\sim s \rightarrow \sim b) \Leftrightarrow \sim s \wedge \sim (\sim b) \Leftrightarrow \sim s \wedge b$. Thus, the translated statement is: You're out of Schlitz and you have beer.

53. Let p: $x - 5 = 4$
   q: $x = 1$
The given statement (symbolically) is: $p \rightarrow q$.
   The negation is $p \wedge \sim q$. Thus, the translated statement is: $x - 5 = 4$ and $x \neq 1$.

55. Let p: $2x + 3y = 8$
   q: $x = 1$
   r: $y = 2$
The given statement (symbolically) is: $(q \wedge r) \rightarrow p$.

Page 56

The negation is $(q \wedge r) \wedge \sim p$. Thus, the translated statement is: $x = 1$ and $y = 2$ and $2x + 3y \neq 8$.

Level 3 Problem Solving, page 106

57. Let d: You purchase your ticket between January 5 and February 15;
f: You fly round trip between February 20 and May 3;
m: You depart on Monday;
t: You depart on Tuesday;
w: You depart on Wednesday;
h: You return on Tuesday;
i: You return on Wednesday;
j: You return on Thursday;
s: You stay over a Saturday night;
e: You obtain 40% off regular fare.

Symbolic statement: $[d \wedge f \wedge (m \vee t \vee w) \wedge (h \vee i \vee j) \wedge s] \rightarrow e$

59. Let ℓ: The tenant lets the premises;
m: The tenant lets a portion of the premises;
s: The tenant sublets the premises;
t: The tenant sublets a portion of the premises;
p: Permission is obtained.

Symbolic statement: $\sim [(\ell \vee m) \vee (s \vee t)]$ unless p. This means:
$\sim p \rightarrow \sim [(\ell \vee m) \vee (s \vee t)]$
or
$\sim p \rightarrow [\sim (\ell \vee m) \wedge \sim (s \vee t)]$.
This can also be written (using De Morgan's law) as
$\sim p \rightarrow [(\sim \ell \wedge \sim m) \wedge (\sim s \wedge \sim t)]$.

3.4 The Nature of Proof, page 107
New Terms Introduced in this Section

Assuming the antecedent	Assuming the consequent (fallacy)	Counterexample
Denying the antecedent (fallacy)	Denying the consequent	Direct reasoning
Fallacy	Fallacy of the converse	Fallacy of the inverse
False chain pattern	Indirect reasoning	Law of detachment
Logical fallacy	Modus ponens	Modus tollens
Syllogism	Theorem	Transitive reasoning

When finished with this section, you should be able to distinguish among direct reasoning, indirect reasoning, and transitive reasoning, as well as give examples of each.

Level 1, page 113

1. *Direct reasoning* consists of two *premises*, or *hypotheses*, and a *conclusion*. The argument form for direct reasoning is $[(p \rightarrow q) \wedge p] \rightarrow q$.

3. *Transitive reasoning* consists of two *premises*, or *hypotheses*, and a *conclusion*. The argument form for transitive reasoning is $[(p \rightarrow q) \wedge (q \rightarrow r)] \rightarrow (p \rightarrow r)$.

5. A *syllogism* is a form of reasoning in which two statements or premises are made and a logical conclusion is drawn from them. In this book, we considered three types of syllogisms: direct reasoning, indirect reasoning, and transitive reasoning.

7. **a**. valid; indirect reasoning
   b. invalid; fallacy of the converse

9. **a**. invalid; fallacy of the inverse
   b. valid; direct

11. This is valid, since it is an example of direct reasoning.

13. Rewrite the argument in if-then format:
    If it is a snark, then it is a fribble. Symbolic: $s \rightarrow f$
    If it is a fribble, then it is ugly. $f \rightarrow u$
    Therefore, if it is a snark, then it is ugly. $\therefore s \rightarrow u$
    It is valid, since it is transitive reasoning.

15. This is valid, since it is an example of direct reasoning.

17. This is valid, since it is an example of direct reasoning.

19. This is invalid, since it is an example of false chain reasoning.

21. This is valid, since it is an example of indirect reasoning.

23. This is valid, since it is an example of indirect reasoning.

25. This is invalid, since it is an example of fallacy of the converse.

27. This is valid, since it is an example of transitive reasoning.

29. Rewrite the augment in if-then format:
    If the person is a mathematician, then the person is an eccentric.   Symbolic: $m \rightarrow e$
    If the person is an eccentric, then the person is rich. $e \rightarrow r$
    Therefore, if the person is a mathematician, then the person is rich. $\therefore m \rightarrow r$
    This is valid, since it is an example of transitive reasoning.

31. This is valid, since it is an example of indirect reasoning.

33. Let c: The crime occurred after 4:00 A.M.
         s: Smith did it.
         j: Jones did it.

w: The crime involved two persons.
Translate the argument into symbolic form:

(1) $c \rightarrow \sim s$ Rearrange the premises:

(2') $j \rightarrow c$	contrapositive of (2)
(1) $c \rightarrow \sim s$	given
(4) $j \rightarrow \sim s$	transitive law
(3') $\sim w \rightarrow j$	contrapositive of (3)
$\therefore \sim w \rightarrow \sim s$	transitive law
$\therefore s \rightarrow w$	contrapositive

(2) $\sim c \rightarrow \sim j$

(3) $\sim j \rightarrow w$

Translation: If Smith committed the crime, then it involved two persons. Valid reasoning.

Level 2, page 114

35. We show this is a fallacy by constructing a truth table:

p	q	r	$p \rightarrow q$	$p \rightarrow r$	$q \rightarrow r$	$(p \rightarrow q) \wedge (p \rightarrow r)$	$[(p \rightarrow q) \wedge (p \rightarrow r)] \rightarrow (q \rightarrow r)$
T	T	T	T	T	T	T	T
T	T	F	T	F	F	F	T
T	F	T	F	T	T	F	T
T	F	F	F	F	T	F	T
F	T	T	T	T	T	T	T
F	T	F	T	T	F	T	F
F	F	T	T	T	T	T	T
F	F	F	T	T	T	T	T

$\uparrow$
Not all Ts

Since the result does not show all Ts, it is a fallacy. This is the false chain pattern.

37. If you learn mathematics, then you understand human nature. (*transitive*)

39. We do not go to the concert. (*indirect*)

41. $b = 0$ (*excluded middle*)

43. We do not interfere with the publication of false information. (*indirect*)

45. I will not eat that piece of pie. (*indirect*)

47. Let p: We win first prize;

     e: We will go to Europe;

     i: We are ingenious.

Given argument: (1) $p \to e$ Rearrange the premises: (2) $i \to p$ *given*
            (2) $i \to p$ (1) $p \to e$ *given*
            (3) i $i \to e$ *transitive law*
                                     (3) i *given*
                                        e *direct reasoning*

Conclusion: We will go to Europe.

Level 3, page 115
49. Let b: This is a baby;
          l: This is an illogical creature;
          d: This is a despised creature;
          c: This is one who can manage a crocodile.
Given argument: (1) $b \to l$ Then we have: (1) $b \to l$ *given*
            (2) $d \to \sim c$ (3) $l \to d$ *given*
            (3) $l \to d$ $b \to d$ *transitive law*
                              (2) $d \to \sim c$ *given*
                              $b \to \sim c$ *transitive law*

Conclusion: If it is a baby, then it cannot manage a crocodile. Or, babies cannot manage crocodiles.

51. Let d: It is a duck;
          w: It waltzes;
          f: It is an officer;
          p: It is my poultry.
Given argument: (1) $d \to \sim w$
                 (2) $f \to w$
                 (3) $p \to d$
Then we have: (3) $p \to d$ *given*
             (1) $d \to \sim w$ *given*
                $p \to \sim w$ *transitive law*
            (2′) $\sim w \to \sim f$ *contrapositive of* (2)
                $p \to \sim f$ *transitive law*

Conclusion: None of my poultry are officers. Here is another possibility: My poultry are not officers.

53. Let g: The government awards the contract to Airfirst Aircraft Company;
    s: Senator Firstair stands to earn a great deal of money;
    f: Airsecond Aircraft Company suffers financial setbacks.

Page 60

Given argument: Given argument: $(1)\ g \to s$

$(2)\ \sim f \to \sim s$

$(3)\ g$

Then we have: | | | |
|---|---|---|
| (1) | $g \to s$ | *given* |
| (3) | g | *given* |
| | s | *direct reasoning* |
| (2) | $\sim f \to \sim s$ | *given* |
| | s | *from above* |
| | f | *indirect reasoning* |

Conclusion: Airsecond Aircraft Company suffers financial setbacks.

Level 3 Problem Solving, page 115

55. The janitor could not have taken the elevator because the building's fuses were blown.

57. Let   *a*: This is a person who appreciates Beethoven;

s: This is a person who keeps silent while the *Moonlight Sonata* is played;

p: This person is a guinea pig;

i: This person is hopelessly ignorant of music. Given argument:

(1) No $a \to \sim s$ or $a \to s$

$(2)\ p \to i$

(3) No $i \to s$ or $i \to \sim s$

Then we have: | | | |
|---|---|---|
| (2) | $p \to i$ | *given* |
| (3) | $i \to \sim s$ | *given* |
| | $p \to \sim s$ | *transitive law* |
| $(1')$ | $\sim s \to \sim a$ | *contrapositive of* (1) |
| | $p \to \sim a$ | *transitive law* |

Thus, the valid conclusion is: Guinea pigs do not appreciate Beethoven.

59. Let *g*: I work a logic problem without grumbling. *u*: It is a logic problem I can understand. *r*: These problems are arranged in regular order like the problems I am used to. *e*: It is an easy problem. *h*: The problem makes my head ache.

Given argument: $(1)\ g \to u$

$(2)\ \sim r$

(3) no e is h or $e \to \sim h$

$(4)\ \sim r \to \sim u$

$(5)\ g$ unless h or $\sim h \to g$

Page 61

Then we have: (3) $e \rightarrow \sim h$ *given*
                (5) $\sim h \rightarrow g$ *given*
                    $e \rightarrow g$ *transitive law*
                (1) $g \rightarrow u$ *given*
                    $e \rightarrow u$ *transitive law*
                (4′) $u \rightarrow r$ *contrapositive of* (4)
                    $e \rightarrow r$ *transitive law*
                (2) $\sim r$ *given*
                    $\sim e$ *indirect reasoning*

Conclusion: This is not an easy problem.

3.5 Problem Solving Using Logic, page 116

Level 1, page 121

1. **a.** The playing board is an undefined term. The directions simply present the board, but do not define it.
   b. This is a stated rule of the game, so we would call this a axiom.
   c. Passing "GO" is an undefined term, because the rules of the game simply refer to GO.
   d. The "chance" cards are part of the definitions, so this is a defined term.
   e. This playing piece is simply given as part of the game, so it qualifies as an undefined term.

3. **a.** The playing pieces in *Sorry* are not specifically defined, so this would be an undefined term.
   b. This is a stated rule of the game, so we would call this a axiom.
   c. "Home" is a defined term in *Sorry*.
   d. This is a stated rule of the game, so we would call this a axiom.
   e. This term is not defined, so it is an undefined term.

5. **a.** This is a stated rule of the game, so we would call this a axiom.
   b. This is a stated rule of the game, so we would call this a axiom.
   c. It is assumed that you know what is meant by a "die," so we take this as an undefined term.
   d. A door's location is precisely located on the playing board, but the notion of a door itself is not defined, so this is an undefined term.
   e. This is a stated rule of the game, so we would call this a axiom.

7. a. While the number, size, and specifications of permitted players are all part of the rules, the word "player" is not defined, so this is an undefined term.
 b. This is a stated rule of the game, so we would call this a axiom.
 c. This is a defined term.
 d. This is a defined term.
 e. This is a consequence of the rules of the game, so this would be considered a theorem.

Level 2, page 122

9.

Statements	Reasons
$MIII$	Given
$MIIIIII$	Doubling (Rule 2)
MUU	Substitution (Rule 3)
M	Deletion (Rule 4)

11.

Statements	Reasons
MI	Given
MII	Doubling (Rule 2)
$MIIII$	Doubling (Rule 2)
$MIIIIIIII$	Doubling (Rule 2)
$MIUIU$	Substitution (Rule 3)

13.

Statements	Reasons
$MIIIUUIIIII$	Given
$MIIIIIIII$	Deletion (Rule 4)
$MIIUU$	Substitution (Rule 3)
MII	Deletion (Rule 4)
$MIIU$	Addition (Rule 1)

15.

Statements	Reasons
EA	Given
AE	Rule 3
AEE	Rule 1
A	Rule 5

17.

Statements	Reasons
AI	Given
AII	Rule 1
$AIII$	Rule 1
AE	Rule 4
AEE	Rule 1
A	Rule 5

19.

Statements	Reasons
AE	Given
AEE	Rule 1
A	Rule 5

21. The given statement is "*either p or q*"   where p: I am a knave; q: He is a knight. Every person would claim to be a knight, so p is a false statement. This forces the speaker to be a knight, so his statement is true. The first is a knight and the second is also a knight. Consider the truth table:

p	q	Either p or q	Comments
T	T	F	Not possible (p must be false).
T	F	T	Not possible (p must be false)
F	T	T	Since the first is not a knave, he must be telling the truth.
F	F	F	Not possible (since the speaker is a knight).

Page 63

23. Since every person would claim to be a knight, we know that Bob's statement is false. In other words, Bob is a knave. Next, we see that Cary's statement is true, so he is a knight.

25. White, since this could happen only at the North Pole.

27. Since there are only two colors, I need to choose three socks.

29. For at least one to be red, there must be 49 red and 1 yellow.

Level 3 Problem Solving, page 123

31. If Alice saw an ace, she would know she was not the only winner, so she would say, "I don't win." Since she did not see an ace, and since she has no information about her own card, she needed to say, "I don't know." This uncertainly tells Ben and Cole that they don't have aces on their heads. The only way they would *know* that they have lost is if they see an ace on Alice's head. The others have lower cards.

33. Follow the reasoning from the previous solution forces Ben and Cole to have kings, with Alice having a lower card.

35. Follow the reasoning from the previous two solutions followed by the conclusive statement by Alice that she loses, means that Ben will also say "I lose" with Cole concluding, "I win" with an queen on his head.

You should take Problems 37-54 together. The key is to begin with Problem 54 and work backward from there. *Begin with the innermost rectangle.*

We see the digits 1, 2, 3, and 4, but we do not see the digits yet to be filled in.

54. There is 1 four.

51. There must be 2 ones (since we have filled in a 1 in Problem 54.)

53. There is 2 threes (since we have filled in a 3 in Problem 52).

52. There are 3 twos (since we have filled in 2s in Problems 51 and 53).

Now we move the next rectangle.

48. There are 5 twos.

50. There are 2 fives.

49. There are 4 threes.

47. There are now 3 ones in this box.

The middle rectangle comes next.

46. There is 1 six.

44. There are 3 fours.

45. There are 3 fives.

43. There are now 7 threes.

Now consider the second-to-the-largest rectangle.

40. Counting, we see there are 5 ones.

Page 64

41. Counting, we see there are 6 twos.

42. There are now 3 sixes.

Finally, we are ready to tackle the largest rectangle.

39. Counting, we see there are 3 sevens.

38. Counting, there are 7 twos,

37. Counting, we see there are 6 ones. We now arrange these answers in the proper order.

37. 6 **39.** 3 **41.** 6 **43.** 7 **45.** 3 **47.** 3 **49.** 4 **51.** 2 **53.** 2

Level 3 Problem Solving, page 123

55. Moe constructed a truth table of all possibilities (underscore means hand is raised):

Harry: B B B W W W B W
Larry: B B W B W B W W
Moe: B W B B B W W W

Moe sees two hands and two black hats and concludes his hat must be black. If Moe's hat were white, then Harry and Larry would each have a solution because they would each see one white hat and one black hat with two hands raised. Thus, Moe knows that his hat must be black.

57. First, let's determine the order of shooting; Consider the following table:

	Order of shooting			
	1	2	3	4
Gary		C3	C1	No; 1
Harry	No; 1	No; 2		No; 3
Iggy	No; 3	C2	C1	
Jack	No; 1		C1	No; 2

Statement 1: Gary shot before Harry and Jack. That is, Gary did not shoot last and Harry and Jack did not shoot 1. These exclusions are shown as No; 1 in the table.

Statement 2: Jack shot before Harry. That is, Jack did not shoot last and Harry did not shoot second (first is already gone). These exclusions are shown as No; 2.

Statement 3: Iggy shot after Harry. That is, Harry did not shoot last, and Iggy did not shoot first. These exclusions are shown as No; 3

Conclusion 1: Harry is not 1, 2, or 4 so Harry shot third. This is shown as ▨. This means that the others can't be third. These exclusions as shown as C1

Conclusion 2: Since there is now one possibility for fourth place we see that Iggy finished fourth. This is shown as ▨. We now see that Iggy can't be second, so we show this as C2

Page 65

Conclusion 3: Finally, we note that Jack must be second, which we fill in as ▢. This means that Gary could not have finished second C3, so that means that Gary shot first.

Next, we form the following table of ages and numbers of marbles.

	Ages				Number of Marbles		
	3	10	17	18	9	12	15
Gary	No; A2	No. 7		No. 9	C5	C5	
Harry	No; A1			No. 9	No. 8		No. 5
Iggy		C4	C4	C4	No. 8		No. 6
Jack	No; 4					No. 8	No; 4

Statement 4: Jack was older than the boy with 15 marbles. Fill in this exclusion as No. 4.

Statement 5: Harry had fewer than 15 marbles. Fill in this exclusion as No. 5.

Assumption 1: Assume that Harry is the youngest.

	Ages				Number of Marbles		
	3	10	17	18	9	12	15
Gary	X				no		
Harry	A1	X	X	X	yes	X	No. 5
Iggy	X	no			no		
Jack	No; 4	yes*			no		No; 4

The result is a contradiction, so Harry is not the youngest.

Assumption 2: Assume that Gary is the youngest.

	Ages				Number of Marbles		
	3	10	17	18	9	12	15
Gary	A2	X	X	X	yes*		X
Harry	No; A1				X	yes	No. 5
Iggy	X						yes
Jack	No; 4						No; 4

The result is a contradiction, so Gary is not the youngest.

Conclusion 4: Iggy is the youngest. Mark this as ▢ and fill in the exclusions as C4.

Statement 6: The youngest boy does not have 15 marbles. Mark this as No. 6.

Page 66

Conclusion 5: This means that Gary has 15 marbles. Mark this as ☐ and then fill in the exclusions as C5 .

Statement 7:   The 10-year-old shot after the 17-year old, so Gary is not the 10-year old (he shot first). Mark this as No. 7

Statement 8:   Gary and Jack together have an even number of marbles. Since Gary has 15 marbles, Jack must have 9. Mark this as ☐ and then fill in the exclusions as No. 8 .

Conclusion 6: Therefore, Harry and Iggy each have 12 marbles. Mark this as ☐.

Statement 9:   Jack is older than the boy with 15 marbles, so Jack must be 18. Mark this as ☐ and fill in the exclusions as No. 9 .

Conclusion 7: This forces Gary to be 17 and Harry to be 10. Mark this as ☐.

Thus, Gary was 17 and had 15 marbles; Harry was 10 and had 12 marbles; Iggy was 3 and had 12 marbles; Jack was 18 and had 9 marbles. They shot in the order of Gary, Jack, Harry, and then Iggy.

59. The problem is that you do not know if the bad coin is either light or heavy. Weigh four coins against four coins, with four coins off the scales. Unbalanced proves that the four unweighed coins are good. The lower pan may have a heavy coin, or the upper pan a light one, but one of them *must* be a bad coin. Now, put two overweight suspect and one underweight suspect in each pan, leaving two underweight suspects off the scales. Unbalance leave the doubtful coins to the two overweight suspects from the descending pan, together with the underweight suspect from the ascending pan. One of these three must be a bad coin. Finally, weight the two remaining overweight suspects against each other. Unbalance identifies a heavy coin in the descending pan. Balance identifies the underweight coin as a light coin. Proceed similarly for other possibilities to conclude that in all cases the maximum number of weighings is three.

3.6 Logic Circuits, page 124
New Terms Introduced in this Section

AND-gate	Circuit	Gate	NOT-gate
OR-gate	Parallel circuit	Series circuit	

Level 1, page 128

1. The light is on when the circuit is complete (does not have a break).   **3.** The light is on when either of the switches are on.   **5.** light   **7.** AND-gate   **9.** NOT-gate
Answers to Problems 10-15 may vary.
11. $\sim p$ **13.** p or q; $p \vee q$ **15.** $p \wedge \sim q$

17.

p	q	$\sim p$	$\sim p \wedge q$
T	T	F	F
T	F	F	F
F	T	T	T
F	F	T	F

19.

p	q	$\sim q$	$p \vee \sim q$	$\sim (p \vee \sim q)$
T	T	F	T	F
T	F	T	T	F
F	T	F	F	T
F	F	T	T	F

Level 2, page 128

21. $p \vee q$

As a circuit: Using gates:

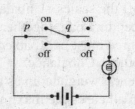

23. $p \wedge \sim q$

As a circuit: Using gates:

25. $\sim (p \wedge q)$

As a circuit: Using gates:

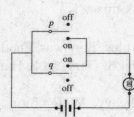

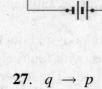

27. $q \rightarrow p$

p	q	$q \rightarrow p$
T	T	T
T	F	T
F	T	F
F	F	T

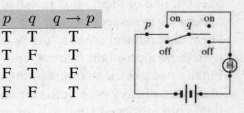

29. $q \rightarrow \sim p$; note, this is the contrapositive of Problem 28, so the truth table and circuits are the same.

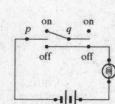

p	q	$\sim p$	$q \rightarrow \sim p$
T	T	F	F
T	F	F	T
F	T	T	T
F	F	T	T

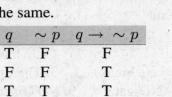

Page 68

31. $\sim (q \rightarrow p)$

p	q	$q \rightarrow p$	$\sim (q \rightarrow p)$
T	T	T	F
T	F	T	F
F	T	F	T
F	F	T	F

33. $(p \wedge q) \vee r$ **35.** $\sim (\sim p \vee q)$ **37.** $(\sim p \wedge q) \vee (p \vee \sim q)$

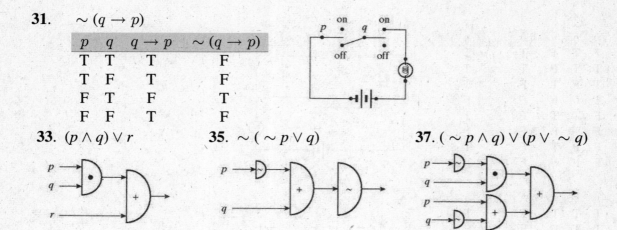

39.

p	q	r	$p \vee q$	$(p \vee q) \vee r$
T	T	T	T	T
T	T	F	T	T
T	F	T	T	T
T	F	F	T	T
F	T	T	T	T
F	T	F	T	T
F	F	T	F	T
F	F	F	F	F

41.

p	q	r	$q \wedge r$	$p \vee (q \wedge r)$
T	T	T	T	T
T	T	F	F	T
T	F	T	F	T
T	F	F	F	T
F	T	T	T	T
F	T	F	F	F
F	F	T	F	F
F	F	F	F	F

Page 69

43.

p	q	r	$p \wedge q$	$(p \wedge q) \vee r$
T	T	T	T	T
T	T	F	T	T
T	F	T	F	T
T	F	F	F	F
F	T	T	F	T
F	T	F	F	F
F	F	T	F	T
F	F	F	F	F

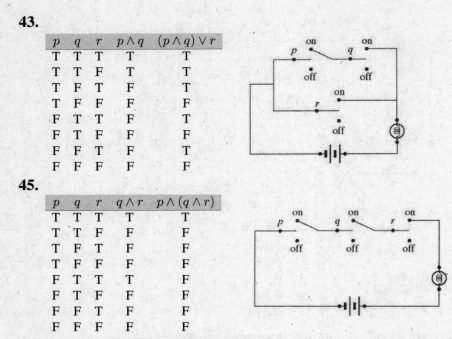

45.

p	q	r	$q \wedge r$	$p \wedge (q \wedge r)$
T	T	T	T	T
T	T	F	F	F
T	F	T	F	F
T	F	F	F	F
F	T	T	T	F
F	T	F	F	F
F	F	T	F	F
F	F	F	F	F

Note that the circuit for this problem is the same as the one for Problem 38. This fact is sometimes called the *associative property for conjunction*. That is,

$$(p \wedge q) \wedge r = p \wedge (q \wedge r)$$

Level 3, page 128

47. Let b_1: the first rest room is occupied;
            b_2: the second rest room is occupied;
            b_3: the third rest room is occupied.

a. $(b_1 \vee b_2) \vee b_3$; note that this is equivalent to $b_1 \vee (b_2 \vee b_3)$.

b.   **c.**

b_1	b_2	b_3	$b_1 \vee b_2$	$(b_1 \vee b_2) \vee b_3$
T	T	T	T	T
T	T	F	T	T
T	F	T	T	T
T	F	F	T	T
F	T	T	T	T
F	T	F	T	T
F	F	T	F	T
F	F	F	F	F

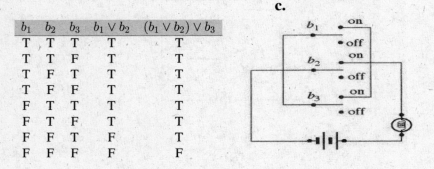

Page 70

49. Let a: the thermostat is set on automatic mode;
          d: all the doors are closed;
          m: the thermostat is set on manual mode;
          c: the proper authorization code has been entered.
   A logical statement might be: $(a \wedge d) \vee (m \wedge c)$.

51.

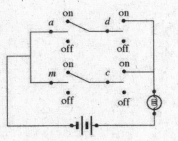

53.

p	q	$p \rightarrow q$	$\sim p$	$\sim p \vee q$	$(p \rightarrow q) \leftrightarrow (\sim p \vee q)$
T	T	T	F	T	T
T	F	F	F	F	T
F	T	T	T	T	T
F	F	T	T	T	T

$(p \rightarrow q) \Leftrightarrow (\sim p \vee q)$

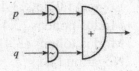

 55. Notice that $p \rightarrow \sim q \Leftrightarrow \sim p \vee \sim q$.

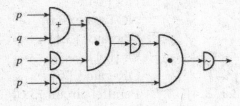

Level 3 Problem Solving, page 129

57. a.   **b.** $140,000

Page 71

59. Let the committee members be a, b, and c, respectively. Light *on* represents a majority. The circuit is shown below. Light is on for a yea majority and light is off otherwise.

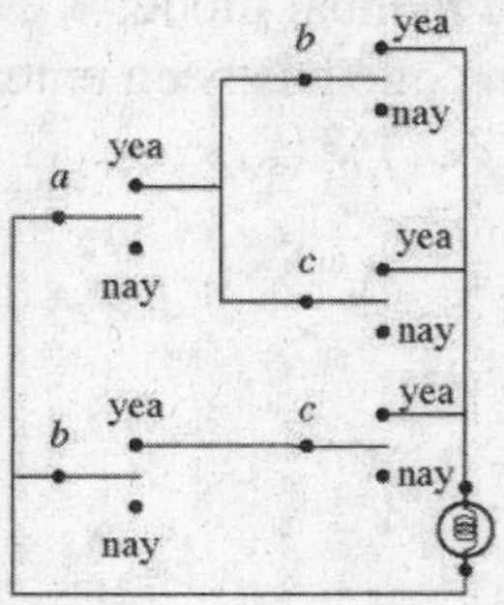

Chapter 3 Review Questions, page 129

STOP

Studying for a chapter examination is a personal process, one which nobody else can do for you. Simply take the time to review what you have done. Here are the new terms in Chapter 3.

And [3.1]	De Morgan's laws [3.3]	Inverse [3.2]
AND-gate [3.6]	Denying the antecedent	Law of contraposition [3.2]
Antecedent [3.2]	(fallacy) [3.4]	Law of detachment [3.4]
Argument [3.1]	Denying the consequent [3.4]	Law of double negation [3.2]
Assuming the antecedent [3.4]	Direct reasoning [3.4]	Law of the excluded middle [3.1]
Assuming the consequent	Disjunction [3.1]	Logic [3.1]
(fallacy) [3.4]	Either ... or [3.1]	Logical equivalence [3.3]
Because [3.3]	Exclusive or [3.1]	Logical fallacy [3.4]
Biconditional [3.3]	Fallacy [3.4]	Modus ponens [3.4]
Circuit [3.6]	Fallacy of the converse [3.4]	Modus tollens [3.4]
Compound statement [3.1]	Fallacy of the inverse [3.4]	Negation [3.1]
Conclusion [3.1]	False chain pattern [3.4]	Negation of a conditional [3.3]
Conditional [3.2]	Fundamental operators [3.2]	Neither ... nor [3.3]
Conjunction [3.1]	Fuzzy logic [3.1]	No p is q [3.3]
Connective [3.1]	Gate [3.6]	Not [3.1]
Consequent [3.2]	Hypothesis [3.1]	NOT-gate [3.6]
Contrapositive [3.2]	Implication [3.3]	Operator [3.1]
Converse [3.2]	Inclusive or [3.1]	Or [3.1]
Counterexample [3.4]	Indirect reasoning [3.4]	OR-gate [3.6]
Deductive reasoning [3.1]	Invalid argument [3.1, 3.4]	Parallel circuit [3.6]

Premise [3.1]	Symbolic logic [3.1]	Truth table [3.2]
Series circuit [3.6]	Tautology [3.3]	Truth value [3.1]
Simple statement [3.1]	Theorem [3.4]	Unless [3.3]
Statement [3.1]	Transitive reasoning [3.4]	Valid argument [3.1]
Syllogism [3.4]	Truth set [3.3]	

If you can describe the term, read on to the next one; if you cannot, then look it up in the text (the section number is shown in brackets). Next, review the types of problems in Chapter 3.

TYPES OF PROBLEMS
Determine whether a sentence is a statement. [3.1]
Write the negation of *all, some,* and *not*. [3.1]
Find truth value of simple and compound statements. [3.1]
Translate statements into symbolic form. [3.1-3.3]
Translate symbolic form into verbal statements. [3.2, 3.3]
Construct a truth table for a given symbolic form. [3.2]
Apply the definition of the conditional. [3.2]
Write the converse, inverse, and contrapositive for a given statement. [3.2]
Determine whether a given symbolic statement is true or false. [3.2]
Decide whether a given statement is a tautology. [3.3]
Write an implication as a disjunction. [3.3]
Write the negation of a compound statement. [3.3]
Given certain real-life premises reach conclusions. [3.3, 3.4]
Be able to recognize, state, and prove valid forms of reasoning, namely direct reasoning, indirect reasoning, and transitive reasoning. [3.4]
Prove logical statements. [3.3, 3.4]
Determine whether a given argument is valid or invalid. [3.4]
Classify valid forms of reasoning and recognize common fallacies. [3.4]
Find a valid conclusion for a given argument. [3.4]
Classify items as an undefined term, defined term, axiom, or theorem. [3.5]
Prove simple theorems using given definitions and axioms. [3.5]
Solve logical puzzles. [3.5]
Design a circuit to simulate truth values of a given logical statement. [3.6]
Use gates to design a circuit. [3.6]

Once again, see if you can verbalize (to yourself) how to do each of the listed types of problems.
    Work all of Chapter 3 Review Questions (whether they are assigned or not). Work through all of the problems before looking at the answers, and *then* correct each of the problems. The

entire solution is shown in the answer section at the back of the text. If you worked the problem correctly, move on to the next problem, but if you did not work it correctly (or you did not know what to do), look back in the chapter to study the procedure, or ask your instructor.

Finally, go back over the homework problems you have been assigned. If you worked a problem correctly, move on to the next problem, but if you missed it on your homework, then you should look back in the book or talk to your instructor about how to work the problem.

If you follow these steps, you should be successful with your review of this chapter.

We give all of the answers to the chapter review questions (not just the odd-numbered questions).

Chapter 3 Review Questions, page 130

1. a. A *logical statement* is a declarative sentence that is either true or false.
    b. A *tautology* is a logical statement in which the conclusion is equivalent to its premise.
    c. $(p \to q) \leftrightarrow (\sim q \to \sim p)$ or, in words: A conditional may always be replaced by its contrapositive without having its truth value affected.

2.

p	q	$\sim p$	$p \wedge q$	$p \vee q$	$p \to q$	$p \leftrightarrow q$
T	T	F	T	T	T	T
T	F	F	F	T	F	F
F	T	T	F	T	T	F
F	F	T	F	F	T	T

3.

p	q	$p \wedge q$	$\sim (p \wedge q)$
T	T	T	F
T	F	F	T
F	T	F	T
F	F	F	T

4.

p	q	$\sim q$	$p \vee \sim q$	$\sim p$	$(p \vee \sim q) \wedge \sim p$	$[(p \vee \sim q) \wedge \sim p] \to \sim q$
T	T	F	T	F	F	T
T	F	T	T	F	F	T
F	T	F	F	T	F	T
F	F	T	T	T	T	T

Page 74

5.

p	q	r	$p \wedge q$	$(p \wedge q) \wedge r$	$[(p \wedge q) \wedge r] \to p$
T	T	T	T	T	T
T	T	F	T	F	T
T	F	T	F	F	T
T	F	F	F	F	T
F	T	T	F	F	T
F	T	F	F	F	T
F	F	T	F	F	T
F	F	F	F	F	T

6.

p	q	$\sim p$	$\sim q$	$p \wedge q$	$\sim(p \wedge q)$	$\sim p \vee \sim q$	$\sim(p \wedge q) \leftrightarrow (\sim p \vee \sim q)$
T	T	F	F	T	F	F	T
T	F	F	T	F	T	T	T
F	T	T	F	F	T	T	T
F	F	T	T	F	T	T	T

7.

$p \to q$
p
_____
$\therefore q$

p	q	$p \to q$	$(p \to q) \wedge p$	$[(p \to q) \wedge p] \to q$
T	T	T	T	T
T	F	F	F	T
F	T	T	F	T
F	F	T	F	T

8. Answers vary.

If you study hard, then you will get an *A*.
You do not get an *A*.
Therefore, you did not study hard.

9. Answers vary; fallacy of the converse, fallacy of the inverse, or false chain pattern; for example:

If you study hard, then you will get an *A*.
You do not study hard, so you do not get an *A*.   *(Fallacy of the inverse)*

10. Yes; it is indirect reasoning.

11. a. Some birds do not have feathers.
b. No apples are rotten.
c. Some cars have two wheels.

© 2012 Cengage Learning. All Rights Reserved. May not be scanned, copied or duplicated, or posted to a publicly accessible website, in whole or in part.

   d. All smart people attend college.
   e. You go on Tuesday and you can win the lottery.

12. a. T   (F → F)   **b.** T   (F → T)   **c.** T   (T → T)   **d.** T   (F → F)   **e.** T   (F → F)

13. a. If P is a prime number, then $P + 2$ is a prime number.
    b. Either P or $P + 2$ is a prime number.

14. a. Let p: This machine is a computer; q: This machine is capable of self-direction.
    $$p \rightarrow \; \sim q$$
    b. Contrapositive:
    $$\sim (\sim q) \; \rightarrow \; \sim p; q \; \rightarrow \; \sim p.$$
    If this machine is capable of self-direction, then it is not a computer.

15. Let p: There are a finite number of primes; q: There is some natural number, greater than 1, that is not divisible by any prime. Then the argument in symbolic form is:
    $$p \rightarrow q$$
    $$\underline{\sim q}$$
    $$\therefore \; \sim p$$

Conclusion: There are infinitely many primes (assuming that "not a finite number" is the same as "infinitely many"). This is indirect reasoning.

16. a.   **b.**

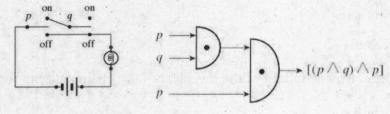

17. Let d: I attend to my duties; r: I am rewarded; ℓ: I am lazy.   Symbolic argument:
   (1) $d \rightarrow r$ Given
   (2) $\ell \rightarrow \; \sim r$ Given
   (3) ℓ Given
   (4) $\therefore \; \sim r$ Direct reasoning (steps 2 and 3)
   (5) $\therefore \; \sim d$ Indirect reasoning (steps 1 and 4)
   Conclusion: I do not attend to my duties.

Page 76

18. Let o: This is organic food; h: This is healthy food; s: This is an artificial sweetener;
p: This is a prune. Symbolic argument:

(1) $o \rightarrow h$ Given
(2) $s \rightarrow \sim h$ Given
(3) $p \rightarrow o$ Given
(4) $\therefore p \rightarrow h$ Transitive (steps 3 and 1)
($2'$) $h \rightarrow \sim s$ Law of contraposition (contrapositive of step 2)
(5) $\therefore p \rightarrow \sim s$ Transitive (steps 4 and $2'$)

Conclusion: No prune is an artificial sweetener.

19. Let s: This is a square; r: This is a rectangle; q: This is a quadrilateral; p: This is a
polygon. Symbolic argument:

(1) $s \rightarrow r$ Given
(2) $r \rightarrow q$ Given
(3) $\therefore s \rightarrow q$ Transitive (steps 1 and 2)
(4) $q \rightarrow p$ Given
(5) $\therefore s \rightarrow p$ Transitive (steps 3 and 4)

Conclusion: All squares are polygons.

20. Let us number the statements:

(1) Josie is the department chairperson.
(2) The members of the department are Josie, Maureen, Terry, and Warren.
(3) They sit at a square table, which we label the sides as follows:

(4) The hobbies are baking, gardening, hiking, and surfing.
(5) Josie sits to Terry's left. There is more than one possibility, but if we disregard a
rotation the table looks like:

(6) Josie's hobby is gardening;

(7) Warren faces Terry.
(8) Warren is not the baker.

Page 77

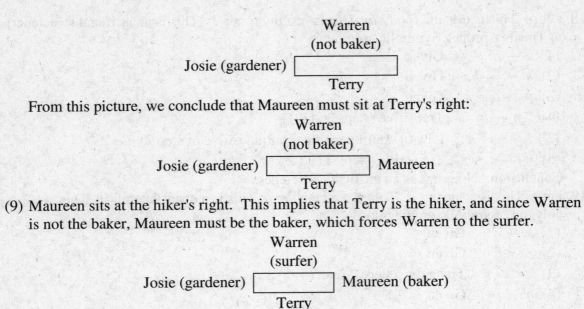

Warren
(not baker)

Josie (gardener) [   ]

Terry

From this picture, we conclude that Maureen must sit at Terry's right:

Warren
(not baker)

Josie (gardener) [   ] Maureen

Terry

(9) Maureen sits at the hiker's right. This implies that Terry is the hiker, and since Warren is not the baker, Maureen must be the baker, which forces Warren to the surfer.

Warren
(surfer)

Josie (gardener) [   ] Maureen (baker)

Terry
(hiker)

Now remember that there are four possibilities (due to rotations), and you can pick any one of them as the correct answer.

North	Warren	Maureen	Terry	Josie
East	Maureen	Terry	Josie	Warren
South	Terry	Josie	Warren	Maureen
West	Josie	Warren	Maureen	Terry

CHAPTER 4 The Nature of Numeration Systems

Chapter Overview
This chapter is concerned with an investigation of the way we have chosen to represent the infinite set of numbers. Remember, the *number* is a concept and the *numeral* is the way we choose to represent that number. For example, there is only one *number* seven, but that number may be represented by the symbol "7" or "111" in different contexts. We conclude this chapter with an optional section which gives an interesting look at the history of computers and calculating devices.

4.1 Early Numeration Systems, page 135
New Terms Introduced in this Section

Addition principle	Counting number	Multiplication principle	Number
Numeral	Numeration system	Positional system	Repetitive system
Simple grouping system	Subtraction principle		

Level 1, page 143

1. A *number* is a concept, whereas a *numeral* is the symbol used to represent a number.

3. Answers vary; among the Egyptian numerations systems's shortcomings are lack of a zero symbol, and the fact that it is difficult and cumbersome to write larger numbers. It is important because it was based on simple groupings (in powers of ten), which is what we use today in our own numeration system.

5. Answers vary; among the Babylonian numerations system's shortcomings is the fact that having only two symbols is difficult, and for most of its history it lacked a zero symbol. It is important because it was a positional numeration system, which is what we use today in our own numeration system. This idea makes it easier to write large numbers.

7. **a**. Egyptian
   b. Roman (because IX and XI represent different numbers), Babylonian
   c. Egyptian, Roman, Babylonian
   d. Egyptian, Roman, Babylonian
   e. Roman, Babylonian
   f. Roman

9. **a**. This is a unique symbol in the Egyptian numeration system; it is $\frac{2}{3}$.
   b. The pollywog represents 100,000 and the heel bone represents 10, so the number is 100,010.

11. a. The two heel bones each represents ten plus four strokes, and a symbol for $\frac{1}{2}$; thus, we have $20 + 4 + \frac{1}{2} = 24\frac{1}{2}$.

   b. The pollywog represents 100 and the oval above makes it a unit fraction; thus

$$\frac{\overset{\frown}{9}}{} = \frac{1}{100}.$$

13. a. The lotus flower represents 1,000 and the scroll represents 100, so the value is 1,100.

   b. Translate the symbols one-by-one and add their values:

1 lotus flower:	$1 \times 1,000 =$	1,000
9 scrolls:	$9 \times 100 =$	900
9 heel bones:	$9 \times 10 =$	90
7 strokes:	$7 \times 1 =$	7
	TOTAL:	$= 1,997$

15. a. $M = 1,000$, $CM = 900$, $XC = 90$, and $IX = 9$, so the number is $1,000 + 900 + 90 + 9 = 1,999$

   b. $M = 1,000$ and $I = 1$, so the number is $1,000 + 1,000 + 1 = 2,001$

17. a. A bar placed over a Roman numeral (or a group of numerals) means 1,000 times the value of the Roman numeral themselves. Thus, since $V = 5$, we have $\overline{V} = 5,000$, $M = 1,000$, $D = 500$, and $C = 100$, so the number is $5,000 + 1,000 + 1,000 + 500 + 100 = 7,600$.

   b. Since $IX = 9$, $D = 500$, $CC = 200$, $X = 10$, and $II = 2$, so we have $\overline{IX}DCCXII = (9 \times 1,000) + 500 + 200 + 10 + 2 = 9,712$.

19. a. $\blacktriangledown \triangleleft \blacktriangledown = (1 \times 60) + 10 + 1 = 71$

   b. $\triangleleft \blacktriangledown \triangleleft \blacktriangledown = [(10 + 1) \times 60] + (10 + 1) = 11 \times 60 + 11 = 660 + 11 = 671$

21. a. Note the use of the Babylonian subtractive symbol:

$$\triangleleft\triangleleft\triangleleft \atop \triangleleft\triangleleft \quad \overline{|\blacktriangledown\blacktriangledown\blacktriangledown} = (10 + 10 + 10 + 10 + 10) - (1 + 1 + 1) = 47$$

   b. $\triangleleft\triangleleft \;\; {\blacktriangledown\,\blacktriangledown \atop \blacktriangledown\,\blacktriangledown} \; \blacktriangledown = (10 + 10) + (1 + 1 + 1 + 1 + 1) = 25$

23. $\cap \cap \cap||||| = 34$ and $\triangleleft\triangleleft\triangleleft\triangleleft\triangleleft \blacktriangledown\blacktriangledown\blacktriangledown\blacktriangledown = 54$, so the total is $45 + 34 + 54 = 133$.

25. One; I am the only one known to be going to St. Ives.

Level 2, page 143

27. $75 = (7 \times 10) + (5 \times 1) = \cap\cap\cap\cap\cap\cap\cap|||||$

29. $521 = (5 \times 100) + (2 \times 10) + 1 = 99999 \cap \cap |$

31. $2,007 = (2 \times 1,000) + (7 \times 1) = $ ⚸ ⚸ $|||||\; ||$

33. $75 = 50 + (2 \times 10) + 5 = $ LXXV

35. $521 = 500 + (2 \times 10) + 1 = $ DXXI

37. $2{,}007 = (2 \times 1{,}000) + 5 + (2 \times 1) = $ MMVII

39. $75 = 60 + 10 + (5 \times 1) = $ ▼◁▼▼▼▼▼

41. $521 = (8 \times 60) + (4 \times 10) + 1 = $ ▼▼▼▼▼▼▼▼◁◁◁◁▼

43. $2{,}007 = (33 \times 60) + (2 \times 10) + (7 \times 1) = $ ◁◁◁▼▼▼◁◁▼▼▼▼▼▼▼

45. Write 𝄞𝄞∩I as 𝄞999999999 ∩ ∩ ∩ ∩ ∩ ∩ ∩ ∩ ∩IIIII IIIII I so that we can subtract the symbols ∩ ∩IIII (as shown). The left-over symbols are: 𝄞999999999∩∩∩∩∩∩∩IIIIII.

47. Add ◁▼▼▼▼▼▼ to ◁◁◁▼▼▼ to obtain
◁▼▼▼▼▼▼▼◁◁◁▼▼▼ = ◁ + ◁ + ◁ + ◁ + (▼▼▼▼▼▼▼▼▼) = ◁◁◁◁◁

49. Write ▼◁▼▼ as ◁◁◁◁◁◁◁▼▼▼▼▼▼▼▼▼▼▼▼ so that we can subtract the symbols ◁◁◁◁▼▼▼▼▼▼. The left-over symbols are: ◁◁▼▼▼▼▼

Note: since the overbar symbol is an operation symbol and not part of the basic set of numerals, you cannot use it in answering this question.

51. VIII $= 8$ **53.** LXXXIX $= 89$ **55.** MMMCMXCIX $= 3{,}999$

Level 3 Problem Solving, page 144

57. a. MDCLXVI $= 1{,}000 + 500 + 100 + 50 + 10 + 5 + 1 = 1{,}666$

Note: since the overbar symbol is an operation symbol and not part of the basic set of numerals, you cannot use it in answering this question.

b. MCDXLIV $= 1{,}000 + (500 - 100) + (50 - 10) + (5 - 1) = 1{,}444$

59.

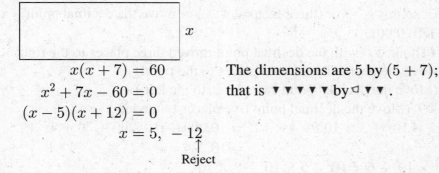

$$x(x + 7) = 60 \qquad \text{The dimensions are 5 by } (5 + 7);$$
$$x^2 + 7x - 60 = 0 \qquad \text{that is ▼▼▼▼▼ by◁▼▼}.$$
$$(x - 5)(x + 12) = 0$$
$$x = 5, \ -12$$
$$\uparrow$$
$$\text{Reject}$$

4.2 Hindu-Arabic Numeration System, page 144

New Terms Introduced in this Section

Decimal numeration system	Decimal point	Expanded notation
Hindu-Arabic numerals	Hundred	Ten

Page 81

Level 1, page 147

1. A *number* is a concept which answers the question "how many" and a *numeral* is the symbol used to represent a number.

3. An abacus is an ancient mechanical device showing the number of ones, tens, hundreds, etc, which is exactly the same thing that we do when we write a number in expanded notation.

5. We represent the numbers by dots separated into groups of ten:

7. $b^n = b \cdot b \cdot \cdots \cdot b$ where there are n factors of b (for any counting number n)

9. The "5" in 805 means 5 units.

11. The "5" in 0.00765 means 5 hundred-thousands.

13. The "5" in 0.00567 means 5 thousandths.

15. a. The exponent is 5, so this is $10 \times 10 \times 10 \times 10 \times 10$ (five factors of 10) or we know the result is "1" followed by five zeros: 100,000. Alternatively, you can view this as 1.0 with the decimal point moved five places to the right.

   b. The exponent is 3, so this is $10 \times 10 \times 10$ (three factors of 10) or move the decimal point three places to the right: 1,000.

17. a. The exponent is -4, so this is $\frac{1}{10^4}$ or (four factors of $\frac{1}{10}$) or move the decimal point four places to the left: 0.0001.

   b. The exponent is -3, so this is $\frac{1}{10^3}$ or (three factors of $\frac{1}{10}$) or move the decimal point three places to the left: 0.001.

19. a. $5 \times 10^3 = 5,000$ (Think: 5.0 with the decimal point moved three places to the right.)

   b. $5 \times 10^2 = 500$ (Move the decimal point two places to the right.)

21. a. $6 \times 10^{-2} = 0.06$ (Move the decimal point two places to the left.)

   b. $9 \times 10^{-5} = 0.00009$ (Move the decimal point five places to the left.)

23. $1 \times 10^4 + 0 \times 10^3 + 2 \times 10^2 + 3 \times 10^1 + 4 \times 10^0 = 10,000 + 0 + 200 + 30 + 4$
$$= 10,234$$

25. $5 \times 10^5 + 2 \times 10^4 + 1 \times 10^3 + 6 \times 10^2 + 5 \times 10^1 + 8 \times 10^0$
$$= 500,000 + 20,000 + 1,000 + 600 + 50 + 8$$
$$= 521,658$$

27. $7 \times 10^6 + 3 \times 10^{-2} = 7{,}000{,}000 + 0.03$
$$= 7{,}000{,}000.03$$

29. $5 \times 10^5 + 4 \times 10^2 + 5 \times 10^1 + 7 \times 10^0 + 3 \times 10^{-1} + 4 \times 10^{-2}$
$$= 500{,}000 + 400 + 50 + 7 + 0.3 + 0.04$$
$$= 500{,}457.34$$

31. $2 \times 10^4 + 6 \times 10^2 + 4 \times 10^{-1} + 7 \times 10^{-3} + 6 \times 10^{-4} + 9 \times 10^{-5}$
$$= 20{,}000 + 600 + 0.4 + 0.007 + 0.0006 + 0.00009$$
$$= 20{,}600.40769$$

33. a. $0.096421 = 0.0 + 0.09 + 0.006 + 0.0004 + 0.00002 + 0.000001$
$$= 9 \times 10^{-2} + 6 \times 10^{-3} + 4 \times 10^{-4} + 2 \times 10^{-5} + 1 \times 10^{-6}$$

   b. $27.572 = 20 + 7 + 0.5 + 0.07 + 0.002$
$$= 2 \times 10^1 + 7 \times 10^0 + 5 \times 10^{-1} + 7 \times 10^{-2} + 2 \times 10^{-3}$$

35. a. $6{,}245 = 6{,}000 + 200 + 40 + 5$
$$= 6 \times 10^3 + 2 \times 10^2 + 4 \times 10^1 + 5 \times 10^0$$

   b. $2{,}305{,}681 = 2{,}000{,}000 + 300{,}000 + 5{,}000 + 600 + 80 + 1$
$$= 2 \times 10^6 + 3 \times 10^5 + 5 \times 10^3 + 6 \times 10^2 + 8 \times 10^1 + 1 \times 10^0$$

37. $0.00000527 = 0.00000\,5 + 0.00000\,02 + 0.00000\,007$
$$= 5 \times 10^{-6} + 2 \times 10^{-7} + 7 \times 10^{-8}$$

39. $893.0001 = 800 + 90 + 3 + 0.0001$
$$= 8 \times 10^2 + 9 \times 10^1 + 3 \times 10^0 + 1 \times 10^{-4}$$

41. $678{,}000.01 = 600{,}000 + 70{,}000 + 8{,}000 + 0.01$
$$= 6 \times 10^5 + 7 \times 10^4 + 8 \times 10^3 + 1 \times 10^{-2}$$

43. $57{,}285.9361 = 50{,}000 + 7{,}000 + 200 + 80 + 5 + 0.9 + 0.03 + 0.006 + 0.0001$
$$= 5 \times 10^4 + 7 \times 10^3 + 2 \times 10^2 + 8 \times 10^1 + 5 \times 10^0$$
$$+ 9 \times 10^{-1} + 3 \times 10^{-2} + 6 \times 10^{-3} + 1 \times 10^{-4}$$

Level 2, page 148

45. Each column on an abacus has meaning, and it might help if you label them:

$10^6 \; 10^5 \; 10^4 \; 10^3 \; 10^2 \; 10^1 \; 10^0$

We see that the number represented is: $3{,}000 + 200 + 1 = 3{,}201$.

47. Consider:

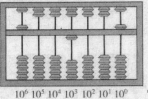

$$10^6 \ 10^5 \ 10^4 \ 10^3 \ 10^2 \ 10^1 \ 10^0$$

The number is: $5,000,000 + 1,000 + 5 = 5,001,005$.

49. Consider:

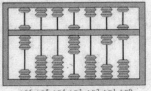

$$10^6 \ 10^5 \ 10^4 \ 10^3 \ 10^2 \ 10^1 \ 10^0$$

$$5,000,000 + 3,000,000 + 5,000 + 4,000 + 20 + 5 + 1 = 8,009,026$$

51. $849 = 8 \times 10^2 + 4 \times 10^1 + 9 \times 10^0$

$$= (5+3) \times 10^2 + 4 \times 10^1 + (5+4) \times 10^0$$

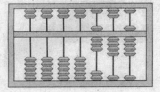

53. $9,387 = 9 \times 10^3 + 3 \times 10^2 + 8 \times 10^1 + 7 \times 10^0$

$$= (5+4) \times 10^3 + 3 \times 10^2 + (5+3) \times 10^1 + (5+2) \times 10^0$$

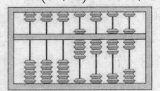

55. $2,001 = 2 \times 10^3 + 1 \times 10^0$

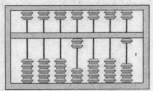

57. $8,007,009 = 8 \times 10^6 + 7 \times 10^3 + 9 \times 10^0$

$$= (5+3) \times 10^6 + (5+2) \times 10^3 + (5+4) \times 10^0$$

Page 84

Level 3, Problem Solving, page 148
59. 0, 1, 2, 3, 4, 10, 11, 12, 13, 14, 20, 21, **22, 23, 24, 30, 31, 32, 33, 34, 40,** $\cdots$.

4.3 Different Numeration Systems, page 148
New Terms Introduced in this Section

Base b numeration system   Binary numeration system   Hexadecimal system
Octal numeration system

Level 1, page 152
1. Write the base eight number in expanded notation, and then carry out the arithmetic.

3. Do repeated division by eight, and then read the remainders down for the base 8 numbers.

5. **a.** There are 9 people shown.
   b. There is one group of five with four left over: 14_{five}
   c. There is one group of nine with no groups of three or any others left over: 100_{three}
   d. There is one group of eight with one left over: 11_{eight}
   e. There is one group of eight, no groups of four or two, but there is one left over: 1001_{two}
   f. There is one group of nine: 10_{nine}

7. $643_{eight} = 6 \times 8^2 + 4 \times 8^1 + 3 \times 8^0$

9. $110111.1001_{two} = 1 \times 2^5 + 1 \times 2^4 + 1 \times 2^2 + 1 \times 2^1 + 1 \times 2^0 + 1 \times 2^{-1} + 1 \times 2^{-4}$

11. $64200051_{eight} = 6 \times 8^7 + 4 \times 8^6 + 2 \times 8^5 + 5 \times 8^1 + 1 \times 8^0$

13. $323000.2_{four} = 3 \times 4^5 + 2 \times 4^4 + 3 \times 4^3 + 2 \times 4^{-1}$

15. $3.40231_{five} = 3 \times 5^0 + 4 \times 5^{-1} + 2 \times 5^{-3} + 3 \times 5^{-4} + 1 \times 5^{-5}$

17. $527_{eight} = 5 \times 8^2 + 2 \times 8^1 + 7 \times 8^0$
$$= 320 + 16 + 7$$
$$= 343$$

19. $25TE_{twelve} = 2 \times 12^3 + 5 \times 12^2 + 10 \times 12^1 + 11 \times 12^0$
$$= 3{,}456 + 720 + 120 + 11$$
$$= 4{,}307$$

21. $431_{five} = 4 \times 5^2 + 3 \times 5^1 + 1 \times 5^0$
$$= 100 + 15 + 1$$
$$= 116$$

23. $1011.101_{two} = 1 \times 2^3 + 0 \times 2^2 + 1 \times 2^1 + 1 \times 2^0 + 1 \times 2^{-1} + 0 \times 2^{-2} + 1 \times 2^{-3}$

$$= 8 + 0 + 2 + 1 + \frac{1}{2} + 0 + \frac{1}{8}$$

$$= 11.625$$

25. $573_{twelve} = 5 \times 12^2 + 7 \times 12^1 + 3 \times 12^0$

$$= 720 + 84 + 3$$

$$= 807$$

27. $2110_{three} = 2 \times 3^3 + 1 \times 3^2 + 1 \times 3^1 + 0 \times 3^0$

$$= 54 + 9 + 3 + 0$$

$$= 66$$

29. $537.1_{eight} = 5 \times 8^2 + 3 \times 8^1 + 7 \times 8^0 + 1 \times \frac{1}{8}$

$$= 320 + 24 + 7 + \frac{1}{8}$$

$$= 351.125$$

31. a.   **b.**

         0 r. 1   0 r. 2

   5)$\overline{1}$ r. 0   4)$\overline{2}$ r. 1

   5)$\overline{5}$ r. 3   4)$\overline{9}$ r. 3

   5)$\overline{2\,8}$ r. 4   4)$\overline{39}$ r. 1

   5)$\overline{144}$ r. 4   4)$\overline{157}$ r. 0

   5)$\overline{724}$ $\leftarrow$ *Start here (work up, read answer down)* $\rightarrow$ 4)$\overline{628}$

   $724 = 10344_{five}$ *Answers* $628 = 21310_{four}$

33. a.   **b.**

         0 r. 3   0 r. 1

   5)$\overline{3}$ r. 1   8)$\overline{1}$ r. 1

   5)$\overline{16}$ r. 2   8)$\overline{9}$ r. 4

   5)$\overline{82}$ r. 2   8)$\overline{76}$ r. 7

   5)$\overline{412}$ $\leftarrow$ *Start here (work up, read answer down)* $\rightarrow$ 8)$\overline{615}$

   $412 = 3122_{five}$ *Answers* $615 = 1147_{eight}$

35. a.   **b.**

$$
\begin{array}{r}
0 \quad \text{r. }1 \\
2\overline{)\ 1}\ \ \text{r. }0 \\
2\overline{)\ 2}\ \ \text{r. }0 \\
2\overline{)\ 4}\ \ \text{r. }0 \\
2\overline{)\ 8}\ \ \text{r. }0 \\
2\overline{)\ 16}\ \ \text{r. }0 \\
2\overline{)\ 32}\ \ \text{r. }0 \\
2\overline{)\ 64}\ \ \text{r. }0 \\
2\overline{)128}\ \ \text{r. }0 \\
2\overline{)256}\ \ \text{r. }0 \\
2\overline{)512}
\end{array}
$$

$$512 = 1000000000_{two}$$

$$
\begin{array}{r}
0 \quad \text{r. }1 \\
3\overline{)\ 1}\ \ \text{r. }0 \\
3\overline{)\ 3}\ \ \text{r. }0 \\
3\overline{)\ 9}\ \ \text{r. }2 \\
3\overline{)\ 29}\ \ \text{r. }1 \\
3\overline{)\ 88}\ \ \text{r. }1 \\
3\overline{)265}\ \ \text{r. }0 \\
3\overline{)795}
\end{array}
$$

← *Start here* →

Answers

$$795 = 1002110_{three}$$

37. a.   **b.**

$$
\begin{array}{r}
0 \ \text{r. }1 \\
8\overline{)\ 1}\ \text{r. }1 \\
8\overline{)\ 9}\ \text{r. }3 \\
8\overline{)\ 75}\ \text{r. }2 \\
8\overline{)602}
\end{array}
$$

$$602 = 1132_{eight}$$

← *Start here (work up, read answer down)* →   *Answers*

$$
\begin{array}{r}
0 \ \text{r. }1 \\
4\overline{)\ 1}\ \text{r. }0 \\
4\overline{)\ 4}\ \text{r. }3 \\
4\overline{)19}\ \text{r. }0 \\
4\overline{)76}
\end{array}
$$

$$76 = 1030_{four}$$

Level 2, page 153

39. We need to change 52 to base 7:

$$52 = 7 \times 7 + 3$$

7 weeks, 3 days

41. We need to change 55 to base 12:

$$55 = 4 \times 12 + 7$$

4 ft, 7 in.

43. We need to change 500 to base 12:

$$500 = 3 \times 144 + 5 \times 12 + 8$$

3 gross, 5 doz, 8 units

45. We need to change 84 to base 5:

$$84 = 3 \times 25 + 1 \times 5 + 4$$

$84 = 314_{five}$ so you would need 8 coins.

Page 87

47. We need to change 954_{twelve} to base 10:
$$954_{twelve} = 9 \times 12^2 + 5 \times 12^1 + 4 \times 12^0$$
$$= 1,296 + 60 + 4$$
$$= 1,360$$
The bookstore ordered a total of 1,360 pencils.

49. $44 = 6 \times 7 + 2$
$= 62_{seven};$ 6 weeks and 2 days

51. $29 = 1 \times 24 + 5$
$= 15_{twenty\text{-}four};$ 1 day and 5 hours

53.   3 years   10 months
<u> 2 years   5 months</u>
  5 years   15 months = 6 years 3 months

55.   6 ft   8 in.
<u> 9 ft   5 in.</u>
  15 ft   13 in. = 16 ft 1 in.

57.   1 gross   9 doz   7 units
<u> 2 gross   8 doz   8 units</u>
  3 gross 17 doz   15 units = 3 gross + (1 gross + 5 doz) + (1 doz + 3 units).
                        = 4 gross + 6 doz + 3 units

59.   3 years   4 months   21 days
<u> 1year   6 months   15 days</u>
  4 years   10 months   36 days = 4 years, 11 months, 6 days

4.4 Binary Numeration Systems, page 154
New Terms Introduced in this Section

ASCII code   Binary numeration system   Bit   Byte
Octal numeration system (problem set)

Level 1, page 157

1. Figure 4.3 is an example of a two-state device which is used to illustrate how computers operate. In particular, if a light is on it represents a "1" and if it is off it represents a "0".

3. $0 \times 2^7 + 0 \times 2^6 + 1 \times 2^5 + 0 \times 2^4 + 0 \times 2^3 + 1 \times 2^2 + 1 \times 2^1 + 1 \times 2^0 = 39$

5. $1 \times 2^7 + 0 \times 2^6 + 1 \times 2^5 + 0 \times 2^4 + 0 \times 2^3 + 1 \times 2^2 + 1 \times 2^1 + 1 \times 2^0 = 167$

7. $1 \times 2^3 + 1 \times 2^2 + 0 \times 2^1 + 1 \times 2^0 = 13$

Page 88

9. $1 \times 2^3 + 0 \times 2^2 + 1 \times 2^1 + 1 \times 2^0 = 11$

11. $1 \times 2^4 + 1 \times 2^3 + 1 \times 2^2 + 0 \times 2^1 + 1 \times 2^0 = 29$

13. $1 \times 2^4 + 1 \times 2^3 + 0 \times 2^2 + 1 \times 2^1 + 1 \times 2^0 = 27$

15. $1 \times 2^6 + 1 \times 2^5 + 0 \times 2^4 + 0 \times 2^3 + 0 \times 2^2 + 1 \times 2^1 + 1 \times 2^0 = 99$

17. $1 \times 2^7 + 0 \times 2^6 + 1 \times 2^5 + 1 \times 2^4 + 1 \times 2^3 + 0 \times 2^2 + 0 \times 2^1 + 0 \times 2^0 = 184$

19. We use repeated division by 2:

$$
\begin{array}{rl}
0 & \text{r. } 1 \\
2\overline{)\ 1\ } & \text{r. } 1 \\
2\overline{)\ 3\ } & \text{r. } 0 \\
2\overline{)\ 6\ } & \text{r. } 1 \\
2\overline{)\ 13\ } & \leftarrow \text{Start here}
\end{array}
$$

Read down for the answer: 1101_{two}.

21. We use repeated division by 2:

$$
\begin{array}{rl}
0 & \text{r. } 1 \\
2\overline{)\ 1\ } & \text{r. } 0 \\
2\overline{)\ 2\ } & \text{r. } 0 \\
2\overline{)\ 4\ } & \text{r. } 0 \\
2\overline{)\ 8\ } & \text{r. } 1 \\
2\overline{)\ 17\ } & \text{r. } 1 \\
2\overline{)\ 35\ } & \leftarrow \text{Start here}
\end{array}
$$

Read down for the answer: 100011_{two}.

23. We use repeated division by 2:

$$
\begin{array}{rl}
0 & \text{r. } 1 \\
2\overline{)\ 1\ } & \text{r. } 1 \\
2\overline{)\ 3\ } & \text{r. } 0 \\
2\overline{)\ 6\ } & \text{r. } 0 \\
2\overline{)\ 12\ } & \text{r. } 1 \\
2\overline{)\ 25\ } & \text{r. } 1 \\
2\overline{)\ 51\ } & \leftarrow \text{Start here}
\end{array}
$$

Read down for the answer: 110011_{two}.

25. We use repeated division by 2:

$$
\begin{array}{rl}
0 & \text{r. } 1 \\
2\overline{)\ 1\ } & \text{r. } 0 \\
2\overline{)\ 2\ } & \text{r. } 0 \\
2\overline{)\ 4\ } & \text{r. } 0 \\
2\overline{)\ 8\ } & \text{r. } 0 \\
2\overline{)\ 16\ } & \text{r. } 0 \\
2\overline{)\ 32\ } & \text{r. } 0 \\
2\overline{)\ 64\ } & \leftarrow \text{Start here}
\end{array}
$$

Read down for the answer: 1000000_{two}.

Page 89

27. We use repeated division by 2:

$$
\begin{array}{r}
0 \quad \text{r. } 1 \\
2)\overline{1} \quad \text{r. } 0 \\
2)\overline{2} \quad \text{r. } 0 \\
2)\overline{4} \quad \text{r. } 0 \\
2)\overline{8} \quad \text{r. } 0 \\
2)\overline{16} \quad \text{r. } 0 \\
2)\overline{32} \quad \text{r. } 0 \\
2)\overline{64} \quad \text{r. } 0 \\
2)\overline{128} \quad \leftarrow \text{ Start here}
\end{array}
$$

Read down for the answer: 10000000_{two}.

29. We use repeated division by 2:

$$
\begin{array}{r}
0 \quad \text{r. } 1 \\
2)\overline{1} \quad \text{r. } 1 \\
2)\overline{3} \quad \text{r. } 0 \\
2)\overline{6} \quad \text{r. } 0 \\
2)\overline{12} \quad \text{r. } 0 \\
2)\overline{24} \quad \text{r. } 1 \\
2)\overline{49} \quad \text{r. } 1 \\
2)\overline{99} \quad \text{r. } 0 \\
2)\overline{198} \quad \text{r. } 1 \\
2)\overline{397} \quad \text{r. } 1 \\
2)\overline{795} \quad \leftarrow \text{ Start here}
\end{array}
$$

Read down for the answer: 1100011011_{two}.

Level 2, page 157

31. Look at Table 4.10: D = 68, O = 79; Note, the O is not shown, so you need to "count down" from "F". The code is 68 79.

33. Look at Table 4.10: E = 69, N = 78 (count down from F), and D = 68. The code is 69 78 68.

35. Look at Table 4.10: 72 = H (count down from F); 65 = A; 86 = V (count down), and 69 = E. The word is HAVE.

37. Look at Table 4.10: 83 = S (count down from F); 84 = T (one more than 83); 85 = U (one more, again), 68 = D, and 89 = Y (count up from Z). The word is STUDY.

39. 11_{two}
$\underline{+\ 10_{two}}$
101_{two}

   Note $1 + 0 = 1;$
   $1 + 1 = 10,$ *carry 1*

41. 110_{two}
$\underline{+\ 111_{two}}$
1101_{two}

   Note $1 + 1 = 1;$
   $1 + 1 = 10,$ *carry 1;*
   $1 + 1 + 1 = 11$ *put down* 1 *and carry 1.*

43. 101_{two}
$\underline{-\ \ 11_{two}}$
10_{two}

   Note $1 - 1 = 0;$
   $10 - 1 = 1$

45. 111_{two}
$\underline{\times\ 101_{two}}$
111_{two}
000_{two}
$\underline{111_{two}}$
100011_{two}

Level 3, page 157

47. The answer is correct because there *must* be either peace or war. If you studied Chapter 2, you will recognize this "or" as *conjunction*. If p: There will be peace, then $\sim p$: There will not be peace, which we can translate as "There will be war." If you look at a truth table for $p \vee \sim p$, you see it is always true.

49. a. From Table 4.10, we see that $5_{eight} = 101_{two}$.
   b. From Table 4.10, we see that $6_{eight} = 110_{two}$.

51. a. Convert digit-by-digit using Table 4.10. We see that $167_{eight} = 001\,110\,111_{two}$.
   b. Convert digit-by-digit using Table 4.10. We see that $624_{eight} = 110\,010\,100_{two}$.

53. a. From Table 4.10, we see that $101_{two} = 5_{eight}$.

    b. From Table 4.10, we see that $100_{two} = 4_{eight}$.

55. Convert digit-by-digit using Table 4.10. We see that
$$000\,000\,111\,111\,101\,000_{two} = 007750_{eight}.$$

57. Convert digit-by-digit using Table 4.10. We see that
$$111\,111\,011\,011\,101\,010\,001_{two} = 773521_{eight}.$$

59. The *Farmer's Manual* is simply a table for base two. Note the column headers are 1, 2, 4, 8, 16, and 32.

4.5 History of Calculating Devices, page 159

New Terms Introduced in this Section

Artificial intelligence	Bulletin board	Chat room	Communications package
Computer abuse	Computer program	Data processing	Database management
Download	e-mail	Hard drive	Hardware
Information retrieval	Input	Input device	Internet
Keyboard	Laptop	Microcomputers	Minicomputers
Modem	Monitor	Mouse	Network
Online	Output	Output device	Password
Pattern recognition	Peripheral	Personal computer	Pixel
Printer	Program	RAM	Resolution
ROM	Simulation	Software	Software package
Spreadsheet	Supercomputer	Upload	Word processing
World Wide Web			

Level 1, page 168

1. Answers vary; fingers, Napier's rods, abacus, slide rule, Pascal's calculator, Leibniz' reckoning machine, Babbage's calculating machines

3. The CPU is the microprocessor the computer uses; for example, INTEL Pentium Pro.

5. ROM is read-only memory, and RAM is random-access memory.

7. Answers vary; writing term papers, using the Internet, chat rooms, shopping, playing games

9. Answers vary. You should express your own opinion backed up by reasons for your opinion. You should write a couple of paragraphs in answering this question.

11. a. With your hands held up and palms facing away from yourself, the middle finger of your left hand should be bent, separating the first two fingers from the remaining seven fingers; the answer shown by your fingers is 27.

  b. The seventh finger from the left should be bent, separating six fingers from the three remaining; the answer shown by your fingers is 63.

  c. The thumb on your right hand should be bent, showing the five fingers on your left hand and the remaining four on your right hand; the answer shown by your fingers is 54.

13. a. With your hands held up and palms facing away from yourself, separate the first two fingers of your left hand (since 27 contains two tens). Then, because 27 has 7 units, bend the seventh finger from the left. To read the answer, 243, note that the first two fingers (on your left hand) indicate the hundreds, the next group shows four fingers, with three fingers at the right.

  b. Separate the first four fingers on your left hand; bend down your eighth finger from the left; read the answer from your fingers by looking at the grouping of fingers; the answer shown is 432.

  c. Separate your left hand away from your right hand, and then bend down other thumb (on your right hand), read the answer from your fingers by looking at the three groups of fingers; the answer shown is 504.

15. The correct arrangement is ENIAC (1942), UNIVAC (1951), Cray (1958), Altair (1975), Apple (1976).

Answers to Problems 16-33 vary.

17. Aristophanes developed a finger-counting systems about 500 BC.

19. Babbage built a calculating machine in the 19th century.

21. Berry helped Atanasoff built the first computer.

23. Cray developed the first supercomputer.

25. Engelbart invented the computer mouse.

27. Jobs was the cofounder of Apple Computer and codesigned the Apple II computer.

29. Leibniz built a calculating device in 1695.

31. Napier invented a calculating device to do multiplication (in 1617).

33. Wozniak co-designed the Apple II computer.

Level 2, page 169

35. Yes, a computer should be used because of its speed and the necessity of carrying out complicated computations in real time.

Page 93

37. Yes, a computer should be used because of its speed and the necessity of carrying our the same steps over and over (repetition).

39. Yes, a computer should be used because of the necessity of repeating the same steps over and over (repetition).

41. Yes; a computer should be used because of its ability to make corrections easily.

43. Yes, a computer should be used because it can help with some of the technical aspects of playing, but it can't do anything to teach you to "feel" the music.

45. Yes, a computer should be used because of its speed, the necessary complicated computations, and the fact that the same processes are repeated. However, because of the number of variables involved, the computer can, at best, offer help in making a prediction, but it cannot actually make a completely accurate forecast.

47. The best choice is E because all of the other choices are good uses for computers.

49. Note that all the possible answers are examples of computer abuse; in this case the abuse is falsifying information, so the correct answer is D.

Level 3, page 169

51. Answers vary; this is a question for which you should have a lot to say! Especially since 9/11 the balance of security and public safety with privacy and the linking of computers is a topic on which much has been written. You should express your opinion, back it up with facts, and present your case. You should probably write about a page on this topic.

53. Answers vary; computers are only as accurate as they have been programmed to be. You should express your opinion, back it up with facts, and present your case   Your answer here should be a couple of paragraphs.

55. Answers vary; before you formulate your answer you might want to look at the previous question. Here are the answers to that question:

   a. Using this criterion, computers can think. In fact, they are superior to man in this category.

   b. Again, the answer is yes.

   c. In this category, computers are not too advanced. However, engineers at the Stanford Research Institute are working on a computer that can make perceptions using a video eye that recognizes objects and patterns.

   d. In this category, computers also fall short of the designation of "thinking." However, it is possible to program a computer so that it can recognize a human face, or one's handwriting.

   e. Computers utterly fail as thinking machines in this category, since they have absolutely no feeling or consideration. However, they can be programmed to simulate all the appropriate responses and reactions that emotions require. They can respond as if they were happy, sad, worried, or angry.

f. According to this criterion, many would say that computers cannot think. However, the answer here is not that clear-cut. Computers can write songs, play original music, write original poems, and draw original designs and paintings. For example, at Stuttgart's Technical College in Germany, a computer was programmed to write original poetry. One of the first tries with the Stuttgart computer produced a staggering result. Fed style and content in the manner of Franz Kafka, the computer turned out a manuscript so Kafkaesque that it fooled language experts. It was, they said, Kafka's writing. In the category of creating, although work is currently being done, we would have to say that a computer cannot (yet!) compete with artists.

g. There is a lot of current research on programming computers to learn by experience. Many computers today do learn by experience, however such efforts are still in the early stages of developments.

Using this as background, formulate your statement of what *you* mean by thinking and reasoning. You should probably write about a page on this topic.

57. Answers vary; you should write about a page on this topic.

Level 3 Problem Solving, page 170

59. Answers vary.

A *syntax error* is an error that involves the way in which you communicate with the computer; that is the exact way (format) the input information is formatted.

A *run-time* error is requesting the computer to do something that is beyond its technical capability.

A *logic error* is when you make a mistake that is logically incorrect either directly or indirectly by asking the computer to do something which leads to a logic impossibility. Put these basic ideas together when writing your response to this question. You should write about a page on this topic.

Chapter 4 Review Questions, page 171

STOP Studying for a chapter examination is a personal process, one which nobody else can do for you. Simply take the time to review what you have done. Here are the new terms in Chapter 4.

Addition principle [4.1]	Base *b* [4.3]	Bit [4.4]
Artificial intelligence [4.5]	Binary numeration	Bulletin board [4.5]
ASCII code [4.4]	system [4.3, 4.4]	Byte [4.4]

Chat rooms [4.5]

Chat rooms [4.5] Internet [4.5] Personal computer [4.5]
Communications package [4.5] Keyboard [4.5] Pixel [4.5]
Computer abuse [4.5] Laptop [4.5] Positional system [4.1]
Computer program [4.5] Microcomputers [4.5] Printer [4.5]
Counting number [4.1] Minicomputers [4.5] Program [4.5]
Data processing [4.5] Modem [4.5] RAM [4.5]
Database management [4.5] Monitor [4.5] Repetitive system [4.1]
Decimal numeration system [4.2] Mouse [4.5] Resolution [4.5]
Multiplication principle [4.1] ROM [4.5]
Decimal point [4.2] Network [4.5] Simple grouping system [4.1]
Download [4.5] Number [4.1] Simulation [4.5]
e-mail [4.5] Numeral [4.1] Software [4.5]
Expanded notation [4.2] Numeration system [4.1] Software package [4.5]
Hard drive [4.5] Octal numeration system [4.3, 4.4] Spreadsheet [4.5]
Hardware [4.5] Subtraction principle [4.1]
Hexadecimal [4.3] Online [4.5] Supercomputer [4.5]
Hindu-Arabic numerals [4.2] Output [4.5] Ten [4.2]
Hundred [4.2] Output device [4.5] Upload [4.5]
Information retrieval [4.5] Password [4.5] Word processing [4.5]
Input [4.5] Pattern recognition [4.5] World Wide Web [4.5]
Input device [4.5] Peripheral [4.5]

If you can describe the term, read on to the next one; if you cannot, then look it up in the text (the section number is shown in brackets). Next, review the types of problems in Chapter 4.

TYPES OF PROBLEMS

Know the principal properties, advantages and disadvantages of the Egyptian, Babylonian, and Roman numeration systems. [4.1]

Write decimal numerals for numbers written in the Egyptian, Babylonian, and Roman numerations systems. [4.1]

Write decimal numerals in the Egyptian, Babylonian, and Roman numeration systems. [4.1]

Perform addition and subtraction in the Egyptian and Babylonian numeration systems. [4.1]

Give the meaning of a particular numeral in the Hindu-Arabic numeration system. [4.2]

Write the decimal representation for a number written in expanded notation. [4.2]

Write a decimal numeral in expanded notation. [4.2]

Use an abacus to illustrate the meaning of a decimal number. [4.2]

Count objects in various number bases. [4.3]
Write numbers in various bases in expanded notation. [4.3]
Change numbers from base *b* to base 10. [4.3]
Change numbers from base 10 to base *b*. [4.3]
Solve applied problems by using number bases. [4.3]

Change a binary numeral to a decimal numeral. [4.4]
Use the binary numeration system to represent a number. [4.4]
Use the binary numeration system and the ASCII code to represent a word. [4.4]
Add, subtract, and multiply using binary numeration systems. [4.4]
Convert from binary to octal and from octal to binary numeration systems. [4.4]

Know some principal events in the history of computers. [4.5]
Know some principal events in the history of the Internet. [4.5]
Know the principal uses for a computer. [4.5]
Know the principal computer abuses. [4.5]

Once again, see if you can verbalize (to yourself) how to do each of the listed types of problems. Work all of Chapter 4 Review Questions (whether they are assigned or not). Work through all of the problems before looking at the answers, and *then* correct each of the problems. The entire solution is shown in the answer section at the back of the text. If you worked the problem correctly, move on to the next problem, but if you did not work it correctly (or you did not know what to do), look back in the chapter to study the procedure, or ask your instructor.

Finally, go back over the homework problems you have been assigned. If you worked a problem correctly, move on to the next problem, but if you missed it on your homework, then you should look back in the book or talk to your instructor about how to work the problem.

If you follow these steps, you should be successful with your review of this chapter.

We give all of the answers to the chapter review questions (not just the odd-numbered questions).

Chapter 4 Review Questions, page 171

1. Answers vary. The position in which the individual digits are listed is relevant; examples will vary.
2. Answers vary. Addition is easier in a simple grouping system; examples will vary.
3. Answers vary. It uses ten symbols; it is positional; it has a placeholder symbol (0); and it uses 10 as its basic unit for grouping.
4. Answers vary. Should include finger calculating, Napier's rods, Pascal's calculator, Leibniz' reckoning machine, Babbage's difference and analytic engines, ENIAC, UNIVAC,

Page 97

Atanasoff's and Eckert and Mauchly's computers, and the dispute they had in proving their position in the history of computers. Should also include the role and impact of the Apple and Macintosh computers, as well as the supercomputers (such as the Cray).

5. Answers vary. Should include illegal (breaking into another's computer, adding, modifying, or destroying information; copying programs without authorization or permission) and ignorance (assuming that output information is correct, or not using software for purposes for which it was intended).

6. Answers vary.
   a. the physical components (mechanical, magnetic, electronic) of a computer system
   b. the routine programs, and associated documentation in a computer system
   c. the process of creating, modifying, deleting, and formatting text and materials
   d. a system for communication between computers using electronic cables, phone lines, or wireless devices
   e. electronic mail — that is, messages sent along computer modems
   f. Random-Access Memory or memory where each location is uniformly accessible, often used for the storage of a program and data being processed
   g. an electronic place to exchange information with others, usually on a particular topic
   h. a computer component that reads and stores data

7. 10^9

8. $4 \times 10^2 + 3 \times 10^1 + 6 \times 10^0 + 2 \times 10^{-1} + 1 \times 10^{-5}$

9. $5 \times 8^2 + 2 \times 8^1 + 3 \times 8^0$

10. $1 \times 2^6 + 1 \times 2^3 + 1 \times 2^2 + 1 \times 2^1$

11. 4,020,005.62

12. $1 \times 2^4 + 1 \times 2^3 + 1 \times 2^2 + 1 \times 2^0 = 29$

13. $1 \times 2^6 + 1 \times 2^5 + 1 \times 2^4 + 1 \times 2^3 + 1 \times 2^1 + 1 \times 2^0 = 123$

14. $1 \times 3^2 + 2 \times 3^1 + 2 \times 3^0 = 17$

15. $8 \times 12^2 + 2 \times 12^1 + 1 \times 12^0 = 1{,}177$

16.
```
       0 r. 1
    2) 1 r. 1
    2) 3 r. 0
    2) 6 r. 0
    2) 12    ←    Start here
    12 = 1100_two   Answer
```

17.
```
       0 r. 1
    2) 1 r. 1
    2) 3 r. 0
    2) 6 r. 1
    2) 13 r. 0
    2) 26 r. 0
    2) 52    ←    Start here
    52 = 110100_two   Answer
```

18.

```
          0  r. 1
   2)      1  r. 1
    2)     3  r. 1
   2)      7  r. 1
   2)     15  r. 1
   2)     31  r. 0
   2)     62  r. 1
   2)    125  r. 0
   2)    250  r. 1
   2)    501  r. 1
    2)1,003  r. 1
   2)2,007          ←      Start here
```

$2{,}007 = 11111010111_{two}$ Answer

19.

```
          0  r. 1
   2)        1  r. 1
   2)        3  r. 1
   2)        7  r. 1
   2)       15  r. 0
   2)       30  r. 1
   2)       61  r. 0
   2)      122  r. 0
   2)      244  r. 0
   2)      488  r. 0
   2)      976  r. 1
   2)    1,953  r. 0
   2)    3,906  r. 0
   2)    7,812  r. 1
   2)   15,625  r. 0
   2)   31,250  r. 0
   2)   62,500  r. 0
   2)  125,000  r. 0
   2)  250,000  r. 0
    2)  500,000  r. 0
   2)1,000,000          ←      Start here
```

$11110100001001000000_{two}$ Answer

20. a.

```
          0  r. 9
   12)     9  r. 2
   12)   110  r. E
   12)1,331          ←      Start here
```

$1{,}331 = 92E_{twelve}$ Answer

b.

```
        0  r. 4
   5)   4  r. 0
   5)  20  r. 0
   5)100          ←      Start here
```

$100 = 400_{five}$ Answer

Page 99

CHAPTER 5 The Nature of Numbers

Chapter Overview
The ideas of a mathematical system and the closure, associative, commutative, and distributive properties are presented in this section. Rather than work with familiar sets of numbers, an abstract system is used so that the students can focus their attention on the *properties*. If we used sets of familiar numbers with ordinary operations, the students might know many of the results *without investigating the properties*. The student should also be encouraged to invent some mathematical systems and then to investigate their properties. The student should be able not only to describe each property but also to check whether a given set satisfies each property for a particular given operation.

5.1 Natural Numbers, page 175

New Terms Introduced in this Section

Addition	Associative property	Closed for addition
Closed for multiplication	Closed set	Closure property
Commutative property	Counting numbers	Distributive property
Multiplication	Natural numbers	Subtraction

Level 1, page 180
Answers to Problems 1-9 may vary.

1. $\mathbb{N} = \{1, 2, 3, 4, 5, \cdots\}$

3. Subtraction is defined in terms of addition: $m - n = x$ means $m = n + x$.

5. For an operation $\circ$ and elements a and b in a set S, $a \circ b = b \circ a$.

7. For operations $\circ$ and $\otimes$ and elements a, b, and c in S, $a \otimes (b \circ c) = (a \otimes b) \circ (a \otimes c)$

9. Associative property changes the grouping and the commutative property changes order.

11. **a.** $3 \cdot 4$ means add 4 three times; $4 + 4 + 4$.
    b. $4 \cdot 3$ means add 3 four times; $3 + 3 + 3 + 3$.
    Note: parts **a** and **b** are not the same.

13. **a.** $2 \cdot 184$ means add 184 two times; $184 + 184$.
    b. $184 \cdot 2$ means add 2 a total of 184 times; $2 + 2 + \cdots + 2$ (a total of 184 terms).
    Note: parts **a** and **b** are not the same.

15. **a.** xy means add y a total of x times; $y + y + y + \cdots + y$ (a total of x terms).
    b. yx means add x a total of y times; $x + x + x + \cdots + x$ (a total of y terms).
    Note: parts **a** and **b** are not the same.

17. Notice that the 3 and the 5 have "switched places"; it illustrates the commutative property.
19. Notice that no switching takes place, but the associations have changed; it illustrates the associative property.
21. The associations have not changed, but the order of the terms has changed; it illustrates the commutative property.
23. The associations have not changed, but the order the terms being added in the second parentheses has changed; it illustrates the commutative property.
25. The associations have not changed, but the order of the terms in the second parentheses has changed; it illustrates the commutative property.
27. The words "sweet" and "dear" have been reversed, so this illustrates the commutative property.
29. None of these are associative.
    a. A "high" school student is not the same as a "high school" student.
    b. A "slow" curve sign is not the same as a "slow curve" sign.
    c. A "bare" facts person is not the same as a "bare facts" person.
    d. A "red" fire engine is not the same as a "red fire" engine.
    e. A "traveling" salesman joke is not the same as a "traveling salesman" joke.
    f. A brown "smoking jacket" is not the same as a "brown smoking" jacket.

Level 2, page 180
31. Dictionary definitions are often circular. You should write at least a paragraph in answering this question.
33. a. $-1 \times i = -i$ b. $i \times i = -1$ c. $-i \times i = 1$ d. $-i \times 1 = -i$ e. $1 \times i = i$
    f. $-i \times (-i) = -1$
35. Yes since every entry in the table is contained in the original set.
37. a. Check to see if $(a \times b) \times c = a \times (b \times c)$ for different choices of a, b, and c.

$$(1 \times i) \times (-i) \overset{?}{=} 1 \times [i \times (-i)] \quad \text{or} \quad [(-1) \times i] \times i \overset{?}{=} (-1) \times (i \times i)$$
$$i \times (-i) \overset{?}{=} 1 \times 1 \qquad\qquad\qquad (-i) \times i \overset{?}{=} (-1) \times (-1)$$
$$1 = 1 \qquad\qquad\qquad\qquad\qquad 1 = 1$$
$$(i \times i) \times (i) \overset{?}{=} i \times (i \times i) \quad \text{or} \quad [i \times (-1)] \times (-1) \overset{?}{=} i \times [(-1) \times (-1)]$$
$$(-1) \times i \overset{?}{=} i \times (-1) \qquad\qquad\qquad -i \times (-1) \overset{?}{=} i \times 1$$
$$-i = -i \qquad\qquad\qquad\qquad\qquad i = i$$

There are other possibilities (you should try some different ones). It appears that the set is associative for the operation of $\times$.

Page 102

b. The table is symmetric with respect to the main diagonal, so it appears the set is commutative for the operation of $\times$ You might wish to try some particular examples showing that $a \times b = b \times a$.

39. Yes, the set of natural numbers ($\mathbb{N}$) is closed for the operations of $\triangleleft$ since the first element listed will always be in the set $\mathbb{N}$.   That is, $a \triangleleft b = a$, which, by definition, is an element of $\mathbb{N}$.

41. a. Find $\square$ at the left of the given table, then find $\triangle$ at the top. The intersection of the row labeled $\square$ and the column headed $\triangle$ contains the entry $\square$; thus, $\square \bullet \triangle = \square$.

b. $\triangle \bullet \circ = \circ$

c. $\circ \bullet \square = \triangle$ and $\square \bullet \circ = \triangle$, so we see $\circ \bullet \square = \square \bullet \circ$.

You might think this shows commutativity, but remember for a set to be commutative it must hold for *all* possibilities. For this operation we see that
$$\circ \bullet \triangle \neq \triangle \bullet \circ$$
Thus, the set is not commutative for $\bullet$.

d. $(\circ \bullet \triangle) \bullet \triangle = \square \bullet \triangle$ and $\circ \bullet (\triangle \bullet \triangle) = \circ \bullet \triangle$
$\qquad\qquad\qquad\quad = \square \qquad\qquad\qquad\qquad\qquad\qquad = \square$

Thus, $(\circ \bullet \triangle) \bullet \triangle = \circ \bullet (\triangle \bullet \triangle)$.

43. Check: $a \downarrow (b \to c) = (a \downarrow b) \to (a \downarrow c)$. Try several examples (we show one):
$$2 \downarrow (3 \to 19) \overset{?}{=} (2 \downarrow 3) \to (2 \downarrow 19)$$
$$2 \downarrow 19 \overset{?}{=} 2 \to 2$$
$$2 = 2$$
It seems to holds.

45. a. $5 \times (100 - 1) = 500 - 5$ **b.** $4 \times (90 - 2) = 360 - 8$
$\qquad\qquad\qquad\qquad\quad = 495 \qquad\qquad\qquad\qquad\qquad\qquad\quad = 352$

c. $8 \times (50 + 2) = 400 + 16$
$\qquad\qquad\qquad\quad = 416$

47. We need to show $(ab)c = a(bc)$ for all possible choices of elements from the set.
$$(0 \cdot 0)0 \overset{?}{=} 0(0 \cdot 0) \quad (0 \cdot 0)1 \overset{?}{=} 0(0 \cdot 1) \quad (0 \cdot 1)1 \overset{?}{=} 0(1 \cdot 1) \quad (1 \cdot 1)1 \overset{?}{=} 1(1 \cdot 1)$$
$$0 \cdot 0 \overset{?}{=} 0 \cdot 0 \qquad\quad 0 \cdot 1 \overset{?}{=} 0 \cdot 0 \qquad\quad 0 \cdot 1 \overset{?}{=} 0 \cdot 1 \qquad\quad 1 \cdot 1 \overset{?}{=} 1 \cdot 1$$
$$0 = 0 \qquad\qquad\quad 0 = 0 \qquad\qquad\quad 0 = 0 \qquad\qquad\quad 1 = 1$$

There are other possibilities, but we note that any choice with any 0 chosen for a, b or c will result in an answer of 0, so we see that the set $\{0, 1\}$ is associative for multiplication.

49. We see that even + even = even, so the set is of even numbers is closed for addition.

51. We see that odd $\times$ odd = odd, so the set of odd numbers is closed for multiplication.

Level 3, page 181

53. Let a and b be any elements in S. Then, $a \odot b = 0 \cdot a + 1 \cdot b = b$ and we know that b is an element of S, so the set S is closed for the operation of $\odot$.

55. It is commutative because we know that $a + b = b + a$ for the set S so
$a \oslash b = 2(a + b)$ and $b \oslash a = 2(b + a)$ so $a \oslash b = b \oslash a$.
Next, check the associative property.

$$[2 \oslash 3] \oslash 4 \overset{?}{=} 2 \oslash [3 \oslash 4]$$
$$[2(2 + 3)] \oslash 4 \overset{?}{=} 2 \oslash [2(3 + 4)]$$
$$10 \oslash 4 \overset{?}{=} 2 \oslash 14$$
$$2(10 + 4) \overset{?}{=} 2(2 + 14)$$
$$2(14) \neq 2(16)$$

To disprove a property, we need find only one counterexample, so we see the set S is not associative for the operation of $\oslash$. You should check the associative property for a different set of numbers.

Level 3 Problem Solving, page 182

57. To fill in the table (Problem 56), you need to try each of the possibilities. Use yourself (or a little toy soldier). The completed table is

$\star$	ℓ	r	a	f
ℓ	a	f	r	ℓ
r	f	a	ℓ	r
a	r	ℓ	f	a
f	ℓ	r	a	f

All of the elements of this table are elements of H, so we see the set is closed for the operations of $\star$.

59. You need to do this by trial and error. One possible answer is:
$$0 + 1 + 2 - 3 - 4 + 5 - 6 + 7 + 8 - 9 = 1$$

5.2 Prime Numbers, page 183

New Terms Introduced in this Section

Canonical form	Composite number	Divides
Divisibility	Divisor	Factor
Factor tree	Factoring	Fundamental theorem of arithmetic
g.c.f.	Greatest common factor	l.c.m. (least common multiple)
Multiple	Prime factorization	Prime number
Relatively prime	Rules of divisibility	Sieve of Eratosthenes

Level 1, page 194

1. A prime number is a counting number with exactly two factors.

3. It is writing a prime factorization of a number so that the factors are written in exponential form in the order of smallest to largest.

5. The abbreviation l.c.m. stands for least common multiple which means the smallest number that all of the given numbers divides into evenly. To find the l.c.m., factor each of the numbers and then take one representative of each different factor, and choose as a representative the one with the largest exponent.

7. **a.** prime; use sieve (Table 5.2).
    b. not prime; $3 \cdot 19$
    c. not prime; only 1 divisor
    d. prime; check primes under 45 since $\sqrt{1,997} \approx 44.69$.

9. **a.** prime; use sieve (Table 5.2).
    b. prime; use sieve.
    c. not prime; $3^2 \cdot 19$
    d. not prime; $3^2 \cdot 223$

11. **a.** $6|48$ because $48 = 6 \cdot 8$; the statement is true (T).
    b. $7\!\!\not|48$ since there does not exist a counting number k so that $48 = 7k$; the statement is false (F).
    c. $8|48$ because $48 = 8 \cdot 6$; the statement is true (T).
    d. $9\!\!\not|48$ since there does not exist a counting number k so that $48 = 9k$; the statement is false (F).

13. **a.** $15\!\!\not|5$ since there does not exist a counting number k so that $5 = 15k$; the statement is false (F).
    b. $5|83,410$ because $83,410 = 5 \cdot 16682$; the statement is true (T).
    c. It is true since $628,174$ is an even number; the statement is true (T).
    d. It is true since $148,729,320$ ends in a zero; the statement is true (T).

15. Use the sieve of Eratosthenes; you only need to cross out the multiples of primes up to 17. The list of primes is: 2, 3, 5, 7, 11, 13, 17, 19, 23, 29, 31, 37, 41, 43, 47, 53, 59, 61, 67, 71, 73, 79, 83, 89, 97, 101, 103, 107, 109, 113, 127, 131, 137, 139, 149, 151, 157, 163, 167, 173, 179, 181, 191, 193, 197, 199, 211, 223, 227, 229, 233, 239, 241, 251, 257, 263, 269, 271, 277, 281, 283, and 293.

17. Use a factor tree to find these factorizations.
    a. $24 = 8 \cdot 3 = 2^3 \cdot 3$
    b. $30 = 6 \cdot 5 = 2 \cdot 3 \cdot 5$
    c. $300 = 3 \cdot 100 = 3 \cdot 4 \cdot 25 = 2^2 \cdot 3 \cdot 5^2$

   d. $144 = 12 \cdot 12 = 4 \cdot 3 \cdot 4 \cdot 3 = 2^4 \cdot 3^2$

19. Use a factor tree to find these factorizations.

   a. $120 = 10 \cdot 12 = 2 \cdot 5 \cdot 4 \cdot 3 = 2^3 \cdot 3 \cdot 5$

   b. $90 = 9 \cdot 10 = 3 \cdot 3 \cdot 2 \cdot 5 = 2 \cdot 3^2 \cdot 5$

   c. $75 = 3 \cdot 25 = 3 \cdot 5^2$

   d. $975 = 25 \cdot 39 = 5 \cdot 5 \cdot 3 \cdot 13 = 3 \cdot 5^2 \cdot 13$

21. 83 is prime (look at Table 5.2).

23. 127 is prime; divide by primes up to 11 (since $\sqrt{127} \approx 11.3$).

25. $377 = 13 \cdot 29$

27. $105 = 5 \cdot 21 = 3 \cdot 5 \cdot 7$

29. 67 is prime (look at Table 5.2).

31. $315 = 5 \cdot 63 = 5 \cdot 9 \cdot 7 = 3^2 \cdot 5 \cdot 7$

33. $567 = 9 \cdot 63 = 3 \cdot 3 \cdot 9 \cdot 7 = 3^4 \cdot 7$

35. $2,869 = 19 \cdot 151$

37. $60 = 2^2 \cdot 3^1 \cdot 5^1$

        $72 = 2^3 \cdot 3^2 \cdot 5^0$

    g.c.f. $= 2^2 \cdot 3^1 \cdot 5^0 = 12$

    l.c.m. $= 2^3 \cdot 3^2 \cdot 5^1 = 360$

39. $12 = 2^2 \cdot 3^1 \cdot 19^0$

        $54 = 2^1 \cdot 3^3 \cdot 19^0$

      $171 = 2^0 \cdot 3^2 \cdot 19^1$

    g.c.f. $= 2^0 \cdot 3^1 \cdot 19^0 = 3$

    l.c.m. $= 2^2 \cdot 3^3 \cdot 19^1 = 2,052$

41. $9 = 2^0 \cdot 3^2 \cdot 7^0$

      $12 = 2^2 \cdot 3^1 \cdot 7^0$

      $14 = 2^1 \cdot 3^0 \cdot 7^1$

    g.c.f. $= 2^0 \cdot 3^0 \cdot 7^0 = 1$

    l.c.m. $= 2^2 \cdot 3^2 \cdot 7^1 = 252$

43. $75 = 2^0 \cdot 3^1 \cdot 5^2$

       $90 = 2^1 \cdot 3^2 \cdot 5^1$

     $120 = 2^3 \cdot 3^1 \cdot 5^1$

    g.c.f. $= 2^0 \cdot 3^1 \cdot 5^1 = 15$

    l.c.m. $= 2^3 \cdot 3^2 \cdot 5^2 = 1,800$

Page 106

Level 2, page 195

45. The least common multiple of 6 and 8 is 24, so the next night off together is in 3 weeks, 3 days.

47. By 6; primes 2, 3, and 5 in the first row are circles and all of the numbers in those column are not primes.

By 7; This is not a particularly useful arrangement; even the fact that the numbers in the column headed by 7 are all crossed out, there does not seem to much advantage to this arrangement.

By 21; The columns headed by 7, 14, and 21, but there does not seem to be any particular advantage to this arrangement. The arrangement by 6 seems to be the best.

49. Carry out the directions for this problem to find the lucky numbers less than 100:
1, 3, 7, 9, 13, 15, 21, 25, 31, 33, 37, 43, 49, 51, 63, 67, 69, 73, 75, 79, 87, 93, and 99

Level 3, page 195

51. Once we move beyond 7, any group of three consecutive odd numbers is certain to include one number that is a multiple of three (and consequently not a prime). To prove this let $2n + 1$, $2n + 3$, and $2n + 5$ be any 3 consecutive odd numbers. If $n = 1$, we have the prime triplets 3, 5, and 7. If $n > 1$, then one of the three numbers must be divisible by 3. Suppose the first is divisible by 3, then they are not prime triplets. If the first is not divisible by 3, then dividing it by 3 leaves a remainder of 1 or 2. If it leaves a remainder of 1, then the middle number is divisible by 3. If it leaves a remainder of 2, then the last one is divisible by 3. In all cases, the numbers will not be prime triplets.

53. a. The set S is closed for the operation $\mathbf{M}$ because any pair of numbers in S will have a least common multiple which is in S.

    b. When finding the least common multiple of a triplet of numbers, the way in which we choose to associate the numbers does not matter. That is, for numbers a, b and c in the set S, $(a\,\mathbf{M}\,b)\,\mathbf{M}\,c = a\,\mathbf{M}\,(b\,\mathbf{M}\,c)$.

    c. When finding the least common multiple of a pair of numbers, the way in which we choose to list the numbers does not matter. That is, for numbers a and b in the set S, $a\,\mathbf{M}\,b = b\,\mathbf{M}\,a$.

Level 3 Problem Solving, page 196

55. One example is

$$2 \cdot 3 \cdot 5 \cdot 7 \cdot 11 \cdot 13 + 1 = 30{,}031$$

which is not prime, since $30{,}031 = 59 \cdot 509$.

57. Some possibilities are:

Page 107

$$1^2 + 1 = 2, \qquad 4^2 + 1 = 17, \qquad 6^2 + 1 = 37, \qquad 10^2 + 1 = 101$$

A computer could help to find some other examples: 197, 257, 401, 577, 677, 1297, 1601, 2917, 3137, 4357, 5477, 7057, 8101, 8837.

59. If $N = 5$, then

$$2^{5-1}(2^5 - 1) = 2^4(2^5 - 1) = 16(31) = 496$$

The proper divisors of 496 are 1, 2, 4, 8, 16, 31, 62, 124, and 248. The sum of these numbers is 496.

5.3 Integers, page 197
New Terms Introduced in this Section

Absolute value	Division	Division by zero	Integers
Opposites	Whole numbers	Zero	

Level 1, page 203

1. If $x = 0$, then $x + 0 = 0 + x = x$. To add nonzero integers x and y, look at the signs of x and y:

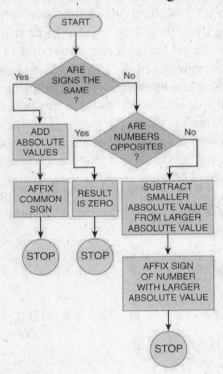

3. If $x = 0$, then $x \cdot 0 = 0 \cdot x = 0$. To multiply nonzero integers x and y, look at the signs of x and y:

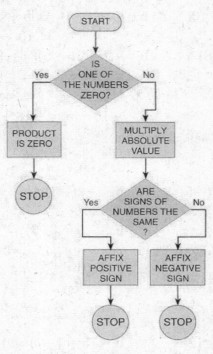

5. $0 \div 5$ means $\frac{0}{5}$ which is equal to 0; on the other hand, $5 \div 0$ is not defined.

7. **a**. $|30|$ is the absolute value of 30; $|30| = 30$ because 30 is a positive number.

    b. $|-30|$ is the absolute value of negative 30; $|-30| = 30$ because -30 is a negative number. Another way of saying this is: $|-30| = -(-30) = 30$.

    c. $-|30|$ is the opposite of the absolute value of 30; $-|30| = -30$

    d. $|30| - |-30| = 30 - 30 = 0$

9. **a**. $5 + 3 = 8$ **b**. $-5 + 3 = -2$

11. **a**. $7 + 3 = 10$ **b**. $-9 + 5 = -4$

13. **a**. $-15 + 8 = -7$ **b**. $|-14 + 2| = |-12| = 12$

15. **a**. $10 - 7 = 3$ **b**. $7 - 10 = 7 + (-10) = -3$

17. **a**. $3(-6) = -18$ **b**. $-5(4) = -20$

19. **a**. $14(-5) = -70$ **b**. $-14(-5) = 70$

21. **a**. $\frac{-12}{4} = -3$ **b**. $\frac{-63}{-9} = 7$

23. **a**. $\frac{-528}{-4} = 132$ **b**. $(-1)^3 = -1$

25. **a**. $7(-8) = -56$ **b**. $-5(15) = -75$

27. **a**. $-2^2 = -(2)(2) = -4$ **b**. $(-2)^2 = (-2)(-2) = 4$

29. a. $-4 - 8 = -4 + (-8) = -12$ **b.** $31 + (-16) = 15$

31. a. $162 + (-12) = 150$ **b.** $-12 + [(-4) + (-3)] = -12 + (-7) = -19$

33. a. $|-5 - (-10)| = |-5 + 10|$ **b.** $|5 - (-5)| = |5 + 5|$
$= 5$ $= 10$

35. a. $|-8| + (-8) = 8 + (-8)$ **b.** $-9 - (4 - 5) = -9 - (-1)$
$= 0$ $= -9 + 1$
$= -8$

37. a. $-6 - (-6) = -6 + 6$ **b.** $-18 - 5 = -18 + (-5)$
$= 0$ $= -23$

Level 2, page 204

39. a. $6 + (-8) - 5 = -2 + (-5)$ **b.** $-2(-3) + (-1)(6) = 6 + (-6)$
$= -7$ $= 0$

41. a. $6 - (-2) = 6 + 2$ **b.** $-5 - (-3) = -5 + 3$
$= 8$ $= -2$

43. a. $\frac{-32}{-8} - 5 - (-7) = 4 + (-5) + 7$ **b.** $\frac{-15}{-5} - 4 - (-8) = 3 + (-4) + 8$
$= -1 + 7$ $= -1 + 8$
$= 6$ $= 7$

45. a. $|-3| - [-(-2)] = 3 - [2]$ **b.** $15 - (-7) = 15 + 7$
$= 3 - 2$ $= 22$
$= 1$

47. a. $[48 \div (-6)] \div (-2) = -8 \div (-2)$ **b.** $48 \div [(-6)] \div (-2)] = 48 \div 3$
$= 4$ $= 16$

49. a. $-5(2) + (-3)(-4) - 6(-7) = -10 + 12 + 42$
$= 2 + 42$
$= 44$

b. $-8(3) - 6(-4) - 2(-8) = -24 + 24 + 16$
$= 16$

Level 3, page 204

51. $1^n = 1$ for all n, so the answer to all these parts is 1: **a.** 1 **b.** 1 **c.** 1 **d.** 1 **e.** 1 **f.** 1

53. $(-1)^n = 1$ if n is even and -1 if n is odd: **a.** 1 **b.** -1 **c.** -1 **d.** 1 **e.** -1 **f.** -1

55. a. An operation $\circ$ is commutative for a set S if $a \circ b = b \circ a$ for all elements a and b in S.
 b. Yes, addition is commutative for the integers (try several examples).
 c. Not commutative for subtraction since $7 - 5 = 2$, but $5 - 7 \neq 2$.

Page 110

d. Yes, multiplication is commutative for integers.

e. Not commutative for division since $8 \div 4 = 2$, but $4 \div 8 \neq 2$.

57. Use the fact that the integers are closed for addition and that the opposite of any integer is also an integer. From this and the definition of subtraction, the desired result follows.

Level 3 Problem Solving, page 204

59. a. 12345668765433; *note:* most calculators will show scientific notation:
   1.234566877 E13

   b. Search for patterns:
$$1 \times 9 = 9$$
$$12 \times 99 = 1188$$
$$123 \times 999 = 122877$$
$$1234 \times 9999 = 12338766 \qquad \text{By patterns: } 12345668765433.$$

5.4 Rational Numbers, page 205

New Terms Introduced in this Section

Denominator	Fundamental property of fractions	Improper fraction
Least common denominator	Numerator	Proper fraction
Rational number	Reduced fraction	Whole number

Level 1, page 210

1. A fraction is reduced if the numerator and denominator have no common factors (except 1 or -1).

3. $\frac{a}{b} \div \frac{c}{d} = \frac{ad}{bc}$ ($c \neq 0$)   You might also look at Example 3 on page 207. This example thoroughly explains the basis for division of rational numbers. You might also want to create several examples to help answer this question.

5.
$$\frac{a}{b} - \frac{c}{d} = \frac{a}{b} + (-1)\frac{c}{d}$$
$$= \frac{a}{b} + \frac{-1}{1} \cdot \frac{c}{d}$$
$$= \frac{a}{b} \cdot \frac{d}{d} + \frac{-c}{d} \cdot \frac{b}{b}$$
$$= \frac{ad}{bd} + \frac{-bc}{bd}$$
$$= \frac{ad - bc}{bd}$$

Page 111

7. a. $\frac{2}{10} = \frac{2 \cdot 1}{2 \cdot 5} = \frac{1}{5}$ **b.** $\frac{3}{12} = \frac{3 \cdot 1}{3 \cdot 4} = \frac{1}{4}$

9. a. $\frac{14}{7} = 2$ (by division)   **b.** $\frac{38}{19} = 2$ (by division)

11. a. $\frac{42}{14} = 3$ (by division)   **b.** $\frac{16}{24} = \frac{8 \cdot 2}{8 \cdot 3} = \frac{2}{3}$

13. a. $\frac{50}{400} = \frac{50 \cdot 1}{50 \cdot 8} = \frac{1}{8}$ **b.** $\frac{140}{420} = \frac{140 \cdot 1}{140 \cdot 3} = \frac{1}{3}$

15. a.
$$78 = 2^1 \cdot 3^1 \cdot 5^0 \cdot 7^0 \cdot 13^1$$
$$455 = 2^0 \cdot 3^0 \cdot 5^1 \cdot 7^1 \cdot 13^1$$
$$\text{g.c.f.} = 2^0 \cdot 3^0 \cdot 5^0 \cdot 7^0 \cdot 13^1 = 13$$
$$\frac{78}{455} = \frac{13 \cdot 6}{13 \cdot 35} = \frac{6}{35}$$

b.
$$75 = 2^0 \cdot 3^1 \cdot 5^2$$
$$500 = 2^2 \cdot 3^0 \cdot 5^3$$
$$\text{g.c.f.} = 2^0 \cdot 3^0 \cdot 5^2 = 25$$
$$\frac{75}{500} = \frac{25 \cdot 3}{25 \cdot 20} = \frac{3}{20}$$

17. a.
$$2{,}431 = 3^0 \cdot 7^0 \cdot 11^1 \cdot 13^1 \cdot 17^1$$
$$3{,}003 = 3^1 \cdot 7^1 \cdot 11^1 \cdot 13^1 \cdot 17^0$$
$$\text{g.c.f.} = 3^0 \cdot 7^0 \cdot 11^1 \cdot 13^1 \cdot 17^0 = 143$$
$$\frac{2{,}431}{3{,}003} = \frac{143 \cdot 17}{143 \cdot 21} = \frac{17}{21}$$

b.
$$47{,}957 = 2^0 \cdot 7^1 \cdot 13^1 \cdot 17^1 \cdot 31^1$$
$$54{,}808 = 2^3 \cdot 7^0 \cdot 13^1 \cdot 17^1 \cdot 31^1$$
$$\text{g.c.f.} = 2^0 \cdot 7^0 \cdot 13^1 \cdot 17^1 \cdot 31^1 = 6{,}851$$
$$\frac{47{,}957}{54{,}808} = \frac{6{,}851 \cdot 7}{6{,}851 \cdot 8} = \frac{7}{8}$$

19. a.
$$35 = 3^0 \cdot 5^1 \cdot 7^1$$
$$15 = 3^1 \cdot 5^1 \cdot 7^0$$
$$\text{l.c.m.} = 3^1 \cdot 5^1 \cdot 7^1 = 105$$
$$\frac{-12}{35} - \frac{8}{15} = \frac{-12}{35} \cdot \frac{3}{3} + \frac{-8}{15} \cdot \frac{7}{7}$$
$$= \frac{-36}{105} + \frac{-56}{105} = \frac{-92}{105}$$

b. $2^{-1} + 3^{-1} = \frac{1}{2} + \frac{1}{3} = \frac{3}{6} + \frac{2}{6} = \frac{5}{6}$

21. a. $\frac{7}{9} - \frac{2}{3} = \frac{7}{9} + \frac{-2}{3} \cdot \frac{3}{3} = \frac{7-6}{9} = \frac{1}{9}$

b. $\frac{4}{7} - \frac{-5}{9} = \frac{4}{7} \cdot \frac{9}{9} + \frac{5}{9} \cdot \frac{7}{7} = \frac{36+35}{63} = \frac{71}{63}$

Page 112

23. a. $\dfrac{5}{3} \div \dfrac{7}{12} = \dfrac{5}{3} \cdot \dfrac{3 \cdot 4}{7}$ **b.** $\dfrac{-2}{9} \div \dfrac{6}{7} = \dfrac{-2}{9} \cdot \dfrac{7}{2 \cdot 3}$

$\qquad\qquad = \dfrac{20}{7}$ $\qquad\qquad = \dfrac{-7}{27}$

25. a. $3^{-1} + 5^{-1} = \dfrac{1}{3} + \dfrac{1}{5}$ **b.** $2^{-1} + 5^{-1} = \dfrac{1}{2} + \dfrac{1}{5}$

$\qquad\qquad = \dfrac{5 + 3}{3 \cdot 5}$ $\qquad\qquad = \dfrac{5 + 2}{2 \cdot 5}$

$\qquad\qquad = \dfrac{8}{15}$ $\qquad\qquad = \dfrac{7}{10}$

27. a. $\dfrac{2}{3} \times \dfrac{5}{7} = \dfrac{10}{21}$ **b.** $\dfrac{-1}{8} \times \dfrac{2}{-5} = \dfrac{-1}{2 \cdot 4} \cdot \dfrac{2}{-5} = \dfrac{1}{20}$

29. a. $7 \times 7^{-1} = 7 \times \dfrac{1}{7} = 1$ **b.** $-14 \times 14^{-1} = -14 \times \dfrac{1}{14} = -1$

31. a. $6 \div 6^{-1} = 6 \div \dfrac{1}{6} = 6 \times 6 = 36$ **b.** $-5 \div 5^{-1} = -5 \div \dfrac{1}{5} = -5 \times 5 = -25$

33. a. $\left(\dfrac{3}{7} \cdot \dfrac{3}{5}\right) \div \dfrac{1}{2} = \dfrac{9}{35} \times \dfrac{2}{1}$ **b.** $\dfrac{2}{-3} \cdot \dfrac{-2}{15} = \dfrac{-4}{-45}$

$\qquad\qquad = \dfrac{18}{35}$ $\qquad\qquad = \dfrac{4}{45}$

35. a. $\dfrac{28}{9} - \dfrac{4}{27} = \dfrac{28}{9} \cdot \dfrac{3}{3} + \dfrac{-4}{27}$ **b.** $\dfrac{1}{-8} + \dfrac{1}{-7} = \dfrac{-1}{8} \cdot \dfrac{7}{7} + \dfrac{-1}{7} \cdot \dfrac{8}{8}$

$\qquad\qquad = \dfrac{84 - 4}{27}$ $\qquad\qquad = \dfrac{-7 - 8}{56}$

$\qquad\qquad = \dfrac{80}{27}$ $\qquad\qquad = \dfrac{-15}{56}$

37. a. $-5 + (-5)^2 = -5 + 25$ **b.** $-3 + (-3)^2 = -3 + 9$

$\qquad\qquad = 20$ $\qquad\qquad = 6$

Level 2, page 210

39. a. $\dfrac{2^{-1} + 3^{-2}}{2^{-1} + 3^{-1}} = \dfrac{\frac{1}{2} + \frac{1}{9}}{\frac{1}{2} + \frac{1}{3}}$ **b.** $\dfrac{6 + 2^{-1}}{\frac{1}{2} + \frac{1}{3}} = \dfrac{\frac{6}{1} + \frac{1}{2}}{\frac{1}{2} + \frac{1}{3}}$

$\qquad\qquad = \dfrac{9 + 2}{2 \cdot 9} \div \dfrac{3 + 2}{2 \cdot 3}$ $\qquad\qquad = \dfrac{12 + 1}{2} \div \dfrac{3 + 2}{6}$

$\qquad\qquad = \dfrac{11}{18} \div \dfrac{5}{6}$ $\qquad\qquad = \dfrac{13}{2} \cdot \dfrac{2 \cdot 3}{5}$

$\qquad\qquad = \dfrac{11}{6 \cdot 3} \cdot \dfrac{6}{5}$ $\qquad\qquad = \dfrac{39}{5}$

$\qquad\qquad = \dfrac{11}{15}$

Page 113

41. a. $\dfrac{-3}{4} \cdot \dfrac{119}{200} + \dfrac{-3}{4} \cdot \dfrac{81}{200} = \dfrac{-3}{4}\left(\dfrac{119}{200} + \dfrac{81}{200}\right)$ **b.** $\dfrac{-3}{4}\left(\dfrac{119}{200} + \dfrac{81}{200}\right) = \dfrac{-3}{4}(1)$

$\qquad\qquad\qquad\qquad\qquad\qquad = \dfrac{-3}{4}\left(\dfrac{200}{200}\right) \qquad\qquad\qquad\qquad\qquad\qquad\qquad = \dfrac{-3}{4}$

$\qquad\qquad\qquad\qquad\qquad\qquad = \dfrac{-3}{4}$

43. $144 = 2^4 \cdot 3^2 \cdot 5^0$

$\qquad 300 = 2^2 \cdot 3^1 \cdot 5^2$

$\qquad 108 = 2^2 \cdot 3^3 \cdot 5^0$

$\qquad \text{l.c.m.} = 2^4 \cdot 3^3 \cdot 5^2 = 10{,}800$

$\dfrac{11}{144} + \dfrac{17}{300} + \dfrac{7}{108} = \dfrac{11}{144} \cdot \dfrac{75}{75} + \dfrac{17}{300} \cdot \dfrac{36}{36} + \dfrac{7}{108} \cdot \dfrac{100}{100}$

$\qquad\qquad\qquad\qquad\quad = \dfrac{11 \cdot 75 + 17 \cdot 36 + 7 \cdot 100}{10{,}800}$

$\qquad\qquad\qquad\qquad\quad = \dfrac{2{,}137}{10{,}800}$

45. $60 = 2^2 \cdot 3^1 \cdot 5^1 \cdot 17^0$

$\qquad 90 = 2^1 \cdot 3^2 \cdot 5^1 \cdot 17^0$

$\qquad 51 = 2^0 \cdot 3^1 \cdot 5^0 \cdot 17^1$

$\qquad \text{l.c.m.} = 2^2 \cdot 3^2 \cdot 5^1 \cdot 17^1 = 3{,}060$

$\dfrac{7}{60} - \dfrac{19}{90} + \dfrac{21}{51} = \dfrac{7}{60} \cdot \dfrac{51}{51} + \dfrac{-19}{90} \cdot \dfrac{34}{34} + \dfrac{21}{51} \cdot \dfrac{60}{60}$

$\qquad\qquad\qquad\qquad = \dfrac{7 \cdot 51 - 19 \cdot 34 + 21 \cdot 60}{3{,}060}$

$\qquad\qquad\qquad\qquad = \dfrac{971}{3{,}060}$

47. $210 = 2^1 \cdot 3^1 \cdot 5^1 \cdot 7^1 \cdot 31^0$

$\qquad 124 = 2^2 \cdot 3^0 \cdot 5^0 \cdot 7^0 \cdot 31^1$

$\qquad 1{,}085 = 2^0 \cdot 3^0 \cdot 5^1 \cdot 7^1 \cdot 31^1$

$\qquad \text{l.c.m.} = 2^2 \cdot 3^1 \cdot 5^1 \cdot 7^1 \cdot 31^1 = 13{,}020$

$\dfrac{143}{210} + \dfrac{15}{124} + \dfrac{11}{1{,}085} = \dfrac{143}{210} \cdot \dfrac{62}{62} + \dfrac{15}{124} \cdot \dfrac{105}{105} + \dfrac{11}{1{,}085} \cdot \dfrac{12}{12}$

$\qquad\qquad\qquad\qquad\qquad = \dfrac{143 \cdot 62 + 15 \cdot 105 + 11 \cdot 12}{13{,}020}$

$\qquad\qquad\qquad\qquad\qquad = \dfrac{10{,}573}{13{,}020}$

Level 3, page 211

49. Given any two rationals $\frac{a}{b}$ and $\frac{c}{d}$; show that $\frac{a}{b} - \frac{c}{d}$ is rational. Now,

$$\frac{a}{b} - \frac{c}{d} = \frac{ad - bc}{bd}$$

by the definition of subtraction. Also, ad and bc are integers and bd is a nonzero integer since a, c are integers and b, d are nonzero integers and the set of integers is closed for

Page 114

multiplication. Finally, $ad - bc$ is an integer because the integers are closed for subtraction. Therefore, $\frac{ad - bc}{bd}$ is a rational by the definition of a rational number.

51. The set of integers, $\mathbb{Z}$, is closed for addition, subtraction, and multiplication, but not for division since $4 \div 5$ is not an integer.

53. We need to show that for any two rational numbers, $\frac{a}{b}$, $\frac{c}{d}$ that $\frac{a}{b} + \frac{c}{d} = \frac{c}{d} + \frac{a}{b}$.
From the definition of addition we have

$$\frac{a}{b} + \frac{c}{d} = \frac{ad + bc}{bd} \qquad \frac{c}{d} + \frac{a}{b} = \frac{cb + da}{db}$$

However, since the integers are commutative, we see that

$$\frac{ad + bc}{bd} = \frac{cb + da}{db}$$

so the set of rationals is commutative.

Level 3 Problem Solving, page 211

55. One possibility is $\frac{3}{4} = \frac{1}{2} + \frac{1}{4}$.

57. One possibility is $\frac{67}{120} = \frac{1}{3} + \frac{1}{8} + \frac{1}{10}$. See if you can find another possibility.

59. Solve $x + \frac{2}{3}x + \frac{1}{2}x + \frac{1}{7}x = 33$

$$\frac{97}{42}x = 33$$

$$x = \frac{1,386}{97}$$

Check papyrus answer: $14 + \frac{1}{4} + \frac{1}{56} + \frac{1}{97} + \frac{1}{194} + \frac{1}{388} + \frac{1}{679} + \frac{1}{776} = \frac{1,386}{97}$; it checks.

5.5 Irrational Numbers, page 211

New Terms Introduced in this Section

e	Hypotenuse	Irrational number	Laws of square root
Leg	Perfect square	π	Positive square root
Pythagorean theorem	Radical form	Radicand	Square number
Square root			

Level 1, page 218

1. For a right $\triangle ABC$, with sides of length a, b, and hypotenuse c, $a^2 + b^2 = c^2$. Also, if $a^2 + b^2 = c^2$ for a triangle with sides a, b, and c, then $\triangle ABC$ is a right triangle.

3. An **exact value** of an irrational number cannot be represented by a terminating or repeating decimal. That is, an irrational number cannot be written as a terminating or a repeating decimal. In particular, an irrational number is usually approximated by 10 or 12 decimal places when using a calculator. The symbol $\sqrt{2}$ is considered an exact representation for the number square root of 2, whereas, if I press $\sqrt{2}$ on my calculator I see a display of 1.414213562 which is considered an approximate representation for the number square root of 2.

5. The square root of 2 is a number which, when squared, is equal to 2. If we look at the last digits of the given number, we see it is 4 and $4 \cdot 4 = 16$, which does not have a last digit of 2 or 0. Consequently, it is not the square root of 2.

7. A dozen is 12, a gross is 144, and a score is 20. Next, we check the limerick by writing it in mathematical notation:

$$\frac{12 + 144 + 20 + 3\sqrt{4}}{7} + 5 \cdot 11 = 9^2 + 0$$

It is a true statement.

9. We use the definition of square root. **a.** 30 **b.** 36 **c.** 807 **d.** 169

11. We use the definition of square root. **a.** a **b.** xy **c.** $4b$ **d.** $40w$

13. Use your calculator to find the approximations of these irrational numbers.
    a. irrational; 3.162 **b.** irrational; 5.477 **c.** irrational; 9.870 **d.** irrational; 7.389

15. Use your calculator to find the approximations of these irrational numbers.
    a. rational; 13 **b.** rational; 20 **c.** irrational; 23.141 **d.** irrational; 22.459

17. Use your calculator to find the approximations of these irrational numbers.
    a. rational; 32 **b.** rational; 44 **c.** irrational; 1.772 **d.** irrational; 1.253

19. **a.** $\begin{aligned} \sqrt{1,000} &= \sqrt{100 \cdot 10} \\ &= \sqrt{10^2} \cdot \sqrt{10} \\ &= 10\sqrt{10} \end{aligned}$ **b.** $\begin{aligned} \sqrt{2,800} &= \sqrt{100 \cdot 28} \\ &= \sqrt{10^2} \cdot \sqrt{2^2} \cdot \sqrt{7} \\ &= (10 \cdot 2)\sqrt{7} \\ &= 20\sqrt{7} \end{aligned}$

   c. $\begin{aligned} \sqrt{2,240} &= \sqrt{64 \cdot 35} \\ &= \sqrt{8^2} \cdot \sqrt{35} \\ &= 8\sqrt{35} \end{aligned}$ **d.** $\begin{aligned} \sqrt{4,410} &= \sqrt{9 \cdot 490} \\ &= \sqrt{3^2} \cdot \sqrt{7^2} \cdot \sqrt{10} \\ &= (3 \cdot 7)\sqrt{10} \\ &= 21\sqrt{10} \end{aligned}$

Page 116

21. a. $\sqrt{\dfrac{1}{2}} = \dfrac{\sqrt{1}}{\sqrt{2}} \cdot \dfrac{\sqrt{2}}{\sqrt{2}}$ **b.** $\sqrt{\dfrac{1}{3}} = \dfrac{\sqrt{1}}{\sqrt{3}} \cdot \dfrac{\sqrt{3}}{\sqrt{3}}$

$\qquad\qquad = \dfrac{\sqrt{2}}{2} \text{ or } \dfrac{1}{2}\sqrt{2}$ $\qquad = \dfrac{\sqrt{3}}{3} \text{ or } \dfrac{1}{3}\sqrt{3}$

c. $\sqrt{\dfrac{3}{5}} = \dfrac{\sqrt{3}}{\sqrt{5}} \cdot \dfrac{\sqrt{5}}{\sqrt{5}}$ **d.** $\sqrt{\dfrac{3}{7}} = \dfrac{\sqrt{3}}{\sqrt{7}} \cdot \dfrac{\sqrt{7}}{\sqrt{7}}$

$\qquad\qquad = \dfrac{\sqrt{15}}{5} \text{ or } \dfrac{1}{5}\sqrt{15}$ $\qquad = \dfrac{\sqrt{21}}{7} \text{ or } \dfrac{1}{7}\sqrt{21}$

23. a. $\dfrac{1}{\sqrt{2}} = \dfrac{1}{\sqrt{2}} \cdot \dfrac{\sqrt{2}}{\sqrt{2}}$ **b.** $\dfrac{-1}{\sqrt{3}} = \dfrac{-1}{\sqrt{3}} \cdot \dfrac{\sqrt{3}}{\sqrt{3}}$

$\qquad\qquad = \dfrac{\sqrt{2}}{2} \text{ or } \dfrac{1}{2}\sqrt{2}$ $\qquad = \dfrac{-\sqrt{3}}{3} \text{ or } -\dfrac{1}{3}\sqrt{3}$

c. $\dfrac{2}{\sqrt{5}} = \dfrac{2}{\sqrt{5}} \cdot \dfrac{\sqrt{5}}{\sqrt{5}}$ **d.** $\dfrac{5}{\sqrt{10}} = \dfrac{5}{\sqrt{10}} \cdot \dfrac{\sqrt{10}}{\sqrt{10}}$

$\qquad\qquad = \dfrac{2\sqrt{5}}{5} \text{ or } \dfrac{2}{5}\sqrt{5}$ $\qquad = \dfrac{5\sqrt{10}}{10}$

$\qquad\qquad\qquad\qquad\qquad\qquad\qquad = \dfrac{\sqrt{10}}{2} \text{ or } \dfrac{1}{2}\sqrt{10}$

25. a. $\sqrt{x^2 + 4}$ is simplified; remember, you are looking for square factors, not square terms.

b. $\sqrt{(x+2)^2} = x + 2$; note here we have a square factor.

27. a. $\sqrt{10^2 - 4(5)(-5)} = \sqrt{100 + 100}$ **b.** $\sqrt{12^2 - 4(3)(12)} = \sqrt{12^2 - 12^2}$

$\qquad\qquad\qquad\qquad\quad = \sqrt{100(1+1)}$ $\qquad\qquad\qquad\qquad = 0$

$\qquad\qquad\qquad\qquad\quad = \sqrt{100}\sqrt{2}$

$\qquad\qquad\qquad\qquad\quad = 10\sqrt{2}$

29. a. $\dfrac{6 + 2\sqrt{5}}{2} = \dfrac{2(3 + \sqrt{5})}{2}$ **b.** $\dfrac{8 - 4\sqrt{3}}{4} = \dfrac{4(2 - \sqrt{3})}{4}$

$\qquad\qquad\qquad = 3 + \sqrt{5}$ $\qquad\qquad\qquad = 2 - \sqrt{3}$

31. a. $\dfrac{3 - 9\sqrt{x}}{3} = \dfrac{3(1 - 3\sqrt{x})}{3}$ **b.** $\dfrac{9 + 3\sqrt{x}}{-3} = \dfrac{-3(-3 - \sqrt{x})}{-3}$

$\qquad\qquad\qquad = 1 - 3\sqrt{x}$ $\qquad\qquad\qquad = -3 - \sqrt{x}$

Page 117

33. a. $\sqrt{\dfrac{4x^2}{25y}} = \dfrac{\sqrt{(2x)^2}}{\sqrt{5^2}\sqrt{y}} \cdot \dfrac{\sqrt{y}}{\sqrt{y}}$ **b.** $\sqrt{\dfrac{5y}{16x}} = \dfrac{\sqrt{5y}}{\sqrt{4^2}\sqrt{x}} \cdot \dfrac{\sqrt{x}}{\sqrt{x}}$

$\qquad\qquad = \dfrac{2x\sqrt{y}}{5y} \text{ or } \dfrac{2x}{5y}\sqrt{y}$ $\qquad\qquad = \dfrac{\sqrt{5xy}}{4x} \text{ or } \dfrac{1}{4x}\sqrt{5xy}$

35. $\dfrac{-(-2) - \sqrt{(-2)^2 - 4(6)(-3)}}{2(6)} = \dfrac{2 - \sqrt{4[1 - 6(-3)]}}{2(6)}$

$\qquad\qquad\qquad\qquad\qquad = \dfrac{2 - 2\sqrt{1 + 18}}{2(6)}$

$\qquad\qquad\qquad\qquad\qquad = \dfrac{2(1 - \sqrt{19})}{2(6)}$

$\qquad\qquad\qquad\qquad\qquad = \dfrac{1 - \sqrt{19}}{6}$

37. $\dfrac{-(-12) + \sqrt{(-12)^2 - 4(1)(-1)}}{2(1)} = \dfrac{12 + \sqrt{144 + 4}}{2}$

$\qquad\qquad\qquad\qquad\qquad\quad = \dfrac{12 + \sqrt{148}}{2}$

$\qquad\qquad\qquad\qquad\qquad\quad = \dfrac{12 + \sqrt{2^2 \cdot 37}}{2}$

$\qquad\qquad\qquad\qquad\qquad\quad = \dfrac{12 + 2\sqrt{37}}{2}$

$\qquad\qquad\qquad\qquad\qquad\quad = \dfrac{2(6 + \sqrt{37})}{2}$

$\qquad\qquad\qquad\qquad\qquad\quad = 6 + \sqrt{37}$

Level 2, page 219

39. Begin with a diagram, and use the Pythagorean theorem. The length of the ladder (26-ft) forms the hypotenuse of a right triangle and one leg of that same triangle is 10 ft.

$$a^2 + b^2 = c^2$$
$$a^2 + 10^2 = 26^2$$
$$a^2 = 676 - 100$$
$$a = \sqrt{576}$$
$$= 24$$

The ladder will reach 24 ft up the wall of the building.

Page 118

41. Saying that the carpenter "wants to be sure that the corner is square" means she wants the sides to meet at a right angle, so we use the Pythagorean theorem with legs 6 ft and 8 ft.

$$a^2 + b^2 = c^2$$
$$6^2 + 8^2 = c^2$$
$$100 = c^2$$
$$10 = c$$

The length of the diagonal must be 10 ft.

43. $c = \sqrt{a^2 + b^2}$
$$= \sqrt{2^2 + 2^2}$$
$$= \sqrt{2^2(1+1)}$$
$$= 2\sqrt{2}$$

The word "exact" means that you do not want to approximate this answer using your calculator. The exact length of the hypotenuse is $2\sqrt{2}$ inches.

45. The diagonal path across the lot is the length of the hypotenuse of a right triangle.

$c = \sqrt{a^2 + b^2}$
$$= \sqrt{400^2 + 300^2}$$
$$= \sqrt{(100 \cdot 4)^2 + (100 \cdot 3)^2}$$
$$= 100\sqrt{16 + 9}$$
$$= 100\sqrt{25}$$
$$= 100(5)$$
$$= 500$$

Walking the length and the width is 700 ft (400 ft + 300 ft) and walking the diagonal is 500 ft, so 200 ft (700 ft − 500 ft) is saved by walking diagonally across the lot. Be sure you answer the question asked and don't stop at the end of the calculation.

47. A sketch may help; in three seconds the balloon rises 12 ft (3 · 4 ft) and it travels horizontally 9 ft (3 · 3 ft). The distance the balloon travels is the length of the hypotenuse of a right triangle, so we use the Pythagorean theorem.

$c = \sqrt{a^2 + b^2}$
$$= \sqrt{12^2 + 9^2}$$
$$= \sqrt{144 + 81}$$
$$= \sqrt{225}$$
$$= 15$$

In three seconds, the balloon travels 15 ft.

Level 3, page 219

49. There are infinitely many possibilities. You can write a decimal which does not terminate or repeat such as $1.2323323332\cdots$. (Can you make up a different one like this?) You may also notice that since $1 = \sqrt{1}$ and $3 = \sqrt{9}$, you can write numbers such as $\sqrt{2}$ or $\sqrt{8} = 2\sqrt{2}$. (Can you make up a different one like these?)

51. There are infinitely many possibilities. You can write a decimal which does not terminate or repeat such as $0.0919919991\cdots$. (Can you make up a different one like this?) You may also notice that since $\frac{1}{10} = \frac{1}{\sqrt{100}}$ and $\frac{1}{11} = \frac{1}{\sqrt{121}}$, you can write numbers such as $\frac{1}{\sqrt{101}}$ or $\frac{1}{\sqrt{120}}$. (Can you make up a different one like these?)

53. Since all the plates are of uniform thickness, the choice need not be based on the volume, but rather area. The solution comes from the Pythagorean theorem, by noting that we need to obtain the sum of the area of the two smaller squares and compare this sum to the area of the one larger square.

    a. The sum of the two smaller squares is $(1 \text{ in.})^2 + (1 \text{ in.})^2 = 2 \text{ in.}^2$.
        The larger square has area $(2 \text{ in.})^2 = 4 \text{ in.}^2$
        Take the 2-in. square.

    b. The sum of the two smaller squares is $(3 \text{ in.})^2 + (4 \text{ in.})^2 = 25 \text{ in.}^2$.
        The larger square has area $(5 \text{ in.})^2 = 25 \text{ in.}^2$
        They are the same, so take either.

    c. The sum of the two smaller squares is $(4 \text{ in.})^2 + (5 \text{ in.})^2 = 41 \text{ in.}^2$.
        The larger square has area $(7 \text{ in.})^2 = 49 \text{ in.}^2$
        Take the 7-in. square.

    d. The sum of the two smaller squares is $(5 \text{ in.})^2 + (7 \text{ in.})^2 = 74 \text{ in.}^2$.
        The larger square has area $(9 \text{ in.})^2 = 81 \text{ in.}^2$
        Take the 9-in. square.

    e. The sum of the two smaller squares is $(10 \text{ in.})^2 + (11 \text{ in.})^2 = 221 \text{ in.}^2$.
        The larger square has area $(15 \text{ in.})^2 = 225 \text{ in.}^2$
        Take the 15-in. square.

Level 3 Problem Solving, page 220

55. You need to copy the diagram and cut out the pieces and then answer this question by trial and error.

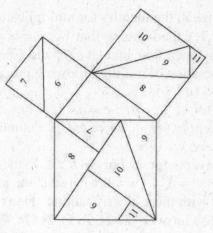

57. It looks like he is saying "Look at this great result!" (Eureka!); any number can be written as the sum of three triangular numbers. He must have meant 3 or fewer triangular numbers, because to represent 2, for example, we write 1 + 1.

59. You may need to look back at Pascal's triangle on page 9. Look at the second diagonal: 1, 3, 6, 10, 15, 21, · · · . If we take the sum of any 2 adjacent of these terms, we obtain square numbers: 4, 9, 16, 25, · · · . (*Note*: Don't forget that we start counting with the 0th diagonal.)

5.6 Groups, Fields, and Real Numbers, page 220

New Terms Introduced in this Section

Abelian group	Additive identity	Additive inverse	Commutative group
Dense set	Field	Group	Identity for addition
Identity for multiplication	Inverse for addition	Inverse for mult.	Multiplicative identity
Multiplicative inverse	Number line	One	Opposites
Real number line	Real numbers	Reciprocal	Repeating decimal
Terminating decimal	Unit distance	Zero	

Level 1, page 228

1. Rational numbers have repeating or terminating decimal representations, and irrational numbers do not.

3. The set $\mathbb{S}$ satisfies the *identity* for $\circ$: There exists a number $I \in \mathbb{S}$ so that $x \circ I = I \circ x = x$ for *every* $x \in \mathbb{S}$. You might also mention the identity for addition in the set of real numbers: There exists in $\mathbb{R}$ a number 0, called **zero**, so that $0 + a = a + 0 = a$ for any

Page 121

$a \in \mathbb{R}$. Finally, you might recall the identity for multiplication in the set of real numbers: There exists in $\mathbb{R}$ a number 1, called **one**, so that $1 \times a = a \times 1 = a$ for any $a \in \mathbb{R}$.

5. Let $\mathbb{S}$ be any set, let $\circ$ be any operation, and let a, b, and c be any elements of $\mathbb{S}$. We say that $\mathbb{S}$ is a **group** for the operation of $\circ$ if the following properties are satisfied:
   1. The set $\mathbb{S}$ is *closed* for $\circ$: $(a \circ b) \in \mathbb{S}$.
   2. The set $\mathbb{S}$ is *associative* for $\circ$: $(a \circ b) \circ c = a \circ (b \circ c)$.
   3. The set $\mathbb{S}$ satisfies the *identity* for $\circ$: There exists a number $I \in \mathbb{S}$ so that
      $x \circ I = I \circ x = x$ for *every* $x \in \mathbb{S}$.
   4. The set $\mathbb{S}$ satisfies the *inverse* for $\circ$: For *each* $x \in \mathbb{S}$, there exists a corresponding
      $x^{-1} \in \mathbb{S}$ so that $x \circ x^{-1} = x^{-1} \circ x = I$, where I is the identity element in $\mathbb{S}$.

7. Use Figure 5.16 to help you with these classifications. Don't forget that you may need to list more than one set for each answer. **a.** $\mathbb{N}, \mathbb{Z}, \mathbb{Q}, \mathbb{R}$ **b.** $\mathbb{Q}, \mathbb{R}$ **c.** $\mathbb{Q}', \mathbb{R}$ **d.** $\mathbb{Q}', \mathbb{R}$ **e.** $\mathbb{Q}, \mathbb{R}$ **f.** $\mathbb{Q}, \mathbb{R}$ **g.** $\mathbb{Q}, \mathbb{R}$ **h.** $\mathbb{N}, \mathbb{Z}, \mathbb{Q}, \mathbb{R}$ **i.** $\mathbb{Z}, \mathbb{Q}, \mathbb{R}$

9. **a.** $\frac{3}{2}$ is $3 \div 2 = 1.5$
   b. $\frac{7}{10}$ is $7 \div 10 = 0.7$
   c. $\frac{3}{5}$ is $3 \div 5 = 0.6$
   d. $\frac{27}{15}$ is $27 \div 15 = 1.8$

11. **a.** $\frac{2}{3}$ is $2 \div 3 = 0.\overline{6}$
    b. $2\frac{2}{13}$ is $2 + 2 \div 13 = 2.\overline{153846}$
    c. $\frac{15}{3}$ is $15 \div 3 = 5$
    d. $\frac{12}{11}$ is $12 \div 11 = 1.\overline{09}$

13. **a.** $0.5 = \frac{5}{10} = \frac{1}{2}$
    b. $0.8 = \frac{8}{10} = \frac{4}{5}$

15. **a.** $0.45 = \frac{45}{100} = \frac{5 \cdot 9}{5 \cdot 20} = \frac{9}{20}$
    b. $0.234 = \frac{234}{1,000} = \frac{2 \cdot 117}{2 \cdot 500} = \frac{117}{500}$

17. **a.** $98.7 = \frac{987}{10}$
    b. $0.63 = \frac{63}{100}$

19. **a.** $15.3 = \frac{153}{10}$
    b. $6.95 = \frac{695}{100} = \frac{5 \cdot 139}{5 \cdot 20} = \frac{139}{20}$

21. Use your calculator, or be sure to align the decimal points if you are not using a calculator.
    a. $6.28 - 3.101 = 3.179$
    b. $-6.824 + 1.32 = -5.504$
    c. $1.36 + 0.541 = 1.901$
    d. $6.31 - 12.62 = -6.31$

23. **a.** $-0.44 \times 0.298 = -0.13112$

Page 122

   b. $-10.5(6.23) = -65.415$

   c. $3.72 \div 0.3 = 12.4$

   d. $(-5.95) \div (-7.00) = 0.85$

25. **a.** $5.2 \times 2.3 - 4.5 = 7.46$; remember to do multiplication before subtraction.

   b. $5.2 - 2.3 \times 4.5 = -5.15$; remember to do multiplication before subtraction.

   c. $8.2 + 2.8 \times 23 = 72.6$; remember to do multiplication before addition.

   d. $8.2 \times 2.8 + 23 = 45.96$; remember to do multiplication before addition.

27. **a.** The order has changed, so it is the commutative property (for addition).

   b. The 3 (outside number) is distributed to each of the inside numbers to change the product to a sum; it is the distributive property.

29. **a.** The order of adding the ingredients has changed, so this is the commutative property.

   b. The order has not changed, but the groupings have changed, so this is an example of the associative property (for addition).

31. Notice that $\frac{\frac{1}{x^2}}{\frac{1}{x^2}} = 1$ and any number can be multiplied by 1 without changing its value according to the identity property (for multiplication).

33. Look at the table and notice that when any symbol is combined with itself, as in 子 ◎ 子 or 丑 ◎ 丑 in each case gives the same symbol; for example 寅 ◎ 寅 = 寅. Another way of saying this is $x \,◎\, x = x$ for any element x in the Chinese zodiac.

35. For an identity element there would need to be an element of Chinese zodiac that could be combined with any element without changing that element; no such identity element exists.

Level 2, page 229

37. $\mathbb{N}$ (the natural numbers) is not a group for subtraction $(-)$ since it is not closed; that is $5 - 8 = -3$, but -3 is not a natural number.

39. $\mathbb{W}$ (the whole numbers) is not a group for $\times$ since the inverse property is not satisfied; that is, the reciprocals of $\{2, 3, 4, \cdots\}$ are not elements of the set of whole numbers.

41. $\mathbb{Z}$ (the integers) is not a group for $\times$ since the inverse property is not satisfied; that is, the reciprocals of the elements in the set are not in the set of integers.

43. $\mathbb{Q}$ (the rational numbers) is not a group for $\times$ since there is no multiplicative inverse for the rational number 0. Be careful here about the definitions. You should distinguish among the inverse property for *multiplication* (page 226), the inverse property for a group (page 226), and the inverse property for a *field* of real numbers (page 228). Notice that the inverse property for a group uses a general operation ∘ so that zero is not excluded, whereas the inverse property for multiplication and the multiplicative inverse property for a field both exclude zero. Therefore, when determining whether a set forms a group for a

given operations we must be certain that *all* the elements of the set have an inverse. *Note*: although the set of rational numbers is not a group for multiplication, the set of *nonzero* rational numbers *is* a group for multiplication.

45. a.

×	1	2	3	4
1	1	2	3	4
2	2	4	6	8
3	3	6	9	12
4	4	8	12	16

b.

*	1	2	3	4
1	2	2	2	2
2	4	4	4	4
3	6	6	6	6
4	8	8	8	8

c. Operation × :

Not closed, since there are numbers in the table that are not in the given set.

Associative, since the set of real numbers is associative for multiplication.

The identity is 1 (from the table).

Except for the number 1, none of the elements have inverses.

The table is symmetric with respect to the principle diagonal, so it is commutative for multiplication.

Operation *:

Not closed, since there are numbers in the table that are not in the given set.

Not associative; for example,

$$(1*2)*3 = 2*3 = 4 \qquad \text{but} \qquad 1*(2*3) = 1*4 = 2$$

The operation of * has no identity, and consequently also has no inverse property.

The table is not symmetric (look at 3*2 and 2*3, as well as others), so the operation of * is not commutative.

Level 3, page 230

Answers for Problems 46-50 may vary.

47. There are infinitely many possibilities.

Two such rational numbers are 2.5 and $\frac{17}{7} = 2.\overline{428571}$; see if you can find another.

Two such irrational numbers are $\sqrt{5}$ and $\frac{2\pi}{3}$. To see how we found these, note that $2 = \sqrt{4}$ and $3 = \sqrt{9}$, so the numbers $\sqrt{5}$, $\sqrt{6}$, $\sqrt{7}$ and $\sqrt{8}$ are between 2 and 3. We also know that π is an irrational number a little larger than three so $\frac{2}{3}(\pi)$ is a little more than 2, but less than 3. You can also write a decimal that does not repeat nor terminate: $2.12112111211112\cdots$. See if you find another.

49. $4.\overline{5}$; $4.545545554\cdots$; see the solution to Problem 47 for an explanation.

51. We begin with halves, then thirds, then fourths (not listing $\frac{2}{4}$ because it is the same as $\frac{1}{2}$ which has been previously listed); then fifths, $\cdots$. Here is the set:

Page 124

$$\left\{ \frac{1}{2}, \frac{1}{3}, \frac{2}{3}, \frac{1}{4}, \frac{3}{4}, \frac{1}{5}, \frac{2}{5}, \frac{3}{5}, \frac{4}{5}, \frac{1}{6}, \frac{5}{6}, \frac{1}{7}, \frac{2}{7}, \cdots \right\}$$

53. It looks like normal multiplication, so we write $a \circ b = ab$.

55. It looks like normal addition, but the answers are one more than they should be, so we write $a \circ b = a + b + 1$.

57. This looks like normal addition, but the answers are twice what they should be, so we write $a \circ b = 2(a + b)$.

59. This one is tough; look at the first numbers only and note that $4^2 = 16$; $5^2 = 25$; $6^2 = 36$; and $2^2 = 4$. What do you see? The answers are one more than the square of the first digit, so we write $a \circ b = a^2 + 1$.

5.7 Discrete Mathematics, page 231
New Terms Introduced in this Section

Algebra	Congruent modulo m	Discrete mathematics	Multiplication
Subtraction	Zero multiplication		

Level 1, page 238

1. Clock arithmetic refers to a modulo system. A twelve-hour clock is a modulo 12 system (with $12 \equiv 0$) and a seven-hour clock is a modulo 7 system.

3. Two numbers are congruent modulo m if the remainders are the same when divided by m.

5. a. $9 + 6 = ?$ Think of this in terms of a 12-hour clock: "If it is 9:00 AM now, at what time will it be 6 hours from now?" We see $9 + 6 \equiv 3$, (mod 12).

 b. $5 - 7 = x$ means $7 + x = 5$. "If it is 7 AM now, in how many hours will it be 5 PM? We see $5 - 7 \equiv 10$, (mod 12).

 c. $5 \times 3 = 5 + 5 + 5 = ?$ Think of a 12-hour clock. "If it is 5:00 AM now, in 5 hours it is 10 AM and then 5 hours still later it is 3 PM. Thus, $5 \times 3 \equiv 3$, (mod 12).

 d. $2 \times 7 = 7 + 7 = ?$ Think of a 12-hour clock. "If it is 7:00 AM now, in 7 hours it is 2 PM. Thus, $2 \times 7 \equiv 2$, (mod 12).

7. a. $5 + 7 = ?$ Think of this in terms of a 12-hour clock: "If it is 5:00 AM now, at what time will it be 7 hours from now?" We see $5 + 7 \equiv 12$, (mod 12; remember on a 12-hour clock we use 12 instead of 0 in mod 12).

 b. $4 - 8 = x$ means $8 + x = 4$. "If it is 8 AM now, in how many hours will it be 4 PM? We see $4 - 8 \equiv 8$, (mod 12).

 c. $2 - 6 = x$ means $6 + x = 2$. "If it is 6 AM now, in how many hours will it be 2 PM? We see $2 - 8 \equiv 8$, (mod 12).

 d. $9 \times 3 = 9 + 9 + 9 = ?$ Think of a 12-hour clock. "If it is 9:00 AM now, in 9 hours it is 6 AM and then 9 hours still later it is 3 PM. Thus, $9 \times 3 \equiv 3$, (mod 12).

9. a. $3 \times 5 - 7 = 15 - 7 = 8$

   b. $7 + 3 \times 2 = 7 + 6 = 13 \equiv 1, \ (\text{mod } 12)$

11. a. $5 + 8 = 13 \equiv 1, \ (\text{mod } 6)$, so the statement is true (T).

   b. $4 + 5 = 9 \equiv 2, \ (\text{mod } 7)$, so the statement is false (F).

13. a. $47 \equiv 2, \ (\text{mod } 5)$, so the statement is true (T).

   b. $108 \equiv 4, \ (\text{mod } 8)$ and $12 \equiv 4, \ (\text{mod } 8)$, so the statement is true (T).

15. a. $2{,}007 \equiv 0, \ (\text{mod } 2{,}007)$, so the statement is true (T).

   b. $246 \equiv 0, \ (\text{mod } 6)$ and $150 \equiv 0, \ (\text{mod } 6)$, so the statement is true (T).

17. a. $9 + 6 = 15 \equiv 0, \ (\text{mod } 5)$

   b. $7 - 11 \equiv 19 - 11 = 8, \ (\text{mod } 5)$

   c. $4 \times 3 = 12 \equiv 2, \ (\text{mod } 5)$

   d. $1 \div 2 = x$ means that $2x = 1$. We know, $2 \times 0 = 0$, $2 \times 1 = 2$, $2 \times 2 = 4$,
$2 \times 3 = 6 \equiv 1, \ (\text{mod } 5)$, $2 \times 4 \equiv 8 \equiv 3, (\text{mod } 5)$.

19. a. $4 + 3 = 7 \equiv 2, \ (\text{mod } 5)$

   b. $6 - 12 \equiv 14 - 12 = 2, \ (\text{mod } 8)$

21. a. $7 + 41 = 48 \equiv 3, \ (\text{mod } 5)$

   b. $4 - 5 \equiv 15 - 5 = 10, \ (\text{mod } 11)$

23. a. We check all possibilities. $\quad x = 0: \quad 0 + 3 \equiv 3, \qquad (\text{mod } 7)$ Ans: $x \equiv 4, \ (\text{mod } 7)$
$x = 1: \quad 1 + 3 \equiv 4, \qquad (\text{mod } 7)$
$x = 2: \quad 2 + 3 \equiv 5, \qquad (\text{mod } 7)$
$x = 3: \quad 3 + 3 \equiv 6, \qquad (\text{mod } 7)$
$x = 4: \quad 4 + 3 = 7 \equiv 0, \quad (\text{mod } 7)$
$x = 5: \quad 5 + 3 = 8 \equiv 1, \quad (\text{mod } 7)$
$x = 6: \quad 6 + 3 = 9 \equiv 2, \quad (\text{mod } 7)$

   b. $x = 0: \quad 4(0) \equiv 0, \qquad (\text{mod } 5)$ Ans: $x \equiv 4, \ (\text{mod } 5)$
$x = 1: \quad 4(1) \equiv 4, \qquad (\text{mod } 5)$
$x = 2: \quad 4(2) = 8 \equiv 3, \quad (\text{mod } 5)$
$x = 3: \quad 4(3) = 12 \equiv 2, \quad (\text{mod } 5)$
$x = 4: \quad 4(4) = 16 \equiv 1, \quad (\text{mod } 5)$

25. a. We check all possibilities. $\quad x = 0: \quad 0 - 2 \equiv 6 - 2 = 4, \quad (\text{mod } 6)$ Ans: $x \equiv 5, \ (\text{mod } 6)$
$x = 1: \quad 1 - 2 \equiv 7 - 2 = 5, \quad (\text{mod } 6)$
$x = 2: \quad 2 - 2 \equiv 0, \qquad (\text{mod } 6)$
$x = 3: \quad 3 - 2 \equiv 1, \qquad (\text{mod } 6)$
$x = 4 \quad 4 - 2 \equiv 2, \qquad (\text{mod } 6)$
$x = 5: \quad 5 - 2 \equiv 3, \qquad (\text{mod } 6)$

b. $x = 0$: $3(0) \equiv 0$,   (mod 7) Ans: $x \equiv 3$, (mod 7)

   $x = 1$: $3(1) \equiv 3$,   (mod 7)

   $x = 2$: $3(2) \equiv 6$,   (mod 7)

   $x = 3$: $3(3) = 9 \equiv 2$,   (mod 7)

   $x = 4$ $3(4) = 12 \equiv 5$,   (mod 7)

   $x = 5$ $3(5) = 15 \equiv 1$,   (mod 7)

   $x = 6$ $3(6) = 18 \equiv 4$,   (mod 7)

27. a. We check all possibilities. $x = 0$: $0^2 \equiv 0$,   (mod 4) Ans: $x \equiv 1, 3$ (mod 4)

   $x = 1$: $1^2 \equiv 1$,   (mod 4)

   $x = 2$ $2^2 = 4 \equiv 0$,   (mod 4)

   $x = 3$: $3^2 = 9 \equiv 1$,   (mod 4)

   b. $x \div 4 \equiv 5$ can be rewritten as $\frac{x}{4} \equiv 5$ which means $x \equiv 4 \cdot 5 = 20 \equiv 2$, (mod 9).

29. a. Given that k represents any natural number, $4k$ represents 4, 8, 12, 16, $\cdots$. Since all of these numbers are multiples of 4, it follows that $4k \equiv 0$, (mod 4). Thus, the solution to the given equation is $x \equiv 0$, (mod 4).

   b. Given that k represents any natural number, $4k + 2$ represents 6, 10, 14, 18, $\cdots$. Since all of these numbers are congruent to 2, (mod 4), it follows that $4k + 2 \equiv 2$, (mod 4). Thus, the solution to the given equation is $x \equiv 2$, (mod 4).

31. We check all possibilities. $x = 0$: $5(0)^3 - 3(0)^2 + 70 = 70 \equiv 0$,   (mod 2)

   $x = 1$: $5(1)^3 - 3(1)^2 + 70 = 72 \equiv 0$,   (mod 2)

$x \equiv 0, 1$ (mod 2). Note, this is the set of all numbers, (mod 2).

Level 2, page 239

33. Since Friday is day 6, we see that Sunday must be day 1.

   a. $6 + 30 = 36 \equiv 1$, (mod 7); this is Sunday.

   b. $6 + 195 = 201 \equiv 5$, (mod 7); this is Thursday.

   c. $6 + 390 = 396 \equiv 4$, (mod 7); this is Wednesday.

   d. $6 + 3 \times 365 = 1,101 \equiv 2$, (mod 7); this is Monday.

35. Let $x =$ the number of miles to your aunt's house. Then the total mileage for the six round trips is $12x \equiv 8$, (mod 10).

$$x = 0: \quad 12(0) \equiv 0, \qquad (\text{mod } 10)$$
$$x = 1: \quad 12(1) = 12 \equiv 2, \quad (\text{mod } 10)$$
$$x = 2: \quad 12(2) = 24 \equiv 4, \quad (\text{mod } 10)$$
$$x = 3 \quad 12(3) = 36 \equiv 6 \quad (\text{mod } 10)$$
$$x = 4: \quad 12(4) = 48 \equiv 8, \quad (\text{mod } 10)$$
$$x = 5 \quad 12(5) = 60 \equiv 0, \quad (\text{mod } 10)$$
$$x = 6 \quad 12(6) = 72 \equiv 2, \quad (\text{mod } 10)$$
$$x = 7 \quad 12(7) = 84 \equiv 4, \quad (\text{mod } 10)$$
$$x = 8 \quad 12(8) = 96 \equiv 6 \quad (\text{mod } 10)$$
$$x = 9 \quad 12(9) = 108 \equiv 8, \quad (\text{mod } 10)$$

We see that $x \equiv 4, 9, (\text{mod } 10)$. The possible distances are 4, 14, 24, 34, $\cdots$ or 9, 19, 29, 39, $\cdots$.

37. Look at the possible answers in Problem 35. The one that is between 10 and 15 miles is 14, so she lives 14 miles from your house.

39. Let $A = \{0, 1, 2, 3, 4, 5, 6\}$. We use the definition of a group to check the appropriate properties.

Closure for $+$: The set A is closed for addition since any two elements of A is in the set A.

Associative for $+$: We check $(a + b) + c \equiv a + (b + c)$ for several possibilities:
$$(4 + 5) + 6 = 9 + 6 = 15 \equiv 1, \text{ mod } 7; \text{ also}$$
$$4 + (5 + 6) = 4 + 11 = 15 \equiv 1, \text{ mod } 7.$$
The associate property is satisfied for $a = 4$, $b = 5$, and $c = 6$.
Next, let $a = 4$, $b = 1$, and $c = 5$:
$$(4 + 1) + 5 = 10 \equiv 3, \text{ mod } 7; \text{ also}$$
$$4 + (1 + 5) = 4 + 6 = 10 \equiv 3; \text{ satisfied.}$$
Try another couple possibilities.

Identity for $+$: The identity element for addition is 0, since (in mod 7):

$$0 + 0 \equiv 0, 1 + 0 \equiv 1, 2 + 0 \equiv 2, 3 + 0 \equiv 3, 4 + 0 \equiv 4, 5 + 0 \equiv 5, \text{ and } 6 + 0 \equiv 6$$

This requirement is satisfied.

Inverse for +: The inverse for addition can be checked by finding the inverse of each element (mod 7).

Page 128

elements in the set   identity

The inverse of 0 is 0: $0 + 0 \equiv 0$

The inverse of 1 is 6: $1 + 6 \equiv 0$

The inverse of 2 is 5: $2 + 5 \equiv 0$

The inverse of 3 is 4: $3 + 4 \equiv 0$

The inverse of 4 is 3: $4 + 3 \equiv 0$

The inverse of 5 is 2: $5 + 2 \equiv 0$

The inverse of 6 is 1: $6 + 1 \equiv 0$

inverses

Since each element has an inverse for addition, we say the inverse property for addition is satisfied.

The set A is a group for addition modulo 7.

41.

+	0	1	2	3	4	5	6	7	8	9	10
0	0	1	2	3	4	5	6	7	8	9	10
1	1	2	3	4	5	6	7	8	9	10	0
2	2	3	4	5	6	7	8	9	10	0	1
3	3	4	5	6	7	8	9	10	0	1	2
4	4	5	6	7	8	9	10	0	1	2	3
5	5	6	7	8	9	10	0	1	2	3	4
6	6	7	8	9	10	0	1	2	3	4	5
7	7	8	9	10	0	1	2	3	4	5	6
8	8	9	10	0	1	2	3	4	5	6	7
9	9	10	0	1	2	3	4	5	6	7	8
10	10	0	1	2	3	4	5	6	7	8	9

×	0	1	2	3	4	5	6	7	8	9	10
0	0	0	0	0	0	0	0	0	0	0	0
1	0	1	2	3	4	5	6	7	8	9	10
2	0	2	4	6	8	10	1	3	5	7	9
3	0	3	6	9	1	4	7	10	2	5	8
4	0	4	8	1	5	9	2	6	10	3	7
5	0	5	10	4	9	3	8	2	7	1	6
6	0	6	1	7	2	8	3	9	4	10	5
7	0	7	3	10	6	2	9	5	1	8	4
8	0	8	5	2	10	7	4	1	9	6	3
9	0	9	7	5	3	1	10	8	6	4	2
10	0	10	9	8	7	6	5	4	3	2	1

43. No, it is not a group for multiplication modulo 11 because 0 does not have a multiplicative inverse.

45. No, because it is not a group (see Problem 43), it cannot be a commutative group.

47. The $(19, 2)$ design requires us to mark off 18 points (labeled 1 to 18) on the circumference of a circle. Draw line segments connecting 1 and 2, 2 and 4, 3 and 6, 4 and 8, 5 and 10, 6 and 12, 7 and 14, 8 and 16, 9 and 18, 10 and 1, 11 and 3, 12 and 5, 13 and 7, 14 and 9, 15 and 11, 16 and 13, 17 and 15, and, finally, 18 and 17. Next, color alternative regions. One possibility is shown.

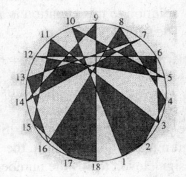

Page 129

49. The $(21, 5)$ design requires us to mark off 20 points
(labeled 1 to 20) on the circumference of a circle.
Draw line segments connecting 1 and 5, 2 and 10,
3 and 15, 4 and 20, 5 and 4, 6 and 9, 7 and 14,
8 and 19, 9 and 3, 10 and 8, 11 and 13, 12 and 18,
13 and 2, 14 and 7, 15 and 12, 16 and 17, 17 and 1,
18 and 6, 19 and 13, and, finally, 20 and 17. Next,
color alternative regions. One possibility is shown.

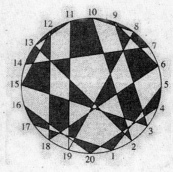

51. The $(65, 2)$ design requires us to mark off 64 points
(labeled 1 to 64) on the circumference of a circle.
Draw line segments connecting the appropriate
points and then color alternative regions.
One possibility is shown.

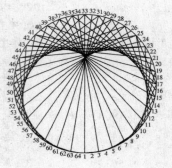

Level 3 Problem Solving, page 240

53. a. The sum is

$$10(0) + 9(5) + 8(3) + 7(4) + 6(1) + 5(3) + 4(7) + 3(2) + 2(8) = 168 \equiv 3, \text{ (mod 11)}$$

The check digit x must satisfy the equation $3 + x \equiv 0$, (mod 11). The one digit
solution to this equation is $x = 8$. Thus, the check (last) digit is 8.

b. The sum is

$$10(0) + 9(6) + 8(9) + 7(1) + 6(0) + 5(2) + 4(3) + 3(5) + 2(6) = 182 \equiv 6, \text{ (mod 11)}$$

The check digit x must satisfy the equation $6 + x \equiv 0$, (mod 11). The one digit
solution to this equation is $x = 5$. Thus, the check (last) digit is 5.

55. The possible check digits for ISBNs are $\{0, 1, 2, 3, 4, 5, 6, 7, 8, 9, 10\}$. Since the check
digit must be a single digit, the X must stand for 10.

57. Assign eleven teams the numbers 1 to 11. On the first day, all teams whose sum is 1, (mod
11) play; day 2, matches those whose sum is 2, (mod 11); ... The last team plays the left

Page 130

over numbered team so each team plays every day. The schedule follows:

Day 1:	1-11;	2-10;	3-9;	4-8;	5-7;	6-12
Day 2:	2-11;	3-10;	4-9;	5-8;	6-7;	1-12
Day 3:	3-11;	4-10;	5-9;	6-8;	1-2;	7-12
Day 4:	4-11;	5-10;	6-9;	7-8;	1-3;	2-12
Day 5:	5-11;	6-10;	7-9;	1-4;	2-3;	8-12
Day 6;	6-11;	7-10;	8-9;	1-5;	2-4;	3-12
Day 7:	7-11;	8-10;	1-6;	2-5;	3-4;	9-12
Day 8:	9-10;	1-7;	2-6;	3-5;	8-11;	4-12
Day 9:	9-11;	1-8;	2-7;	3-6;	4-5;	10-12
Day 10:	10-11;	1-9;	2-8;	3-7;	4-6;	5-12
Day 11:	1-10;	2-9;	3-8;	4-7;	5-6;	11-12

Note: The answer is not unique. In fact, there are 39,916,800 possible solutions.

59. Let x represent the total number of doubloons. There are three equations.

(1) Since x coins divide equally among 17 pirates with 3 coins left over, we have $x \equiv 3$, (mod 17).

(2) After 6 pirates were killed in a fight, 11 pirates were left. Since dividing the doubloons among these remaining pirates resulted in a remainder of 4, we have $x \equiv 4$, (mod 11).

(3) After the shipwreck, 5 more pirated perished, leaving 6 pirates in all. When the doubloons were equally divided among these pirates, the remains was 5 coins. Thus, $x \equiv 5$, (mod 6).

Thus, the questions amounts to finding the smallest possible solution to the following system of congruence equations:

$$\begin{cases} x \equiv 3, \text{(mod 17)} \\ x \equiv 4, \text{(mod 11)} \\ x \equiv 5, \text{(mod 6)} \end{cases}$$

There are methods for solving systems such as this, called *Diophantine* equations. This topic is usually introduced in number theory, which is beyond the scope of this course. Instead, we will use a spreadsheet, computer program, or trial and error to find the smallest value satisfying these equations is 785. At first, there were 46 coins for each pirate plus 3 for the cook. After 6 were killed, there are 71 coins for each of the 11 pirates plus 4 for the cook. Finally, the 6 remaining pirates received 130 coins each with 5 for the cook.

5.8 Cryptography, page 240
New Terms Introduced in this Section

Ciphertext	Cryptography	Decoding key	Encoding key
Encript	Modular codes		

Level 1, page 244

1. Use the inner ring of Figure 5.24 to help you with this encoding:
14-5-22-5-18-29-19-1-25-29-14-5-22-5-18-28

3. Use the inner ring of Figure 5.24 to help you with this encoding:
25-15-21-29-2-5-20-29-25-15-21-18-29-12-9-6-5-28

5. Use the inner ring of Figure 5.24 to help you with this encoding:
ARE WE HAVING FUN YET

7. Use the inner ring of Figure 5.24 to help you with this encoding:
FAILURE TEACHES SUCCESS

9. The decoding key is the inverse of the encoding key, so the inverse of multiply by 8 is to divide by 8.

11. The inverse of multiplying by 20 and then adding 2 is to subtract 2 and then divide by 20.

13. The inverse of multiplying by 4, adding 2, and then doubling the result is to divide by 2, then subtract 2, then divide by 4.

Level 2, page 244

15. The process of encoding a message consists of four steps:
- Replace each letter, space, or punctuation mark with its numerical value;
- Apply the encoding key to these numerical values;
- Modulate (modulo 29); use Figure 5.24.
- Rewrite these modulated numbers in terms of letters/spaces/punctuation.

We began this process in Problem 1.

NEVER SAY NEVER	*Given*
14-5-22-5-18-29-19-1-25-29-14-5-22-5-18-28	*Original message*
42-15-66-15-54-87-57-3-75-87-42-15-66-15-54-84	*Encode (multiply by 3)*
13-15-8-15-25-29-28-3-17-29-13-15-8-15-25-26	*Modulate*
MOHOY .CQ MOHOYZ	*Coded message*

17. See the solution to Problem 15 above for the procedure. This problem is a continuation of Problem 3.

YOU BET YOUR LIFE.	*Given*
25-15-21-29-2-5-20-29-25-15-21-18-29-12-9-6-5-28	*Original*
55-35-47-63-9-15-45-63-55-35-47-41-63-29-23-17-15-61	*Encode*
26-6-18-5-9-15-16-5-26-6-18-12-5-29-23-17-15-3	*Modulate*
ZFREIOPEZFRLE WQOC	*Coded message*

19. To decode this message, you must reverse the previously described four-step process described in the solution to Problem 15.

XEJSBQ LEJSBQ ,. C RCGG UEQZ   Given
24-5-10-19-2-17-29-12-5-10-19-2-17-29-27-28-29-3-29-18-3-7-7-29-21-5-17-26

The decoding key is to divide by 3. Before doing this, we use Figure 5.24 to replace all the numbers on our list that are not divisible by 3 with numbers (congruent modulo 29) that are divisible by 3:

24-63-39-48-60-75-87-12-63-39-48-60-75-87-27-57-87-3-87-18-3-36-87-21-63-75-84

Now we apply the decoding key:

8-21-13-16-20-25-29-4-21-13-16-20-25-29-9-19-29-1-29-6-1-12-12-29-7-21-25-28

Finally use Figure 5.24 to replace each of these numbers with its corresponding symbol:

HUMPTY DUMPTY IS A FALL GUY.

21. To decode this message, you must reverse the previously described four-step process described in the solution to Problem 15.

SKAXBVIXAVXVGKQDV.WSYQCLT   Given
19-11-1-24-2-22-9-24-1-22-24-22-7-11-17-4-22-28-23-19-25-17-3-12-20

The decoding key is to add 7 and then divide by 2. First we add 7:
26-18-8-31-9-29-16-31-8-29-31-29-14-18-24-11-29-35-30-26-32-24-10-19-27

Next, replace all the numbers on our list that are not divisible by 2 with numbers (congruent modulo 29) that are divisible by 2:
26-18-8-60-38-58-16-60-8-58-60-58-14-18-24-40-58-64-30-26-32-24-10-48-56

Now divide each number by 2:
13-9-4-30-19-29-8-30-4-29-30-29-7-9-12-20-29-32-15-13-16-12-5-24-28

Finally use Figure 5.24 to replace each of these numbers with its corresponding symbol:

MIDAS HAD A GILT COMPLEX.

23. See the solution to Problem 15 above for the procedure.

Page 133

SHOW ME A DROPOUT FROM A DATA
PROCESSING SCHOOL AND I WILL
SHOW YOU A NINCOMPUTER.
19-8-15-23-29-13-5-29-1-29-4-18-15-16-15-21-20-29-6-18-15-13-29-1-29-4-1-20-1-29-
16-18-15-3-5-19-19-9-14-7-29-19-3-8-15-15-12-29-1-14-4-29-9-29-23-9-12-12-29-
19-8-15-23-29-25-15-21-29-1-29-14-9-14-3-15-13-16-21-20-5-18-28
Apply the two-step encoding key by multiplying by 3 and then adding 5:
62-29-50-74-92-44-20-92-8-92-17-59-50-53-50-68-65-92-23-59-50-44-92-8-92-17-8-65-8-92-
53-59-50-14-20-62-62-32-47-26-92-62-14-29-50-50-41-92-8-47-17-92-32-92-74-32-41-41-92
62-29-50-74-92-80-50-68-92-8-92-47-32-47-14-50-44-53-68-65-20-59-89

Modulate:
4-29-21-16-5-15-20-5-8-5-17-1-21-24-21-10-7-5-23-1-21-15-5-8-5-17-8-7-8-5-
24-1-21-14-20-4-4-3-18-26-5-4-14-29-21-21-12-5-8-18-17-5-3-5-16-3-12-12-5-
4-29-21-16-5-22-21-10-5-8-5-18-3-18-14-21-15-24-10-7-20-1-2

Finally use Figure 5.24 to replace each of these numbers with its corresponding symbol:
D UPEOTEHEQAUXUJGEWAUOEHEQHGHEXAUNTDDCRZEDN
UULEHRQECEPCLLED UPEVUJEHERCRNUOXJGTAB

Level 3 Problem Solving, page 245
25. First use the hint to conclude that the three-letter word GHV represents AND. Assuming
that G is A, H in N, and V is D, we fill in those letters of the puzzle:

AN N   A   A
GHJWHY NDW LOGKL TGRLBK,
     A D AND   N
UBLRGXV, GHV XYOZLD WH DZL
        D   A   AN
DWR VWS ZL RXBOJ G UGH MWX
A   A   N N
GOO LYGLWHZHSL.

Now notice the two-letter word WH. Since H is N, W must be a vowel. It can't be an A,
though, because G is A. Consequently, W must be either I or O. Based on the first two
words (GHJWHY and NDW), it seems more likely that the W is an O.

AN ON   O   A   A
GHJWHY NDW LOGKL TGRLBK,
     A D AND   ON
UBLRGXV, GHV XYOZLD WH DZL

Page 134

```
 O  DO        A    AN  O
DWR VWS ZL RXBOJ G  UGH MWX
A    A  ON  N
GOO LYGLWHZHSL.
```

A reasonable guess for the first word (GHJWHY) is ANYONE. If that is correct, the second word (NDW) is probably WHO. So let's assume that J is Y, Y is E, N is W, and D is H. If this is correct, the coded message becomes:

```
ANYONE WHO   A    A
GHJ WHY  NDW LOGKL TGRLBK,
      A  D AND  E   H ON H
UBLRGXV, GHV XYOZLD WH  DZL
HO  DO        YA  AN  O
DWR VWS ZL RXBOJ G  UGH MWX
A    EA  ON  N
GOO LYGLWHZHSL.
```

Now, the two-letter word ZL suggests Z is one of the two remaining vowels, either I or U. If Z is U, then L must be P, but the three-letter word DZL (along with the high number of Ls) makes us doubt the likelihood of this line of reasoning. So, it is more likely that Z is I and L is S or T. After testing both possibilities, we go with L being S, which gives:

```
ANYONE WHO S A S  A S
GHJ WHY  NDW LOGKL TGRLBK,
    S  A D AND  E  ISH ON HIS
UBLRGXV, GHV XYOZLD WH  DZL
HO  DO  IS      YA  AN  O
DWR VWS ZL RXBOJ G  UGH MWX
A    SEASONIN S
GOO LYGLWHZHSL.
```

It is now clear that S is G. Then R (from DWR) is probably T. Also, since the word GOO begins with an A and has two repeated letter, O must be L. All of this leads to the following translation:

```
ANYONE WHO SLA S  ATS
GHJ WHY  NDW LOGKL TGRLBK,
    STA D AND  ELISH ON HIS
UBLRGXV, GHV XYOZLD WH  DZL
HOT DOG IS T   LY A  AN  O
DWR  VWS ZL RXBOJ G UGH MWX
ALL SEASONINGS
GOO LYGLWHZHS L.
```

Finally, make some trial-and-error guesses. Using the context of the known letters, we guess that X must be R, and then UBLRGXV is MUSTARD. Replacing X with R, U with M, B with U, and R with T gives:

ANYONE WHO SLA S ATSU
GHJ WHY NDW LOGKL TGRLBK,
MUSTARD AND RELISH ON HIS
UBLRGXV, GHV XYOZLD WH DZL
HOT DOG IS TRULY A MAN   OR
DWR VWS ZL RXBOJ G UGH MWX
ALL SEASONINGS
GOO LYGLWHZHS L.

We can now complete the deciphered message:

ANYONE WHO SLAPS CATSUP,
MUSTARD, AND RELISH ON HIS HOT
DOG IS TRULY A MAN FOR
ALL SEASONINGS.

27. We begin with the hint to identify the vowels.
Focus on the two-letter word HR and three-letter word RHZ we guess that the letter H is the vowel O. After the H, the most frequent letter is W, so we guess it stands for either E or A. Trying each, we conclude it most likely stands for A, so we replace W by A.

```
      A     A   A    O
MZDGBWV-MVCWZ WTZ HR
   O    O   A     O
ZHMAD XHKEBWTEG NZHLTUCG
  A      O      O
TUCWV CELTZHEXCEB RHZ
  A      A   O
PCWBSCZ GBWBTHE
```

The next most frequently occurring vowel (from the hint) is the letter C, so it seems reasonable to assume this letter is E.

```
      A    EA   A    O
MZDGBWV-MVCWZ WTZ HR
   O    O    A     O   E
ZHMAD XHKEBWTEG NZHLTUCG
```

```
   EA  E     O   E     O
TUCWV CELTZHEXCEB RHZ
   EA  E      A   O
PCWBSCZ GBWBTHE
```

Now, it is tough and you will need to study what you have. Look again at HR and RHZ, and guess that the letter R represents F. This leads us to replace Z by R.

```
   R    A     EAR A R OF
MZDGBWV-MVCWZ WTZ HR
RO    O    A    RO    E
ZHMAD XHKEBWTEG NZHLTUCG
   EA  E   RO  E   FOR
TUCWV CELTZHEXCEB RHZ
   EA  ER    A  O
PCWBSCZ GBWBTHE
```

Now, look again at the vowels and the word WTZ to replace the T by I, so the remaining vowel K is U.

```
   R    A     EAR A IR OF
MZDGBWV-MVCWZ WTZ HR
RO    OU  AI    RO I  E
ZHMAD XHKEBWTEG NZHLTUCG
 I EA  E   IRO  E   FOR
TUCWV CELTZHEXCEB RHZ
   EA  ER    A  IO
PCWBSCZ GBWBTHE
```

The word MVCWZ looks like the word CLEAR, so we replace the M with C and the V with L.

```
CR    A   CLEAR A IR OF
MZDGBWV-MVCWZ WTZ HR
ROC    OU  AI    RO I  E
ZHMAD XHKEBWTEG NZHLTUCG
 I EALE   IRO  E   FOR
TUCWV CELTZHEXCEB RHZ
   EA  ER    A  IO
PCWBSCZ GBWBTHE
```

What hyphenated word goes with CLEAR? We conclude it must be CRYSTAL, so we replace the D with Y, G with S, B with T, and V with L:

CRYSTAL CLEAR A IR OF
MZDGBWV-MVCWZ WTZ HR
ROC Y OU TAI S RO I ES
ZHMAD XHKEBWTEG NZHLTUCG
I EAL E IRO E T FOR
TUCWV CELTZHEXCEB RHZ
EAT ER STATIO
PCWBSCZ GBWBTHE

We are almost there now; the A in ZHMAD must be K, and the following word must then be MOUNTAINS, so we replace the X with M, and the E with N.

CRYSTAL CLEAR A IR OF
MZDGBWV-MVCWZ WTZ HR
ROCKY MOUNTA INS RO I ES
ZHMAD XHKEBWTEG NZHLTUCG
I EAL EN IRONMENT FOR
TUCWV CELTZHEX CEB RHZ
EAT ER STATION
PCWBSCZ GBWBTHE

We can now complete the deciphered message:
CRYSTAL-CLEAR AIR OF
ROCKY MOUNTAINS PROVIDES
IDEAL ENVIRONMENT FOR
WEATHER STATION.

Answers for Problems 28-30 are not unique.

29. You need to arrive at these numerals by trial and error. One possible answer is:

$$
\begin{array}{r}
3915 \\
15 \\
+\ 4826 \\
\hline
8756
\end{array}
$$

Chapter 5 Review Questions, page 246

Studying for a chapter examination is a personal process, one which nobody else can do for you. Simply take the time to review what you have done. Here are the new terms in Chapter 5.

Abelian group [5.6]	Addition [5.1]	Additive inverse [5.6]
Absolute value [5.3]	Additive identity [5.6]	Algebra [5.7]

Associative property [5.1]
Canonical form [5.2]
Ciphertext [5.8]
Closed for addition [5.1]
Closed for multiplication [5.1]
Closed set [5.1]
Closure property [5.1]
Commutative group [5.6]
Commutative property [5.1]
Composite number [5.2]
Congruent modulo m [5.7]
Counting numbers [5.1]
Cryptography [5.8]
Decoding key [5.8]
Denominator [5.4]
Dense set [5.6]
Discrete mathematics [5.7]
Distributive property [5.1]
Divides [5.2]
Divisibility [5.2]
Division [5.3; 5.7]
Division by zero [5.3]
Divisor [5.2]
e [5.5]
Encoding key [5.8]
Encrypt [5.8]
Factor [5.2]
Factor tree [5.2]
Factoring [5.2]
Field [5.6]

Fundamental property of
    fractions [5.4]
Fundamental theorem of
    arithmetic [5.2]
g.c.f. [5.2]
Greatest common factor [5.2]
Group [5.6]
Hypotenuse [5.5]
Identity for addition [5.6]
Identity for multiplication [5.6]
Improper fraction [5.4]
Integers [5.3]
Inverse for addition [5.6]
Inverse for multiplication [5.6]
Irrational number [5.5]
Laws of square roots [5.5]
l.c.m. [5.2]
Least common
        denominator [5.4]
Least common
        multiple [5.2]
Leg of a triangle [5.5]
Modular codes [5.8]
Modulo 5 [5.7]
Multiple [5.2]
Multiplication [5.1; 5.7]
Multiplicative identity [5.6]
Multiplicative inverse [5.6]
Natural numbers [5.1]
Number line [5.6]

Numerator [5.4]
One [5.6]
Opposites [5.6]
Perfect square [5.5]
π [5.5]
Positive square root [5.5]
Prime factorization [5.2]
Prime number [5.2]
Proper fraction [5.4]
Pythagorean theorem [5.5]
Radical form [5.5]
Radicand [5.5]
Rational number [5.4]
Real number line [5.6]
Real numbers [5.6]
Reciprocal [5.6]
Reduced fraction [5.4]
Relatively prime [5.2]
Repeating decimal [5.6]
Rules of divisibility [5.2]
Sieve of Eratosthenes [5.2]
Square number [5.5]
Square root [5.5]
Subtraction [5.1; 5.7]
Terminating decimal [5.6]
Unit distance [5.6]
Whole numbers [5.4]
Zero [5.6]
Zero multiplication [5.7]

If you can describe the term, read on to the next one; if you cannot, then look it up in the text (the section number is shown in brackets).   Next, review the types of problems in Chapter 5.

TYPES OF PROBLEMS
Demonstrate the definition of multiplication. [5.1]
Determine whether a given set with a given operation is closed. [5.1]
Recognize and distinguish the commutative and associative properties.[5.1]

Apply the distributive property with a variety of operations. [5.1]

Determine whether a given number is prime or not. [5.2]

Tell whether one number divides another number. [5.2]

Find the prime factorization and write the answer in canonical form. [5.2]

Find the least common multiple of a set of numbers. [5.2]

Find the greatest common factor of a set of numbers. [5.2]

Show that there is no largest prime number. [5.2]

Use problem-solving techniques to solve applied problems. [5.2-5.8]

Find the absolute value of a number. [5.3]

Carry out operations with integers. [5.3]

Reduce fractions using the fundamental property of fractions. [5.4]

Carry out operations with fractions. [5.4]

Use the definition of square root to simplify radical expressions. [5.5]

Classify numbers as rational or irrational. [5.5]

Determine into which of the following sets that a given number belongs: $\mathbb{N}$ (natural numbers),
$\mathbb{Z}$ (integers), $\mathbb{Q}$ (rational numbers), $\mathbb{Q}'$ (irrational numbers), or $\mathbb{R}$ (real numbers). [5.6]

Express a rational number as a decimal. [5.6]

Express a terminating decimal as a fraction. [5.6]

Use the order of operations to simplify real numbers. [5.6]

Find a rational number or an irrational number between each of a given pair of numbers. [5.6]

Recognize and distinguish examples of the closure, associative, commutative, identity, and
inverse properties.[5.6]

Know and be able to describe each of the field properties. [5.6]

Carry out operations in modular arithmetic. [5.7]

Solve modular equations. [5.7]

Decide if a given set and operation forms a group. [5.7]

Decide if a given set and two operations forms a field. [5.7]

Encode and decode simple phrases.[5.8]

Break a simple code. [5.8]

Once again, see if you can verbalize (to yourself) how to do each of the listed types of problems.

Work all of Chapter 5 Review Questions (whether they are assigned or not). Work through all of the problems before looking at the answers, and *then* correct each of the problems. The entire solution is shown in the answer section at the back of the text. If you worked the problem correctly, move on to the next problem, but if you did not work it correctly (or you did not know what to do), look back in the chapter to study the procedure, or ask your instructor.

Page 140

Finally, go back over the homework problems you have been assigned. If you worked a problem correctly, move on to the next problem, but if you missed it on your homework, then you should look back in the book or talk to your instructor about how to work the problem.

If you follow these steps, you should be successful with your review of this chapter.

We give all of the answers to the chapter review questions (not just the odd-numbered questions).

Chapter 5 Review Questions, page 246

1. $-4 + 5(-3) = -4 - 15$
$$= -19$$

2. $\dfrac{4}{7} \cdot \dfrac{9}{9} + \dfrac{5}{9} \cdot \dfrac{7}{7} = \dfrac{36}{63} + \dfrac{35}{63}$
$$= \dfrac{71}{63}$$

3.
$$30 = 2^1 \cdot 3^1 \cdot 5^1 \cdot 7^0 \cdot 11^0$$
$$42 = 2^1 \cdot 3^1 \cdot 5^0 \cdot 7^1 \cdot 11^0$$
$$99 = 2^0 \cdot 3^2 \cdot 5^0 \cdot 7^0 \cdot 11^1$$
$$\text{l.c.m.} = 2^1 \cdot 3^2 \cdot 5^1 \cdot 7^1 \cdot 11^1$$
$$= 6{,}930$$

$$\frac{7}{2 \cdot 3 \cdot 5} \cdot \frac{3 \cdot 7 \cdot 11}{3 \cdot 7 \cdot 11} = \frac{1{,}617}{2 \cdot 3^2 \cdot 5 \cdot 7 \cdot 11}$$
$$\frac{5}{2 \cdot 3 \cdot 7} \cdot \frac{3 \cdot 5 \cdot 11}{3 \cdot 5 \cdot 11} = \frac{825}{2 \cdot 3^2 \cdot 5 \cdot 7 \cdot 11}$$
$$\frac{5}{3^2 \cdot 11} \cdot \frac{2 \cdot 5 \cdot 7}{2 \cdot 5 \cdot 7} = \frac{350}{2 \cdot 3^2 \cdot 5 \cdot 7 \cdot 11}$$
$$\frac{2{,}792}{2 \cdot 3^2 \cdot 5 \cdot 7 \cdot 11} = \frac{1{,}396}{3{,}465}$$

4. $-\sqrt{10} \cdot \sqrt{10} = -10$

5. $\left(\dfrac{11}{12} + 2 \right) + \dfrac{-11}{12} = 2 + \left(\dfrac{11}{12} + \dfrac{-11}{12} \right)$
$$= 2$$

6. $\dfrac{3^{-1} + 4^{-1}}{6} = \dfrac{\frac{1}{3} + \frac{1}{4}}{6}$
$$= \dfrac{\frac{7}{12}}{6}$$
$$= \dfrac{7}{12} \times \dfrac{1}{6}$$
$$= \dfrac{7}{72}$$

7. $\dfrac{-7}{9} \cdot \dfrac{99}{174} + \dfrac{-7}{9} \cdot \dfrac{75}{174} = \dfrac{-7}{9}\left(\dfrac{99}{174} + \dfrac{75}{174}\right)$

$\qquad\qquad\qquad\qquad\qquad = \dfrac{-7}{9}(1)$

$\qquad\qquad\qquad\qquad\qquad = -\dfrac{7}{9}$

8. $\dfrac{-3 + \sqrt{3^2 + 4(2)(3)}}{2(2)} = \dfrac{-3 + \sqrt{33}}{4}$

9. $\frac{8}{3}$ (Note; $2\frac{2}{3}$ is mixed-number form, but $\frac{8}{3}$ is reduced.)

10. $\dfrac{16}{18} = \dfrac{2 \cdot 8}{2 \cdot 9}$

$\qquad\quad = \dfrac{8}{9}$

11. $\dfrac{100}{825} = \dfrac{25 \cdot 4}{25 \cdot 33}$

$\qquad\quad = \dfrac{4}{33}$

12. $\dfrac{184}{207} = \dfrac{23 \cdot 8}{23 \cdot 9}$

$\qquad\quad = \dfrac{8}{9}$

13. $\dfrac{1,209}{2,821} = \dfrac{3 \cdot 13 \cdot 31}{7 \cdot 13 \cdot 31}$

$\qquad\quad = \dfrac{3}{7}$

14. 0.375, rational since the decimal terminates

15. $2.\overline{3}$, rational since the decimal repeats

16. $0.\overline{428571}$, rational, since the decimal repeats

17. 6.25, rational, since the decimal terminates

18. $0.\overline{230769}$, rational, since the decimal repeats

19. The number 89 is not divisible by 2, 3, 5, or 7 (primes up to $\sqrt{89}$), so 89 is prime.

20. The number 101 is not divisible by 2, 3, 5, or 7 (primes up to $\sqrt{101}$), so 101 is prime.

21. The number 349 is not divisible by 2, 3, 3, 5, 7, 11, 13, or 17 (primes up to $\sqrt{349}$), so 349 is prime.

22. $1,001 = 7 \cdot 11 \cdot 13$ (use a factor tree)

23. $6,825 = 3 \cdot 5^2 \cdot 7 \cdot 13$ (use a factor tree)

24. $\frac{x}{5} \equiv 2$, (mod 8) means $x = 5 \cdot 2 = 10 \equiv 2$, (mod 8)

Page 142

25. $2x \equiv 3$, (mod 7); consider the set $\{0, 1, 2, 3, 4, 5, 6\}$ and we check all possibilities.

$x = 0$: $2(0) = 0 \equiv 0$, (mod 7)

$x = 1$: $2(1) = 2 \equiv 2$, (mod 7)

$x = 2$: $2(2) = 4 \equiv 4$, (mod 7)

$x = 3$: $2(3) = 6 \equiv 6$, (mod 7)

$x = 4$ $2(4) = 8 \equiv 1$ (mod 7)

$x = 5$: $2(5) = 10 \equiv 3$, (mod 7)

$x = 6$: $2(6) = 12 \equiv 5$, (mod 7)

We see $x \equiv 5$, (mod 7).

26. $2x^2 + 7x + 1 \equiv 0$, (mod 2); consider the set $\{0, 1\}$ and try each:

$x = 0$ $2(0)^2 + 7(0) + 1 = 1 \equiv 1$, (mod 2) Answer: $x \equiv 1$, (mod 2)

$x = 1$ $2(1)^2 + 7(1) + 1 = 10 \equiv 0$, (mod 2)

27. $49 = 7^2 \cdot 11^0 \cdot 13^0$

$1{,}001 = 7^1 \cdot 11^1 \cdot 13^1$

$2{,}401 = 7^4 \cdot 11^0 \cdot 13^0$

g.c.f. $= 7^1 \cdot 11^0 \cdot 13^0$

$= 7$

28. $49 = 7^2 \cdot 11^0 \cdot 13^0$

$1{,}001 = 7^1 \cdot 11^1 \cdot 13^1$

$2{,}401 = 7^4 \cdot 11^0 \cdot 13^0$

l.c.m. $= 7^4 \cdot 11^1 \cdot 13^1$

$= 343{,}343$

29. $(1 \odot 2) \odot (3 \odot 4) = (1 \times 2 + 1 + 2) \odot (3 \times 4 + 3 + 4)$

$= 5 \odot 19$

$= 5 \times 19 + 5 + 19$

$= 95 + 5 + 19$

$= 119$

30. $(1 \downarrow 2) \uparrow (2 \downarrow 3) = 1 \uparrow 2$

$= 2$

31. For $a \neq 0$, then multiplication is defined as: $a \times b$ means $\underbrace{b + b + b + \cdots + b}_{a \text{ addends}}$. If $a = 0$, then $0 \times b = 0$.

32. $a - b = a + (-b)$

33. $\frac{a}{b} = m$ means $a = bm$ where $b \neq 0$.

34. If we use the definition of division, $\frac{x}{0} = m$, then $0 \cdot m = x$. If $x \neq 0$, then there is no solution because $0 \cdot m = 0$ for all numbers m. Also, if $\frac{0}{0} = m$, then $0 = 0 \cdot m$ which is true for *every* number m. Thus, $\frac{0}{0} = 0$ checks, and $\frac{0}{0} = 1$ checks, and since two numbers equal to the same number must also be equal, we obtain the statement $0 = 1$; thus we say $\frac{0}{0}$ is indeterminate.

35. Answers vary; $34.1011011101111011\cdots$

36. A **field** is a set $\mathbb{R}$, with two operations $+$ and $\times$ satisfying the following properties for any elements $a, b, c \in \mathbb{R}$:

	Addition, $+$	Multiplication, $\times$
Closure	1. $(a + b) \in \mathbb{R}$	2. $ab \in \mathbb{R}$
Associative	3. $(a + b) + c = a + (b + c)$	4. $(a \times b) \times c = a \times (b \times c)$
Identity	5. There exists $0 \in \mathbb{R}$ so that $0 + a = a + 0 = a$ for every element a in $\mathbb{R}$.	6. There exists $1 \in \mathbb{R}$ so that $1 \times a = a \times 1 = a$ for every element a in $\mathbb{R}$.
Inverse	7. For each $a \in \mathbb{R}$, there is a unique number $(-a) \in \mathbb{R}$ so that $a + (-a) = (-a) + a = 0$	8. For each $a \in \mathbb{R}$, $a \neq 0$, there is a unique number $\frac{1}{a} \in \mathbb{R}$ so that $a \times \frac{1}{a} = \frac{1}{a} \times a = 1$
Commutative	9. $a + b = b + a$	10. $ab = ba$
Distributive	11. $a \times (b + c) = a \times b + a \times c$	

37. Let a, b, and c be any elements in $\mathbb{N}$.

Closure: $a \, D \, b \in \mathbb{N}$ because the g.c.f. is the product of factors and $\mathbb{N}$ is closed for multiplication.

Associative: $(a \, D \, b) \, D \, c = a \, D \, (b \, D \, c)$ because the g.c.f. is the product of factors and $\mathbb{N}$ is associative for multiplication.

Identity: Look for I so that $a \, D \, I = I \, D \, a = a$, but there is no such number.

Inverse: Since there is no identity, there can be no inverse property.

$\mathbb{N}$ is not a group for D, and therefore cannot be a commutative group.

38. $\text{BMI} = \dfrac{703w}{h^2}$ **a.** $\dfrac{703(165)}{65^2} \approx 27.5$ **b.** $\dfrac{703(185)}{72^2} \approx 25.1$

    c. Solve $25 = \frac{703w}{(5 \times 12 + 6)^2}$

$$25 = \frac{703w}{(5 \times 12 + 6)^2}$$ The desired weight is 155 lb.

$$25(5 \times 12 + 6)^2 = 703w$$

$$w = \frac{25(5 \times 12 + 6)^2}{703} \approx 154.9$$

39. You can use a sieve or prime factorizations. A door will be opened or closed by a tenant only if the door number can be divided evenly by the number of that tenant. For example, door 9 will be touched (opened or closed) by tenants 1, 3, and 9; door 10 by tenants 1, 2, 5, and 10. Thus, the only doors left open are those with an odd number of divisors. The open doors are the perfect squares: 1, 4, 9, 16, 25, 36, $\cdots$, 841, 900, 961. There are 31 doors left open.

40. 8 ft = 96 in. and $\frac{96}{8} = 12$, so 12 stairs are necessary. The total length of the segments is $12 + 8 = 20$ ft, and the length of the diagonal is $\sqrt{12^2 + 8^2} = \sqrt{208} = 4\sqrt{13}$.

CHAPTER 6 The Nature of Algebra

Chapter Overview
Because of the prerequisites for this course, you should have had some exposure to the material of this chapter. However, this chapter will provide a good review of algebra, as well as present a new perspective on many familiar ideas.

6.1 Polynomials, page 251
New Terms Introduced in this Section

Algebra	Binomial	Binomial theorem	Degree	Expand
FOIL	Like terms	Linear	Monomial	Numerical coefficient
Polynomial	Quadratic	Similar terms	Simplify	Term
Trinomial	Variable			

Level 1, page 257

1. A *polynomial* is a term or a sum of terms.

3. Look for similar terms to add and subtract. You should add a few words of justification showing how this is really an application of the distributive property. For example,
$$8x^2 - 5x^2 = (8 - 5)x^2 \text{ by the distributive property}$$

5. You should explain in your own words what we mean by FOIL. Begin by looking at page 253 and then put the mnemonic into words. You might conclude with a few words about why you think this process is important.

7. For any positive integer n,

$$(a + b)^n = \binom{n}{0}a^n + \binom{n}{1}a^{n-1}b + \binom{n}{2}a^{n-2}b^2 + \cdots + \binom{n}{n-1}ab^{n-1} + \binom{n}{n}b^n$$

where $\binom{n}{r}$ is the number in the nth row, rth diagonal of Pascal's triangle. You should conclude your response to this question by explaining in your own words when you would use this theorem and what all this notation is saying.

9. $(2x - 4) - (3x + 4) = 2x - 4 - 3x - 4$ This is a 1st-degree binomial.
$$= (2x - 3x) + (-4 - 4)$$
$$= -x - 8$$

11. $(x - y - z) + (2x - 5y - 3z) = (x + 2x) + (-y - 5y) + (-z - 3z)$
$$= 3x - 6y - 4z$$

This is a 1st-degree trinomial.

13. $(2x^2 - 5x + 4) + (3x^2 - 2x - 11) = (2x^2 + 3x^2) + (-5x - 2x) + (4 - 11)$
$$= 5x^2 - 7x - 7$$
This is a 2nd-degree trinomial.

15. $(x^2 + 4x - 3) - (2x^2 + 9x - 6) = x^2 + 4x - 3 - 2x^2 - 9x + 6$
$$= (x^2 - 2x^2) + (4x - 9x) + (-3 + 6)$$
$$= -x^2 - 5x + 3$$
This is a 2nd-degree trinomial.

17. $3(x - 5) - 2(x + 8) = 3x - 15 - 2x - 16$　This is a 1st degree binomial.
$$= x - 31$$

19. $3(2x^2 + 5x - 5) + 2(5x^2 - 3x + 6) = 6x^2 + 15x - 15 + 10x^2 - 6x + 12$
$$= 16x^2 + 9x - 3$$
This is a 2nd-degree trinomial.

21. $2(x + 3) - 3(x^2 - 3x + 1) + 4(x - 5) = 2x + 6 - 3x^2 + 9x - 3 + 4x - 20$
$$= -3x^2 + 15x - 17$$
This is a 2nd-degree trinomial.

23. a. $(x + 3)(x + 2) = x^2 + 5x + 6$
　　b. $(y + 1)(y + 5) = y^2 + 6y + 5$
　　c. $(z - 2)(z + 6) = z^2 + 4z - 12$
　　d. $(s + 5)(s - 4) = s^2 + s - 20$

25. a. $(c + 1)(c - 7) = c^2 - 6c - 7$
　　b. $(z - 3)(z + 5) = z^2 + 2z - 15$
　　c. $(2x + 1)(x - 1) = 2x^2 - x - 1$
　　d. $(2x - 3)(x - 1) = 2x^2 - 5x + 3$

27. a. $(x + y)(x + y) = x^2 + 2xy + y^2$
　　b. $(x - y)(x - y) = x^2 - 2xy + y^2$
　　c. $(x + y)(x - y) = x^2 - y^2$
　　d. $(a + b)(a - b) = a^2 - b^2$

29. a. $(x + 4)^2 = x^2 + 8x + 16$
　　b. $(y - 3)^2 = y^2 - 6y + 9$
　　c. $(s + t)^2 = s^2 + 2st + t^2$
　　d. $(u - v)^2 = u^2 - 2uv + v^2$

Level 2, page 257

31. $(2x - 1)(3x^2 + 2x - 5) = (2x - 1)3x^2 + (2x - 1)2x + (2x - 1)(-5)$
$$= 6x^3 - 3x^2 + 4x^2 - 2x - 10x + 5$$
$$= 6x^3 + x^2 - 12x + 5$$

Page 146

33. $(5x+1)(x^3 - 2x^2 + 3x) = (5x+1)x^3 + (5x+1)(-2x^2) + (5x+1)3x$
$$= 5x^4 + x^3 - 10x^3 - 2x^2 + 15x^2 + 3x$$
$$= 5x^4 - 9x^3 + 13x^2 + 3x$$

35. $3(3x^2 - 5x + 2) - 4(x^3 - 4x^2 + x - 4) = 9x^2 - 15x + 6 - 4x^3 + 16x^2 - 4x + 16$
$$= -4x^3 + 25x^2 - 19x + 22$$

37. $(2x-3)(3x+2) + (x+2)(x+3) = 6x^2 - 5x - 6 + x^2 + 5x + 6$
$$= 7x^2$$

39.

$x+4$

	x^2	x	x	x	x
x	x	1	1	1	1
x	x	1	1	1	1

$x+2$

$(x+2)(x+4) = x^2 + 6x + 8$

41.

$x+4$

	x^2	x	x	x	x
x	x	1	1	1	1
x	x	1	1	1	1
x	x	1	1	1	1

$x+3$

$(x+3)(x+4) = x^2 + 7x + 12$

43.

$3x+2$

	x^2	x^2	x^2	x	x
	x^2	x^2	x^2	x	x
	x	x	x	1	1
	x	x	x	1	1
	x	x	x	1	1

$2x+3$

$(2x+3)(3x+2) = 6x^2 + 13x + 6$

45. $(x-1)^3 = x^3 + 3x^2(-1) + 3x(-1)^2 + (-1)^3$
$$= x^3 - 3x^2 + 3x - 1$$

47. $(x+y)^6 = x^6 + 6x^5y + 15x^4y^2 + 20x^3y^3 + 15x^2y^4 + 6xy^5 + y^6$

Page 147

49. $(x - y)^8 = x^8 + 8x^7(-y) + 28x^6(-y)^2 + 56x^5(-y)^3 + 70x^4(-y)^4 + 56x^3(-y)^5$
$$+ 28x^2(-y)^6 + 8x(-y)^7 + (-y)^8$$
$$= x^8 - 8x^7y + 28x^6y^2 - 56x^5y^3 + 70x^4y^4 - 56x^3y^5 + 28x^2y^6 - 8xy^7 + y^8$$

51. $(2x - 3y)^4 = (2x)^4 + 4(2x)^3(-3y) + 6(2x)^2(-3y)^2 + 4(2x)(-3y)^3 + (-3y)^4$
$$= 16x^4 - 96x^3y + 216x^2y^2 - 216xy^3 + 81y^4$$

Level 3, page 258

53. You should look at the fourteenths row of Figure 6.1 for the coefficients.
$$(x + y)^{14} = \cdots + 91x^2y^{12} + 14xy^{13} + y^{14}$$

55. Since $(6x + 2)(51x - 7) = 306x^2 + 60x - 14$ and since the number of seats in an auditorium is the product of the number of seats in each row times the number of rows, we see that there are $306x^2 + 60x - 14$ seats in the auditorium.

57. Since $(6b + 15)(10 - 2b) = 150 + 30b - 12b^2$ and since the distance traveled is $d = rt$ for a rate of r and time t we see that the distance traveled by the boat is $150 + 30b - 12b^2$.

Level 3 Problem Solving, page 258

59. $(10x + y)(10x + z) = 100x^2 + 10xz + 10xy + yz$
$$= 100x^2 + 10x(z + y) + yz$$
$$= 100x^2 + 10x(10) + yz$$
$$= 100x^2 + 100x + yz$$
$$= 100(x^2 + x) + yz$$
$$= 100[x(x + 1)] + yz$$

6.2 Factoring, page 258
New Terms Introduced in this Section

Common factor   Completely factored   Difference of squares   Factor

Level 1, page 263

1. (1) look for common factor, then
(2) factor a difference of squares, and then
(3) use FOIL to factor if it is a trinomial.

3. $10xy - 6x = 2x(5y - 3)$ common factor

5. $8xy - 6x = 2x(4y - 3)$ common factor

7. $x^2 - 4x + 3 = (x - 3)(x - 1)$ FOIL

9. $x^2 - 5x + 6 = (x - 3)(x - 2)$ FOIL

Page 148

11. $x^2 - 7x + 12 = (x - 4)(x - 3)$ FOIL
13. $x^2 - x - 30 = (x - 6)(x + 5)$ FOIL
15. $x^2 - 2x - 35 = (x - 7)(x + 5)$ FOIL
17. $3x^2 + 7x - 10 = (3x + 10)(x - 1)$ FOIL
19. $2x^2 - 7x + 3 = (2x - 1)(x - 3)$ FOIL
21. $3x^2 - 5x - 2 = (3x + 1)(x - 2)$ FOIL
23. $2x^2 + 9x + 4 = (2x + 1)(x + 4)$ FOIL
25. $3x^2 + x - 2 = (3x - 2)(x + 1)$ FOIL
27. $5x^3 + 7x^2 - 6x = x(5x^2 + 7x - 6)$ common factor
$\qquad\qquad\qquad = x(5x - 3)(x + 2)$ FOIL
29. $7x^4 - 11x^3 - 6x^2 = x^2(7x^2 - 11x - 6)$ common factor
$\qquad\qquad\qquad = x^2(7x + 3)(x - 2)$ FOIL
31. $x^2 - 64 = (x - 8)(x + 8)$ difference of squares
33. $25x^2 + 50 = 25(x^2 + 2)$ common factor
35. $x^4 - 1 = (x^2 - 1)(x^2 + 1)$ difference of squares
$\qquad\quad = (x - 1)(x + 1)(x^2 + 1)$ difference of squares again

Level 2, page 263

37.

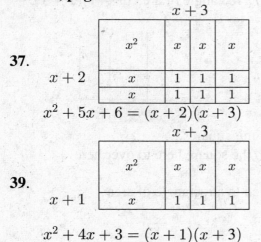

$x^2 + 5x + 6 = (x + 2)(x + 3)$

39.

$x^2 + 4x + 3 = (x + 1)(x + 3)$

Page 149

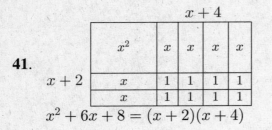

41.

$$x^2 + 6x + 8 = (x+2)(x+4)$$

In Problems 43-48, cut out the darker shaded portions and move the lightly shaded pieces.

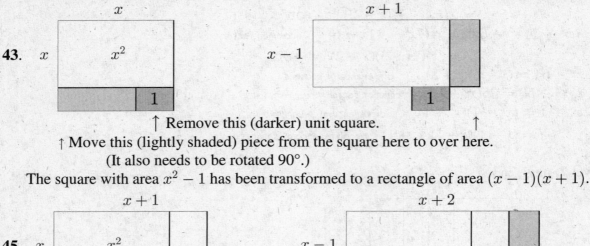

43.

↑ Remove this (darker) unit square.   ↑

↑ Move this (lightly shaded) piece from the square here to over here.

(It also needs to be rotated 90°.)

The square with area $x^2 - 1$ has been transformed to a rectangle of area $(x-1)(x+1)$.

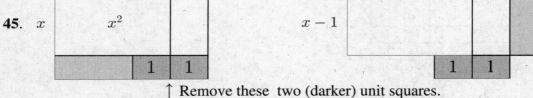

45.

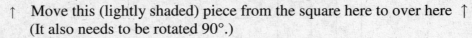

↑ Remove these two (darker) unit squares.

↑ Move this (lightly shaded) piece from the square here to over here ↑

(It also needs to be rotated 90°.)

The square with area $x^2 + x - 2$ has been transformed to a rectangle of area $(x-1)(x+2)$.

Page 150

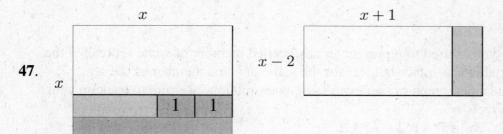

47.

First remove a rectangle from the bottom with width 1 unit and length x units. Now the width of the original rectangle is $x - 1$. Next, remove two unit squares (shown with darker shading). Finally, move the lightly shaded portion over as shown. The original was a square of area x^2 with a rectangle of length x and a height of 1 unit removed, as well as two unit squares cut off. What we are left with is a rectangle of length $x + 1$ and height $x - 2$. Thus, $x^2 - x - 2 = (x + 1)(x - 2)$.

Level 3, page 263

49. Since $x^2 - 2x - 143 = (x - 13)(x + 11)$ and since the formula for the area of a rectangle is $A = \ell w$, we see the dimensions of the rectangle are $x - 13$ feet by $x + 11$ feet.

51. Since $6x^2 + 5x - 4 = (3x + 4)(2x - 1)$ and since the formula for the distance traveled is $d = rt$, we are given that the rate is $3x + 4$ which means the time is $2x - 1$ hours.

53. $(x + 2)(x + 4) + (5x + 6)(x - 1) = x^2 + 6x + 8 + 5x^2 + x - 6$
$$= 6x^2 + 7x + 2$$
$$= (3x + 2)(2x + 1)$$

55. $x^6 - 13x^4 + 36x^2 = x^2(x^4 - 13x^2 + 36)$
$$= x^2(x^2 - 9)(x^2 - 4)$$
$$= x^2(x - 3)(x + 3)(x - 2)(x + 2)$$

57. $20x^2y^2 + 17x^2yz - 10x^2z^2 = x^2(20y^2 + 17yz - 10z^2)$
$$= x^2(4y + 5z)(5y - 2z)$$

Level 3, Problem Solving, page 263

59. Let $x - 1, x$, and $x + 1$ be the three integers. Then $(x - 1)(x + 1) = x^2 - 1$ so the square of the middle integer is 1 more than the product of the first and third.

6.3 Evaluation, Applications, and Spreadsheets, page 264

New Terms Introduced in this Section

Cell   Evaluate   Replication   Spreadsheet

Page 151

Level 1, page 271

1. A *variable* is a symbol used to represent an unspecified member of some set (called the domain). A variable is a "place holder" for the name of some member of the set.

3. You should write a paragraph or two expressing your opinion. Be sure to back up your statements with facts.

5. **a.** +(2/3)*A1^2   **b.** +5*A1^2 − 6*A2^2

7. **a.** +12*(A1^2+4)   **b.** +(15*A1+7)/2

9. **a.** +(5 − A1)*(A1+3)^2   **b.** +6*(A1+3)*(2*A1 − 7)^2

11. **a.** +(1/4)*A1^2 − (1/2)*A1+12   **b.** +(2/3)*A1^2+(1/3)*A1 − 17

13. **a.** $4x + 3$ **b.** $5x^2 − 3x + 4$

15. **a.** $\frac{5}{4}x + 14^2$ **b.** $(\frac{5}{4}x + 14)^2$

17. **a.** $\frac{x}{y}(z)$ **b.** $\frac{x}{yz}$

19. $A = x + z + 8$
$$= 1 + 4 + 8$$
$$= 13$$

21. $C = 10 − w$
$$= 10 − 2$$
$$= 8$$

23. $E = 25 − y^2$
$$= 25 − 2^2$$
$$= 25 − 4$$
$$= 21$$

25. $G = 5x + 3z + 2$
$$= 5(1) + 3(4) + 2$$
$$= 5 + 12 + 2$$
$$= 19$$

27. $I = 5y − 2z$
$$= 5(2) − 2(4)$$
$$= 10 − 8$$
$$= 2$$

29. $K = wxy$
$$= 2(1)(2)$$
$$= 4$$

Page 152

31. $M = (x + y)^2$
$ = (1 + 2)^2$
$ = 3^2$
$ = 9$

33. $P = y^2 + z^2$
$ = 2^2 + 4^2$
$ = 4 + 16$
$ = 20$

35. $R = z^2 - y^2 - x^2$
$ = 4^2 - 2^2 - 1^2$
$ = 16 - 4 - 1$
$ = 11$

37. $T = x^2 + y^2 z$
$ = 1^2 + 2^2(4)$
$ = 1 + 16$
$ = 17$

39. $V = \dfrac{3wyz}{x}$
$ = \dfrac{(3)(2)(2)(4)}{1}$
$ = 48$

41. $X = (x^2 z + x)^2 z$
$ = [1^2(4) + 1]^2(4)$
$ = [5]^2(4)$
$ = 25(4)$
$ = 100$

Level 2, page 272

43. In order to complete this problem you should first work Problems 19-42 (both even and odd problems), but as you get close to the end you will be able to guess parts of the saying. You can use this as a double-check on the solutions to Problems 19-42.

T	H	E		F	O	U	R		M	A	I	N
P	R	O	C	E	S	S	E	S		O	F	
A	L	G	E	B	R	A		A	R	E		
F	A	C	T	O	R	,	S	O	L	V	E	,
S	I	M	P	L	I	F	Y	,		A	N	D
E	V	A	L	U	A	T	E	.				

45. The given values are $A1 = 6$, $B1 = -4$, $C1 = 2$, and $D1 = 3$.

a. $+ A1 - B1/C1 = 6 - \dfrac{-4}{2}$

$$= 6 + 2$$
$$= 8$$

b. $+ (A1 - B1)/C1 = \dfrac{6 - (-4)}{2}$

$$= \dfrac{10}{2}$$
$$= 5$$

Note how the parentheses changes the order of operations in parts **a** and **b**.

c. $+A1/C1*D1 = \dfrac{6}{2} \cdot 3$

$$= 3 \cdot 3$$
$$= 9$$

d. $+ (A1 + \cdots + D1)/2 = \dfrac{6 + (-4) + 2 + 3}{2}$

$$= \dfrac{7}{2}$$
$$= 3.5$$

47. Remember that the $ symbol (as in A1) means that the reference to A1 should remain constant as it is replicated.

a.
b.

	A	B	C	D	E
1	1	3	4	5	6
2	1	3	3	3	3
3					
4					
5					

49. a. Enter 0 in cell B1. Then, the formula in cell C3 takes this 0 from B1 and adds 1, causing cell C3 to have the value of 1. Then, cell A2 references cell C3 and adds 1 to that value to be 2. Cell A3 uses the value of 2 (from cell A2) to compute its value

Page 154

$(2 + 1 = 3)$; be sure to continue in the same fashion, in order to obtain the following result:

	A	B	C
1	7	0	5
2	2	4	6
3	3	8	1

b. Follow the steps shown in part **a** except begin with the value 100 instead of 0. The result is shown.

	A	B	C
1	107	100	105
2	102	104	106
3	103	108	101

51.

	A	B	C	D	E
1	NAME	SALES	COST	PROFIT	COMMISSION
2				+B2-C2	+.08*D2
3	Replicate Row	2 for Rows 3 to 21.			
4					
5					

53. This problem is similar to Example 4. For this problem, $p = 0.65$ and $q = 0.35$.
Genotype:
black (BB): $p^2 = (0.65)^2 = 0.4225$ or 42.25%

black (recessive brown, Bb): $2pq = 2(0.65)(0.35) = 0.455$ or 45.5%.

brown (bb): $q^2 = (0.35)^2 = 0.1225$ or 12.25%

Phenotype (this refers to the physical appearance):
black: $0.4225 + 0.455 = 0.8775$ or 87.75%

brown: 12.25%

55. See Example 4; in this problem, the genotypes and phenotypes are the same where
$p = 0.20$ and $q = 0.80$.
red (rr): $p^2 = (0.20)^2 = 0.04$ or 4%

pink (rw or wr): $2pq = 2(0.20)(0.80) = 0.32$ or 32%

white (ww): $q^2 = (0.8)^2 = 0.64$ or 64%

Level 3 Problem Solving, page 273
57. $(B + b)^2 = B^2 + 2Bb + b^2$; if $b^2 = 0.25$, then $b = 0.5$, so $B = 1 - 0.5 = 0.5$. Thus, $B = 50\%$
and $b = 50\%$.

59. The answers for your class vary, of course, on your class. However, you might be interested to know that for the general population, FF is 49%, Ff is 42%, and ff is 9%; Since what you observe is the phenotype, the 91% of general population would have free hanging ear lobes with 9% attached.. For the general population we would estimate that since FF is 49% then *F* is 70%, and since ff is 9% so then *f* is 30%.

6.4 Equations, page 274
New Terms Introduced in this Section

Addition property	Division property	Equation	Equation properties
Equivalent equations	Linear equation	Multiplication property	Multiplicity
Quadratic equation	Quadratic formula	Root	Satisfy
Solution	Solve an equation	Subtraction property	
Symmetric property of equality		Zero-product rule	

Level 1, page 281

1. The procedure for solving a first-degree equation is to isolate the variable on one side of the equation. We do this by using the equation properties and the idea of opposite operations to decide which one of these properties to use.

3. a.
$x - 5 = 10$ *Given linear equation*
$x - 5 + 5 = 10 + 5$ *Add 5 to both sides.*
$x = 15$ *Simplify.*

b. $6 = x - 2$ *Given linear equation*
$6 + 2 = x - 2 + 2$ *Add 2 to both sides.*
$8 = x$ *Simplify.*

c. $8 + x = 4$ *Given linear equation*
$8 + x - 8 = 4 - 8$ *Sub. 8 from both sides.*
$x = -4$ *Simplify.*

d. $18 + x = 10$ *Given linear equation*
$18 + x - 18 = 10 - 18$ *Subtract 18 from both sides.*
$x = -8$ *Simplify.*

5. a. $13x = 0$ *Given linear equation*
$\frac{13x}{13} = \frac{0}{13}$ *Divide both sides by 13.*
$x = 0$ *Simplify.*

b. $-\frac{1}{2}x = 0$ *Given linear equation*
$-2\left(-\frac{1}{2}x\right) = -2(0)$ *Multiply both sides by −2.*
$x = 0$ *Simplify.*

c. $2x - 5 = 11$ *Given linear equation*
$2x = 16$ *Add 5 to both sides.*
$x = 8$ *Divide both sides by 2.*

d. $8 - 3x = 2$ *Given linear equation*
$-3x = -6$ *Subtract 8 from both sides.*
$x = 2$ *Divide both sides by −3*

Page 156

7. a. $n = 0.1111$ **b.** $n = 0.5222\cdots$
$\quad\ \ = \dfrac{1{,}111}{10{,}000}$ $\quad 10n = 5.2222\cdots$
$\qquad\qquad\qquad\qquad\ \ 9n = 4.7$
$\qquad\qquad\qquad\qquad\quad\ n = \dfrac{4.7}{9}\cdot\dfrac{10}{10} = \dfrac{47}{90}$

9. a. $\quad n = 0.3939\cdots$ **b.** $n = 0.622\cdots$
$\quad 100n = 39.3939\cdots$ $\quad 10n = 6.22\cdots$
$\quad\ \ 99n = 39$ $\qquad 9n = 5.6$
$\qquad\quad n = \dfrac{39}{99} = \dfrac{13}{33}$ $\qquad\ = \dfrac{5.6}{9}\cdot\dfrac{10}{10}$
$\qquad\qquad\qquad\qquad\qquad\qquad\ \ = \dfrac{56}{90} = \dfrac{28}{45}$

By now you should be getting used to doing the arithmetic steps in your head, so we will now begin showing less of the arithmetic steps. Also, as long as we are adding, subtracting, multiplying, or dividing both sides by the same number, we will not give a written reason. In other words, what we show here is a model of the way your work should look.

11. a. $-5X = -1$ **b.** $6 = C - 4$
$\qquad\quad X = \dfrac{1}{5}$ $\qquad 10 = C$

13. a. $16F - 5 = 11$ **b.** $6 = 5G - 24$
$\qquad\ \ 16F = 16$ $\qquad 30 = 5G$
$\qquad\qquad F = 1$ $\qquad\ \ 6 = G$

15. a. $\dfrac{J}{5} = 3$ **b.** $\dfrac{2K}{3} = 6$
$\qquad\ J = 15$ $\qquad 2K = 18$
$\qquad\qquad\qquad\qquad\qquad\ K = 9$

17. a. $\quad 7 = \dfrac{2N}{3} + 11$ **b.** $-5 = \dfrac{2P+1}{3}$
$\qquad -4 = \dfrac{2N}{3}$ $\qquad -15 = 2P + 1$
$\qquad -12 = 2N$ $\qquad -16 = 2P$
$\qquad\ \ -6 = N$ $\qquad\ \ -8 = P$

Page 157

19.
$$5(6S - 81) = -3(15 + 5S)$$
$$30S - 405 = -45 - 15S$$
$$45S - 405 = -45$$
$$45S = 360$$
$$S = 8$$

21.
$$3(U - 3) - 2(U - 12) = 18$$
$$3U - 9 - 2U + 24 = 18$$
$$U + 15 = 18$$
$$U = 3$$

23.
$$5(W + 3) - 6(W + 5) = 0$$
$$5W + 15 - 6W - 30 = 0$$
$$-W - 15 = 0$$
$$-15 = W$$

25.
$$5(Z - 2) - 3(Z + 3) = 9$$
$$5Z - 10 - 3Z - 9 = 9$$
$$2Z - 19 = 9$$
$$2Z = 28$$
$$Z = 14$$

Level 2, page 282

27.

$x^2 = 10x$	Given quadratic equation
$x^2 - 10x = 0$	Subtract $10x$ from both sides to get a 0 on one side.
$x(x - 10) = 0$	Factor nonzero side.
$x = 0$ or $x - 10 = 0$	Apply zero-product rule.
$x = 10$	Solve each of the resulting linear equations.

The solution is $x = 0$, $x = 10$.

29.

$5x + 66 = x^2$	Given quadratic equation
$0 = x^2 - 5x - 66$	Subtract $5x$ and 66 from both sides to get a 0 on one side.
$0 = (x - 11)(x + 6)$	Factor nonzero side.
$x - 11 = 0$ or $x + 6 = 0$	Apply zero-product rule.
$x = 11$ or $x = -6$	Solve each of the resulting linear equations.

The solution is $x = 11$, $x = -6$.

Page 158

31.

$$x^3 = 4x \qquad \text{Given (cubic) equation}$$

$$x^3 - 4x = 0 \qquad \text{Subtract } 4x \text{ from both sides to get a 0 on one side.}$$

$$x(x^2 - 4) = 0 \qquad \text{Factor nonzero side.}$$

$$x(x - 2)(x + 2) = 0$$

$$x = 0 \text{ or } x - 2 = 0 \text{ or } x + 2 = 0 \qquad \text{Apply zero-product rule.}$$

$$x = 2 \quad \text{or} \quad x = -2 \qquad \text{Solve each of the resulting linear equations.}$$

The solution is $x = 0$, $x = 2$, $x = -2$.

33.

$$4x(x - 9) = 9(1 - 4x) \qquad \text{Given quadratic equation}$$

$$4x^2 - 36x = 9 - 36x \qquad \text{Simplify.}$$

$$4x^2 - 9 = 0 \qquad \text{Add } 36x \text{ and subtract 9 from both sides.}$$

$$(2x - 3)(2x + 3) = 0 \qquad \text{Factor nonzero side.}$$

$$2x - 3 = 0 \text{ or } 2x + 3 = 0 \qquad \text{Apply zero-product rule.}$$

$$x = \frac{3}{2} \quad \text{or} \quad x = -\frac{3}{2} \qquad \text{Solve each of the resulting linear equations.}$$

The solution is $x = \frac{3}{2}$, $x = -\frac{3}{2}$. You can also write this as $x = \pm \frac{3}{2}$.

35. $x^2 + 7x + 2 = 0$ \qquad Given quadratic equation

$$x = \frac{-7 \pm \sqrt{(7)^2 - 4(1)(2)}}{2(1)} \qquad \text{Quadratic formula } a = 1,\ b = 7,\ c = 2.$$

$$= \frac{-7 \pm \sqrt{49 - 8}}{2} \qquad \text{Simplify.}$$

$$= \frac{-7 \pm \sqrt{41}}{2}$$

37. $x^2 - 5x - 3 = 0$ \qquad Given quadratic equation

$$x = \frac{-(-5) \pm \sqrt{(-5)^2 - 4(1)(-3)}}{2(1)} \qquad \text{Quadratic formula } a = 1,\ b = -5,\ c = -3.$$

$$= \frac{5 \pm \sqrt{25 + 12}}{2} \qquad \text{Simplify.}$$

$$= \frac{5 \pm \sqrt{37}}{2}$$

As with linear equations, we now show the work as your work will look (without the reasons for each step).

Page 159

39. $x^2 - 6x + 7 = 0$

$$x = \frac{6 \pm \sqrt{36 - 4(1)(7)}}{2(1)}$$

$$= \frac{6 \pm \sqrt{8}}{2}$$

$$= \frac{6 \pm 2\sqrt{2}}{2}$$

$$= 3 \pm \sqrt{2}$$

41. $3x^2 + 5x - 4 = 0$

$$x = \frac{-5 \pm \sqrt{25 - 4(3)(-4)}}{2(3)}$$

$$= \frac{-5 \pm \sqrt{25 + 48}}{6}$$

$$= \frac{-5 \pm \sqrt{73}}{6}$$

43. $\qquad 4x^2 + 2x = -5$

$4x^2 + 2x + 5 = 0$

$$x = \frac{-2 \pm \sqrt{4 - 4(4)(5)}}{2(4)}$$

$$= \frac{-2 \pm \sqrt{4 - 80}}{8}$$

$$= \frac{-2 \pm \sqrt{-76}}{8}$$

There are no real values because there is a negative under the square root.

45. $\qquad 3x^2 = 11x + 4$

$3x^2 - 11x - 4 = 0$

$(3x + 1)(x - 4) = 0$

$$x = -\frac{1}{3}, 4$$

Page 160

47.
$$6x^2 = 5x$$
$$6x^2 - 5x = 0$$
$$x(6x - 5) = 0$$
$$x = 0, \frac{5}{6}$$

49. $\qquad x^2 + 4 = 3\sqrt{2}x \quad a = 1, \ b = -3\sqrt{2}, \text{ and } c = 4.$
$$x^2 - 3\sqrt{2}x + 4 = 0$$
By calculator, $x \approx 1.41, 2.83$ (rounded to the nearest hundredth).

51. Since $\sqrt{2}x^2 + 2x - 3 = 0$ is given we see $a = \sqrt{2}, \ b = 2, \text{ and } c = -3$
By calculator, $x \approx 0.91, -2.33$ (rounded to the nearest hundredth).

53. Since $0.02x^2 + 0.831x + 0.0069 = 0$ is given we see $a = 0.02, \ b = 0.831, \text{ and }$
$c = 0.0069$. By calculator, $x \approx -41.54, -0.01$ (rounded to the nearest hundredth).

Level 3, page 282

55. a. $\text{CHILD'S DOSE} = \dfrac{\text{AGE OF CHILD}}{\text{AGE OF CHILD} + 12} \times \text{ADULT DOSE}$
$$= \frac{10}{10 + 12} \times (100 \text{ mg})$$
$$\approx 45.5 \text{ mg}$$
b. $\text{CHILD'S DOSE} = \dfrac{\text{AGE OF CHILD}}{\text{AGE OF CHILD} + 12} \times \text{ADULT DOSE}$
$$10 \text{ mg} = \frac{12}{12 + 12} \times \text{ADULT DOSE}$$
$$(10 \text{ mg})2 \approx \text{ADULT DOSE}$$
$$20 \text{ mg} \approx \text{ADULT DOSE}$$

Level 3 Problem Solving, page 282

57. On the last step, dividing by $a + b - c$ is not allowed because $a + b = c$ so $a + b - c = 0$; cannot divide by 0.

59. a. $\qquad x^2 + 10x = 39$
$$x^2 + 10x - 39 = 0$$
$$(x + 13)(x - 3) = 0$$
$$x = -13, 3$$

    b. It is a geometric proof (which, by the way, gives only one of the answers).

        Al-Khwârizmî starts with a square of side x, which therefore represents x^2, as shown at the below.

Page 161

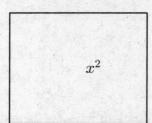

To this square we must add $10x$, and this is done by adding four rectangles each of breadth 10/4 the length x to the square.

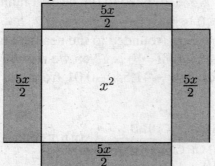

The figure how has area $x^2 + 10x$ which is equal to 39. Now complete the square by adding four little squares each of area $(5/2)^2 = 25/4$. Thus, the outside square has area $4(25/4) + 39 = 64$. The side of the square is therefore 8. But the length $5/2 + x + 5/2$, so $x + 5 = 8$, which gives $x = 3$.

6.5 Inequalities, page 283
New Terms Introduced in this Section

Addition property of inequality	Comparison property	Inequality
Inequality symbol	Multiplication property of inequality	Solve an inequality

Page 162

Level 1, page 287

1. For any two numbers x and y, exactly one of the following is true: 1. $x = y$ (x is equal to y; that is, *the same as*). 2. $x > y$ (x is greater than y; that is, *larger than*). 3. $x < y$ (x is less than y; that is, *smaller than*).

3. The solution to $x < 5$ is all values of x to the left of 5 on a number line; open circle for the endpoint.

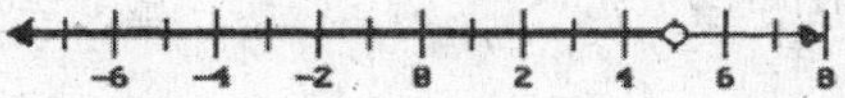

5. The solution to $x \geq -3$ is all values of x to the right of -3 on a number line; closed circle for the endpoint.

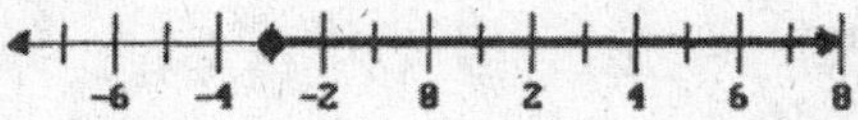

7. Rewrite the inequality $4 \geq x$ so the variable is on the left: $x \leq 4$. The solution to $x \leq 4$ is all values of x to the left of 4 on a number line; closed circle for the endpoint.

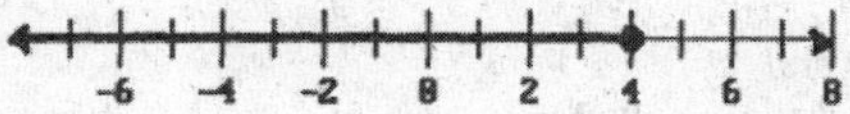

9. Rewrite the inequality $\frac{x}{2} > 3$ as $x > 6$. The solution to $x > 6$ is all values of x to the right of 6 on a number line; open circle for the endpoint.

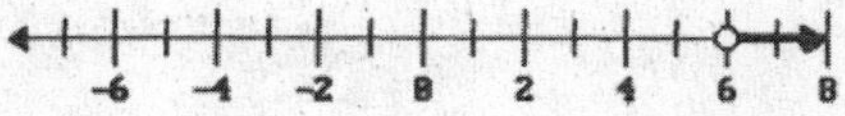

11. $-2 > \dfrac{x}{-4}$

$8 < x$

$x > 8$

The solution is all values to the right of 8 on a number line; open circle for the endpoint.

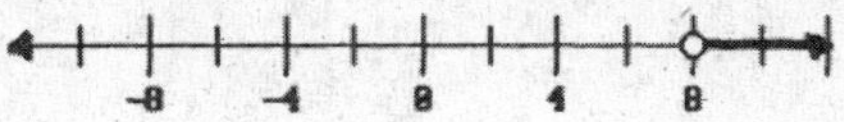

13. The solution to $x \geq 100$ is all values of x to the right of 100 on a number line; closed circle for the endpoint.

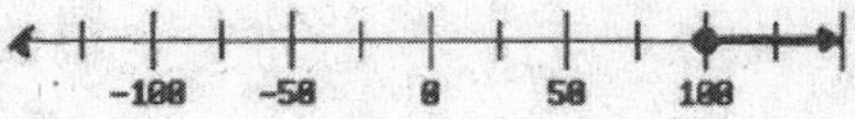

15. The solution to $x \leq -75$ is all values of x to the left of -75 on a number line; closed circle for the endpoint.

17. $x + 7 \geq 3$ **19.** $x - 2 \geq -4$

   $\quad x \geq -4$ Subtract 7 from both sides. $\quad x \geq -2$ Add 2 to both sides.

21. $-4 < 2 + y$

   $\quad -6 < y$ Subtract 2 from both sides.

   $\quad y > -6$ Variable on the left; reverse inequality and order.

23. $2 > -s$

   $\quad s > -2$ Add s to both sides and

   $\qquad$ subtract 2 from both sides.

25. $-m > -5$

   $\quad m < 5$ Divide both sides by -1, reverse order.

27. $3 > 2 - x$

   $\quad x > -1$ Add x to both sides and

   $\qquad$ subtract 3 from both sides.

29. $2x + 6 \leq 8$

   $\quad 2x \leq 2$

   $\quad\ x \leq 1$

31. $\quad 3 > s + 9$

   $\quad -6 > s$

   $\quad\ s < -6$

33. $4 \leq a + 2$

   $\quad 2 \leq a$

   $\quad a \geq 2$

35. $3s + 2 > 8$

   $\quad 3s > 6$

   $\quad\ s > 2$

37. $7u - 5 \leq 9$

   $\quad 7u \leq 14$

   $\quad\ u \leq 2$

39. $5 - 3w > 8$

   $\quad -3w > 3$

   $\quad\ w < -1$

Page 164

Level 2, page 287

41. $7 - 5A < 2A + 7$
$\quad -7A < 0$
$\quad\quad A > 0$

43. $3C > C + 19$
$\quad 2C > 19$
$\quad\quad C > \dfrac{19}{2}$

45. $5E - 4 < 3E - 6$
$\quad 2E < -2$
$\quad\quad E < -1$

47. $4G - 1 > 3(G + 2)$
$\quad 4G - 1 > 3G + 6$
$\quad\quad G > 7$

49. $2 - 3I > 7(1 - I)$
$\quad 2 - 3I > 7 - 7I$
$\quad\quad 4I > 5$
$\quad\quad\quad I > \dfrac{5}{4}$

51. $7(\text{NUMBER}) + 35 > 0$
$\quad\quad 7N + 35 > 0$
$\quad\quad\quad\quad 7N > -35$
$\quad\quad\quad\quad\quad N > -5$

The number is any number
greater than -5.

53. $3N + 12 < 0$
$\quad\quad 3N < -12$
$\quad\quad\quad N < -4$

The number is any number less than -4.

Level 3 Problem Solving, page 287

55. $x = -x + 4$
$\quad 2x = 4$
$\quad\quad x = 2$

The number is 2.

57. $x < (-x) + 4$

$2x < 4$

$x < 2$

The number is less than 2.

59.
$$\text{LENGTH} + \text{GIRTH} \leq 130$$
$$6(\text{SHORTER SIDE}) + 4(\text{SHORTER SIDE}) \leq 130$$
$$6s + 4s \leq 130$$
$$10s \leq 130$$
$$s \leq 13$$

The height and width must each be 13 in. or less and the length is 78 in. or less.

6.6 Algebra in Problem Solving, page 288

New Term Introduced in this Section

Consecutive integer	difference	evolve	product
quotient	solve	substitution	sum
translate			

Level 1, page 297

1. First: You have to *understand the problem*. This means read the problem and note what it is all about. Focus on processes rather than numbers. You cannot work a problem you do not understand. A sketch may help in understanding the problem. Second: *Devise a plan.* Write down a verbal description of the problem using operation signs and an equal or inequality sign. Third: *Carry out the plan.* In the context of word problems, we *translate, evolve, solve,* and *answer.* Fourth: *Look back.* Be sure your answer makes sense by checking it with the original question in the problem. **Remember to answer the question that was asked.**

Translate means to begin with a verbal statement and then let it *evolve* (which means use substitution) into a symbolic statement. The word *solve* applies to finding the replacements that make an equation true.

---

3. **a.** $3 + 2 \times 4 = 11$ **b.** $3(2 + 4) = 18$
5. **a.** $8 \times 9 + 10 = 82$ **b.** $8(9 + 10) = 152$
7. **a.** $3^2 + 2^3 = 17$ **b.** $3^3 - 2^2 = 23$
9. **a.** $4^2 + 9^2 = 97$ **b.** $(4 + 9)^2 = 169$

Page 166

11. a. $3(n+4) = 16$ open, conditional equation

$$3n + 12 = 16$$
$$3n = 4$$
$$n = \frac{4}{3}$$

b. $5(n+1) = 5n + 5$ Recognize this as the distributive property, so this is an open equation which is an identity.

13. a. $3^2 + 4^2 = 25$; true equation

b. $1^2 + 2^2 + 3^2 = n$; $n = 14$; open, conditional equation

15. a. $6n + 12 = 6(n+2)$ Recognize this as the distributive property, so this an open equation which is an identity

b. $n(7 + n) = 0$ open, conditional equation

$$n = 0, -7$$

17. $A = bh$

19. $A = \frac{1}{2}pq$

21. $V = s^3$

23. $V = \frac{1}{3}\pi r^2 h$

25. $V = \frac{4}{3}\pi r^3$

Level 2, page 298

27. $2(\text{NUMBER}) - 12 = 6$ The number is 9.

$$2n - 12 = 6 \qquad \text{Let } n = \text{NUMBER}$$
$$2n = 18$$
$$n = 9$$

29. $3(\text{NUMBER}) - 6 = 2(\text{NUMBER})$ The number is 6.

$$3n - 6 = 2n \qquad \text{Let } n = \text{NUMBER}$$
$$n = 6$$

31. EVEN INTEGER + NEXT EVEN INTEGER = 94

$$n + (n + 2) = 94 \qquad \text{Let } n = \text{EVEN INTEGER}$$
$$2n + 2 = 94$$
$$2n = 92$$
$$n = 46$$
$$n + 2 = 48$$

The numbers are 46 and 48.

Page 167

33. INTEGER + NEXT INTEGER + 3RD CONSECUTIVE INTEGER + 4TH CONSECUTIVE INTEGER = 74

$n + (n + 1) + (n + 2) + (n + 3) = 74$ Let n = INTEGER

$4n + 6 = 74$

$4n = 68$

$n = 17$

$n + 1 = 18; n + 2 = 19; n + 3 = 20$

The numbers are 17, 18, 19, and 20.

35. PRICE OF FIRST CABINET + PRICE OF SECOND CABINET = 4,150

PRICE OF FIRST CABINET + 4(PRICE OF FIRST CABINET) = 4,150

Let p = PRICE OF THE FIRST (cheaper) CABINET

$p + 4p = 4,150$

$5p = 4,150$

$p = 830$

$4p = 3,320$

The cabinets cost $830 and $3,320.

37. 4(AMOUNT OF MONEY TO START) $- 72 = 48$

$4m - 72 = 48$ Let m = AMT OF MONEY TO START

$4m = 120$

$m = 30$

He started with $30.

39. $8^2 + 14^2 = (\text{LENGTH OF BRACE})^2$

$260 = b^2$ Let b = LENGTH OF BRACE

$b = \sqrt{260}$

$b = 2\sqrt{65}$

The exact length of the base is $2\sqrt{65}$, which is approximately 16 ft.

Page 168

41.

DIST FROM N TO M + DIST FROM M TO C + DIST FORM C TO D = TOTAL DISTANCE

DIST FROM N TO M + (DIST FROM N TO M + 90) + (DIST FROM N TO M − 150) = 1,140

Let DIST FROM N TO M $= d$ $d + (d + 90) + (d − 150) = 1,140$

$$3d − 60 = 1,140$$
$$3d = 1,200$$
$$d = 400$$
$$d − 150 = 250$$

It is 250 miles from Cincinnati to Detroit.

43. DIST OF SLOWER RUNNER + HEAD START = DIST OF FASTER RUNNER

(RATE OF SLOWER RUNNER)(TIME) + 50 = (RATE OF FASTER RUNNER)(TIME)

$$6(\text{TIME}) + 50 = 10(\text{TIME}) \qquad \text{Let } t = \text{TIME}$$
$$6t + 50 = 10t$$
$$50 = 4t$$
$$12.5 = t$$

It will take the faster runner 12.5 seconds to catch the slower runner.

45. DIST OF CAR + HEAD START = DIST OF POLICE CAR

(RATE OF CAR)(TIME) + 2 = (RATE OF POLICE CAR)(TIME)

$$80(\text{TIME}) + 2 = 100(\text{TIME}) \qquad \text{Let } t = \text{TIME}$$
$$80t + 2 = 100t$$
$$2 = 20t$$
$$0.1 = t$$

Since the rates are in miles per hour, the time is in hours so $\frac{1}{10}$ hr = 6 min. It will take the police car 6 minutes to catch the speeder.

Page 169

47.

DIST OF 1ST JOGGER + DIST OF 2ND JOGGER = TOTAL DISTANCE

$(\text{RATE OF 1st JOGGER})(\text{TIME}) + (\text{RATE OF 2nd JOGGER})(\text{TIME}) = 21$

$(\text{SLOWER JOGGER'S RATE} + 2)(1.5) + (\text{SLOWER JOGGER'S RATE})(1.5) = 21$ Let r = SLOWER RATE

$$(r + 2)(1.5) + r(1.5) = 21$$
$$2r + 2 = 14$$
$$2r = 12$$
$$r = 6$$
$$r + 2 = 8$$

The faster person jogged 8 miles per hour and the other at 6 miles per hour.

Level 3, page 299

49. Area = $\frac{1}{2}bh$ = 17.5 cm^2. Let x = length of shortest side.

$$\frac{1}{2}(x + 2)x = 17.5$$
$$x^2 + 2x = 35$$
$$x^2 + 2x - 35 = 0$$
$$(x + 7)(x - 5) = 0$$
$$x = -7, 5$$

The length of the shorter side is 5 cm.

51. Area = $\frac{1}{2}bh$; let x = height of the triangle.

$$3 = \frac{1}{2}(x + 2)x$$
$$6 = x^2 + 2x$$
$$x^2 + 2x - 6 = 0$$
$$x = \frac{-2 \pm \sqrt{2^2 - 4(1)(-6)}}{2}$$
$$\approx 1.646, -3.646 \quad \textit{Reject} - 3.646 \textit{ since x is a distance.}$$

The base is 3.6 ft and the height is 1.6 ft.

53. Find r when $A = 2$, so solve:

$$(1 + r)^2 = 2$$
$$1 + r = \sqrt{2} \quad \textit{Disregard negative values.}$$
$$r = \sqrt{2} - 1$$
$$\approx 0.4142$$

Need to obtain about 41.4% interest.

Level 3 Problem Solving, page 299

55. Recall, 1 mile = 5,280 ft = 63,360 in.; $\frac{1}{2}$ mi = 31,680 in.

$$(\text{SIDE})^2 + (\text{HEIGHT})^2 = (\text{HYPOTENUSE})^2$$

$$(31{,}680)^2 + (\text{HEIGHT})^2 = (0.5 \text{ mi} + \frac{1}{4} \text{ in.})^2$$

$$(31{,}680)^2 + (\text{HEIGHT})^2 = (31{,}680 + 0.25)^2 \qquad \textit{Let } h = \text{HEIGHT}$$

$$31{,}680^2 + h^2 = 31{,}680.25^2$$

$$h^2 = 15{,}840.063$$

$$h \approx 125.86$$

The height is approximately 10.5 ft.

57. $\quad x - (\frac{1}{3}x + \frac{1}{5}x + \frac{1}{6}x + \frac{1}{4}x) = 6$

$$60x - (20x + 12x + 10x + 15x) = 360$$

$$3x = 360$$

$$x = 120$$

There are 120 lilies.

59. FIRST MONKEY'S DISTANCE = SECOND MONKEY'S DISTANCE

$$100 + 200 = x + y$$

$$300 = x + \sqrt{(100 + x)^2 + 200^2}$$

$$300 - x = \sqrt{(100 + x)^2 + 200^2}$$

$$300^2 - 600x + x^2 = (100 + x)^2 + 200^2$$

$$300^2 - 600x + x^2 = 100^2 + 200x + x^2 + 200^2$$

$$-800x = 100^2 + 200^2 - 300^2$$

$$x = 50$$

The second monkey jumped 50 cubits into the air.

6.7 Ratios, Proportions, and Problem Solving, page 300
New Terms Introduced in this Section

Property of proportions   Proportion   Ratio   Solve a proportion

Level 1, page 306

1. A *ratio* is a quotient of two numbers; a *proportion* is a statement of equality between ratios.

Page 171

3. It saves a few steps; with equations it is necessary to multiply both sides by the least common denominator, and then to simplify, whereas the property of proportions does the same thing in one step.

5. $\dfrac{60}{3} = \dfrac{20}{1}$ The ratio of cement to water is 20 to 1.

7. $\dfrac{106}{100} = \dfrac{53}{50}$ The ratio of males to females is 53 to 50.

9. $\dfrac{279}{15\frac{1}{2}} = \dfrac{279}{\frac{31}{2}}$

$\qquad = \dfrac{2 \cdot 279}{31}$

$\qquad = \dfrac{18}{1}$

The ratio of miles to gallons is 18 to 1; sometimes this is abbreviated as 18 mpg.

11. a. Yes, since $1 \cdot 21 = 7 \cdot 3$.

b. Yes, since $8 \cdot 9 = 6 \cdot 12$.

c. Yes, since $6 \cdot 5 = 3 \cdot 10$.

13. a. Yes, since $4 \cdot 75 = 3 \cdot 100$.

b. No, since $3 \cdot 67 \neq 2 \cdot 100$

c. No, since $3 \cdot 7\frac{1}{3} \neq 5 \cdot 4$.

15. a. Since $1 \cdot 8 > 6 \cdot 1$, we see that $\frac{1}{6} > \frac{1}{8}$.

b. Since $1 \cdot 3 < 4 \cdot 1$, we see that $\frac{1}{4} < \frac{1}{3}$.

c. Since $1 \cdot 8 > 5 \cdot 1$, we see that $\frac{1}{5} > \frac{1}{8}$.

17. a. Since $8000 < 8001$ we see that $0.8000 < 0.8001$.

b. Since $8000 > 7999$, we see that $0.8000 > 0.7999$.

c. Since $280 < 281$, we see that $2.80 < 2.81$

d. Since $280 < 288$, we see that $2.80 < 2.88$

19. $\dfrac{5}{1} = \dfrac{A}{6}$

$\quad A = 30$

21. $\dfrac{C}{2} = \dfrac{5}{1}$

$\quad 10 = C$

23. $\dfrac{12}{18} = \dfrac{E}{12}$

$\quad 18E = 144$

$\qquad E = 8$

Page 172

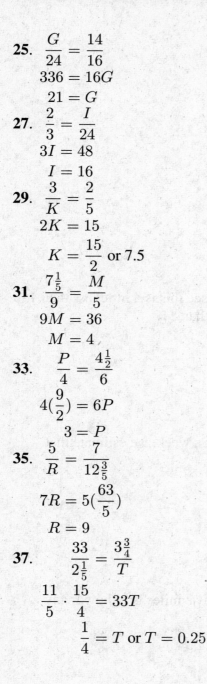

25. $\dfrac{G}{24} = \dfrac{14}{16}$

    $336 = 16G$

    $21 = G$

27. $\dfrac{2}{3} = \dfrac{I}{24}$

    $3I = 48$

    $I = 16$

29. $\dfrac{3}{K} = \dfrac{2}{5}$

    $2K = 15$

    $K = \dfrac{15}{2}$ or 7.5

31. $\dfrac{7\frac{1}{5}}{9} = \dfrac{M}{5}$

    $9M = 36$

    $M = 4$

33. $\dfrac{P}{4} = \dfrac{4\frac{1}{2}}{6}$

    $4\left(\dfrac{9}{2}\right) = 6P$

    $3 = P$

35. $\dfrac{5}{R} = \dfrac{7}{12\frac{3}{5}}$

    $7R = 5\left(\dfrac{63}{5}\right)$

    $R = 9$

37. $\dfrac{33}{2\frac{1}{5}} = \dfrac{3\frac{3}{4}}{T}$

    $\dfrac{11}{5} \cdot \dfrac{15}{4} = 33T$

    $\dfrac{1}{4} = T$ or $T = 0.25$

Page 173

39.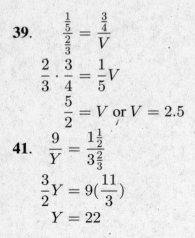

$$\frac{\frac{1}{5}}{\frac{2}{3}} = \frac{\frac{3}{4}}{V}$$

$$\frac{2}{3} \cdot \frac{3}{4} = \frac{1}{5}V$$

$$\frac{5}{2} = V \text{ or } V = 2.5$$

41. $\dfrac{9}{Y} = \dfrac{1\frac{1}{2}}{3\frac{2}{3}}$

$$\frac{3}{2}Y = 9(\frac{11}{3})$$

$$Y = 22$$

Level 2, page 307

43. Begin by estimating an answer. If 4 melons sell for 52¢, we see this is a little less than 15¢ per melon, so we estimate that 7 melons should be a bit less than $1.

$$\frac{4 \text{ melons}}{0.52 \text{ dollars}} = \frac{7 \text{ melons}}{x \text{ dollars}}$$

$$x = \frac{7 \cdot 0.52}{4}$$

$$\approx 0.91$$

Seven melons cost $0.91.

45. Begin by estimating an answer. Since four days is about half a week, we estimate that about half of $3\frac{1}{2}$ gallons or a bit more than $1\frac{3}{4}$ gallons.

$$\frac{7 \text{ days}}{3\frac{1}{2} \text{ gallons}} = \frac{4 \text{ days}}{x \text{ gallons}}$$

$$x = \frac{3\frac{1}{2} \cdot 4}{7}$$

$$= 2$$

The family needs 2 gallons.

47. Rephrase this as a proportion: Three miles is to 40 minutes as 2 miles is to how many minutes?

$$\frac{3 \text{ miles}}{40 \text{ minutes}} = \frac{2 \text{ miles}}{x \text{ minutes}}$$

$$x = \frac{40 \cdot 2}{3}$$

$$= 26\frac{2}{3}$$

It will take Jack 27 minutes. (Round $26\frac{2}{3}$ to the nearest unit.)

49. Rephrase this as a proportion. One hundred forty pounds is to 2,100 calories as 165 pounds is how many calories?

$$\frac{140 \text{ lbs}}{2,100 \text{ calories}} = \frac{165 \text{ lbs}}{x \text{ calories}}$$

$$x = \frac{2,100 \cdot 165}{140}$$

$$= 2,475$$

A 165-lb person needs 2,475 calories per day.

Level 3, page 308

51.
$$\frac{V}{T} = \frac{V'}{T'} \qquad \text{Given equation}$$

$$\frac{175}{300} = \frac{V'}{273} \qquad \text{Substitute known values.}$$

$$V' = \frac{175 \cdot 273}{300} \qquad \text{Property of proportions}$$

$$= 159.25 \qquad \text{Simplify.}$$

53. Since the span given is 24 feet, the half-span is 12 feet.

$$\text{PITCH} = \frac{\text{RISE}}{\text{HALF-SPAN}}$$

$$= \frac{8}{12}$$

$$= \frac{2}{3}$$

55. We formulate a proportion using the ratio of short side to long side lengths:

$$\frac{3}{5} = \frac{2}{x}$$

$$x = \frac{5 \cdot 2}{3}$$

$$= \frac{10}{3}$$

Page 175

The longer side should be $3\frac{1}{3}$ ft or 3 ft, 4 in.

57. We formulate a proportion using the ratio of house value to tax:

$$\frac{180,000}{1,080} = \frac{130,000}{x}$$

$$x = \frac{1,080 \cdot 130,000}{180,000}$$

$$= 780$$

The tax is $780.

Level 3 Problem Solving, page 308

59. Notice that she filled the soft-drink barrels with pure water *only* each day — she *never* added any more pure soft drink. So, if after 4 days only 1 pint of pure soft drink remains, she must have served all 32 gallons except for the 1 remaining pint. Since there are 4 quarts in a gallon and 2 points in a quart, she served 31 gal, 3 qt, and 1 pt.

6.8 Percents, page 308
New Terms Introduced in this Section

Base   Percent   Percentage   Percent problem
Rate

Level 1, page 316
Use Table 6.1, page 310 for Problems 1-21. You may know most of these from memory or previous work.

1. $\frac{3}{4}$; 75%

3. $\frac{2}{5}$; 0.4

5. $0.\overline{3}$; $33\frac{1}{3}$%

7. $\frac{17}{20}$; 85%

9. 0.375; 37.5%

11. $\frac{6}{5}$; 1.2

13. $\frac{1}{20}$; 5%

15. $0.1\overline{6}$; $16\frac{2}{3}$%

17. $\frac{2}{9}$; $0.\overline{2}$

19. $\frac{7}{40}$; 17.5%

Page 176

21. $\frac{1}{400}$; 0.25%

Estimates in Problems 22-28 may vary.

23. a. 10% of 95,000 is $\frac{1}{10}$ of 95,000, which is 9,500.

  b. 10% of 85.6 is $\frac{1}{10}$ of 85.6, which is 8.56.

25. a. 25% of 819 is about $\frac{1}{4}$ of 800, which is 200.

  b. 25% of 790 is about $\frac{1}{4}$ of 800, which is 200.

27. a. 40% of 93 is about 40% of 100, which is 40.

  b. 90% of 8,741 is about 90% of 9,000, which is 8,100.

29. $\dfrac{15}{100} = \dfrac{A}{64}$

$$A = \frac{15 \cdot 64}{100}$$
$$= \frac{960}{100}$$
$$= 9.6$$

31. $\dfrac{14}{100} = \dfrac{21}{W}$

$$W = \frac{100 \cdot 21}{14}$$
$$= 150$$

33. $\dfrac{P}{100} = \dfrac{10}{5}$ P is 200%.

$$P = \frac{100 \cdot 10}{5}$$
$$= 200$$

35. $\dfrac{P}{100} = \dfrac{4}{5}$ P is 80%.

$$P = \frac{100 \cdot 4}{5}$$
$$= 80$$

37. $\dfrac{P}{100} = \dfrac{9}{12}$ P is 75%.

$$P = \frac{100 \cdot 9}{12}$$
$$= 75$$

Page 177

39. $\dfrac{35}{100} = \dfrac{49}{W}$

$\qquad W = \dfrac{100 \cdot 49}{35}$

$\qquad\quad = 140$

41. $\dfrac{120}{100} = \dfrac{16}{W}$

$\qquad W = \dfrac{100 \cdot 16}{120}$

$\qquad\quad = 13\dfrac{1}{3}$

43. Recognize $\dfrac{33\frac{1}{3}}{100}$ to be $\frac{1}{3}$.

$\qquad \dfrac{1}{3} = \dfrac{12}{W}$

$\qquad W = \dfrac{3 \cdot 12}{1}$

$\qquad\quad = 36$

45. $\dfrac{6}{100} = \dfrac{A}{8,150}$ $\qquad$ *A* is \$489.

$\qquad A = \dfrac{6 \cdot 8,150}{100}$

$\qquad\quad = 489$

Level 2, page 316

47. 11% of 210 million is $0.11 \times 210 = 23.1$, so 23,100,000 Americans live in poverty.

49. 6% of \$181 is $0.06 \times \$181 = \$10.86$.

51. What percent of \$3,020 is \$151.

$\qquad \dfrac{P}{100} = \dfrac{151}{3,020}$ $\qquad$ The tax rate is 5%.

$\qquad P = \dfrac{100 \cdot 151}{3,020}$

$\qquad\quad = 5$

53. Since $80 - 48 = 32$. we ask the question, "32 is what percent of 80?"

$$\frac{P}{100} = \frac{32}{80} \qquad \text{The decrease is 40\%.}$$
$$P = \frac{100 \cdot 32}{80}$$
$$= 40$$

55. 32% of $8,200 is $0.32 \times 8,200 = 2,624$, so the tax withheld is $2,624.

57. Ask, "18 is what percent of 20?"
$$\frac{P}{100} = \frac{18}{20} \qquad \text{The percentage correct is 90\%.}$$
$$P = \frac{100 \cdot 18}{20}$$
$$= 90$$

59. Ask, "8% of what amount is $100."
$$\frac{8}{100} = \frac{100}{W} \qquad \text{The old wage was \$1,250 and the new wage is \$1,350.}$$
$$W = \frac{100 \cdot 100}{8}$$
$$= 1,250$$

6.9 Modeling Uncategorized Problems, page 317
Level 1, page 323

1. Answers vary. You should write a couple of paragraphs giving support for your comments.

3. Bus and driver: $150.00
    10 children @ $5.00 each: $ 50.00
    6 teens @ $6.00 each: $ 36.00
    3 adults @ $7.00 each: $ 21.00
    20 people for lunch @ $2.50 each: $ 50.00
    TOTAL: $307.00   Answer is A.

5. One hour (minimum): $20.00
    30 sheets @ $0.10 each: $ 3.00
    binding fee: $ 1.25
    Additional copy @ 7¢/pg: $ 2.10
    TOTAL: $26.35   Answer is A.

7. 15% less is 0.85($125.00): $106.25
    Plus sales tax, 0.07($106.25): $ 7.44
    TOTAL: $113.69   Answer is A.

Page 179

9.   35% of LOAN PAYMENT = $180

$$\text{LOAN PAYMENT} = \frac{\$180}{35\%}$$
$$\approx \$514.29$$

Combined total is $514.29 + $180.00 = $694.29.   Answer is B.

11. Mortgage payment: $2,320 + 0.03($2,320) = $2,320(1 + 0.03)$
$$= \$2,389.60$$

Increase is: $2,389.60 − $2,320.00 = $69.60.   Answer is D.

13. 2 hoses @ 2.50 gal/min = 5.00 gal/min.   This is 60(5) gal/hr = 300 gal/hr

Thus, the number of hours to fill the pool is $\dfrac{25{,}500 \text{ gal}}{300 \text{ gal/hr}} = 85$ hr.   Answer is C.

15. Votes in favor: $\frac{1}{4}(40) = 10$

Votes against: $\frac{3}{5}(40) = 24$

TOTAL: 34

Since there are 40 voters, we see that there are 6 abstentions.   Answer is A.

17. Alice's fare: $350

Bobbie's fare: $278

Cal's face: $322

TOTAL: $950   Answer is A.

19. 184 + 7 = 191 passes incomplete or intercepted. Thus, 541 − 191 = 350 completed

passes. The percent completed: $\frac{350}{541} = 64.7\%$.   Answer is A.

21. Income from 2.5% investment: $0.025(\$2{,}500) = \$$ 62.50

Income from 3.2% investment: $0.032(\$21{,}300) = \$$ 681.60

Income form 3.8% investment: $0.038(\$8{,}540) = \$$ 324.52

TOTAL: $1,068.62

Taxes: $0.38(\$1{,}068.62) \approx \$406.08$

Income (after taxes): $1,068.62 − $406.08 = $662.54   Answer is C.

23. Hourly salary: $83,000 ÷ 52 ÷ 40 ≈ $39.90   Answer is C.

25. The prime numbers less than 100 are: 2, 3, 5, 7, 11, 13, 17, 19, 23, 29, 31, 37, 41, 43, 47, 53, 59, 61, 67, 71, 73, 79, 83, 89, and 97.   Answer is C.

27. Let x be the length of one guy wire. From the Pythagorean theorem,

$$10^2 + 12^2 = x^2$$
$$244 = x^2$$
$$x = \sqrt{244} \approx 15.62$$

The guy wires are $3x \approx 46.86$.   Answer is B.

29. The (distinct) prime factors of 100 are: 2 and 5.   Answer is A.

31. Answers vary; support your comments.

Level 2, page 324

33. FIRST EVEN NUMBER $\times$ NEXT EVEN NUMBER $= 255$

$$n(n + 2) = 255 \qquad \text{Let } n = \text{FIRST EVEN NUMBER.}$$
$$n^2 + 2n = 255$$
$$n^2 + 2n - 255 = 0$$
$$(n - 15)(n + 17) = 0$$
$$n = 15, -17$$
$$n + 2 = 17, -15$$

The numbers are 15 and 17. (Disregard the negative values because the problem tells us the numbers must be positive.)

35. NUMBER $+ 2$(RECIPROCAL) $= 3$

$$n + \frac{2}{n} = 3 \qquad \text{Let } n = \text{NUMBER.}$$
$$n^2 + 2 = 3n$$
$$n^2 - 3n + 2 = 0$$
$$(n - 1)(n - 2) = 0$$
$$n = 1, 2$$

There are two numbers, 1 and 2.

37. a. $251,000 $-$ $112,950 = $138,050 so the percent loss is $\frac{\$138,050}{\$251,000} = 0.55$ or 55%

    b. To gain back $138,050 the percent gain is $\frac{\$138,050}{\$112,950} \approx 1.222$ or about 122%

39.   HEIGHT OF THE STD. OIL BLDG $+$ HEIGHT OF SEARS BLDG $= 2,590$

HEIGHT OF THE STD. OIL BLDG $+ \left(\text{HEIGHT OF THE STD. OIL BLDG} + 318\right) = 2,590$

    Let $h =$ HEIGHT OF THE STD. OIL BLDG $\qquad\qquad h + (h + 318) = 2,590$
$$2h = 2,272$$
$$h = 1,136$$
$$h + 318 = 1,454$$

The height of the Standard Oil Building is 1,136 ft and the height of the Sears Building is 1,454 ft.

Level 3, page 325

41. We look for a pattern. The sum of the first
2 consecutive even numbers: $2 + 4 = 6$ Notice: $6 = 2 \cdot 3$

3 consecutive even numbers: $2 + 4 + 6 = 12$ Notice: $12 = 3 \cdot 4$
4 consecutive even numbers: $2 + 4 + 6 + 8 = 20$ Notice: $20 = 4 \cdot 5$
5 consecutive even numbers: $2 + 4 + 6 + 8 + 10 = 30$ Notice: $30 = 5 \cdot 6$
6 consecutive even numbers: $2 + 4 + 6 + 8 + 10 + 12 = 42$ Notice: $42 = 6 \cdot 7$

It looks the sum of n consecutive even numbers is $n(n+1)$, so the sum of the first 100 consecutive even numbers is $100(101) = 10{,}100$.

43. The number of MBAs who do not also have business degrees is $520 - 450 = 70$. Thus, there are 70 nonbusiness degrees.

45. To fill the first 1/2 of the pool, it will take $\frac{1}{2}(36)=18$ hours. For the last 1/2 of the pool we find the time (in hours):

WORK DONE BY INLET PIPE $-$ WORK DONE BY DRAIN $=$ ONE-THIRD JOB COMPLETED

$$\frac{1}{36}(\text{TIME}) - \frac{1}{40}(\text{TIME}) = \frac{1}{2}$$

$$360\left[\frac{1}{36}(\text{TIME}) - \frac{1}{40}(\text{TIME})\right] = 360\left[\frac{1}{2}\right]$$

$$10(\text{TIME}) - 9(\text{TIME}) = 180$$

$$\text{TIME} = 180$$

It will take 180 additional hours (or $7\frac{1}{2}$ days) to fill the pool.

47. If we eliminate the commercials, the recording time is not ℓ hours, but

$\ell - \frac{c}{60}\ell = \frac{1}{60}(60 - c)\ell$ hours. Using this instead of 30 in Example 2, we find

$$\frac{x}{40} + \frac{\frac{1}{60}(60 - c)\ell - x}{20} = 1$$

$$x + \frac{1}{30}(60\ell - c\ell) - 2x = 40$$

$$x + 2\ell - \frac{1}{30}c\ell - 2x = 40$$

$$2\ell - \frac{1}{30}c\ell - 40 = x$$

49.

$$\text{AMT INVESTED IN SAVINGS} + \text{AMT IN ANNUITIES} = \text{TOTAL INVESTED}$$
$$\text{AMT INVESTED IN SAVINGS} + \text{AMT IN ANNUITIES} = 100,000$$
$$0.035(\text{AMT INVESTED IN SAVINGS}) + 0.0625(\text{AMT IN ANNUITIES}) = \text{TOTAL INCOME}$$
$$\downarrow \qquad\qquad\qquad\qquad\qquad \downarrow$$
$$0.035(\text{AMT INVESTED IN SAVINGS}) + 0.0625(100,000 - \text{AMT IN SAVINGS}) = 4,490$$
$$0.035x + 0.0625(100,000 - x) = 4,490$$
$$0.035x + 6,250 - 0.0625x = 4,490$$
$$1,760 = 0.0275x$$
$$64,000 = x$$
$$100,000 - x = 36,000$$

The investor must put \$64,000 in savings and \$36,000 in annuities.

Level 3 Problem Solving, page 325

51. Draw a picture of a box with sides w, $w - 1$, and $w + 3$. Then,
$$w^2 + (w - 1)^2 + (w + 3)^2 = 17^2$$
$$w^2 + w^2 - 2w + 1 + w^2 + 6w + 9 = 17^2$$
$$3w^2 + 4w - 279 = 0$$
$$(3w + 31)(w - 9) = 0$$
$$w = -\frac{31}{3}, 9 \qquad \textit{Reject negative width.}$$
$$w - 1 = 8$$
$$w + 3 = 12$$

The dimensions of the box are 9, 8, by 12. To lay flat on the bottom the diagonal, d, at the base is $w^2 + (w + 3)^2 = d^2$
$$9^2 + (9 + 3)^2 = d^2 \qquad \textit{Substitute w = 9.}$$
$$225 = d^2$$
$$d = 15, -15 \qquad \textit{Square root property; reject negative length.}$$

Two inches must be cut off the rod.

53. $(\text{RATE WITH THE WIND})(\text{TIME WITH WIND}) = \text{DISTANCE} = 630$
$$(\text{RATE WITH THE WIND})(2.5) = 630$$
$$\text{RATE WITH THE WIND} = 252$$

Similarly,

Page 183

$$(\text{RATE AGAINST THE WIND})(\text{TIME AGAINST WIND}) = \text{DISTANCE} = 630$$
$$(\text{RATE AGAINST THE WIND})(3) = 630$$
$$\text{RATE AGAINST THE WIND} = 210$$

Finally,

$$\text{RATE OF PLANE} + \text{RATE OF WIND} = \text{RATE WITH THE WIND}$$
$$\text{RATE OF PLANE} - \text{RATE OF WIND} = \text{RATE AGAINST THE WIND}$$
$$\text{RATE OF PLANE} + \text{RATE OF WIND} = 252$$
$$\text{RATE OF PLANE} - \text{RATE OF WIND} = 210$$
$$\text{RATE OF PLANE} = 210 + \text{RATE OF WIND}$$
$$(\text{RATE OF PLANE}) + \text{RATE OF WIND} = 252$$
$$\downarrow$$
$$(210 + \text{RATE OF WIND}) + \text{RATE OF WIND} = 252$$
$$(210 + w) + w = 252 \qquad \text{Let } w = \text{RATE OF WIND}$$
$$2w = 42$$
$$w = 21$$

The wind speed is 21 mph.

55. We begin by looking at a pattern.

Number	Cost/person	Income	Expense
100	\$2,500	$100(\$2,500) = \$250,000$	$\$240,500 + 500(100) = \$290,500$
101	\$2,495	$101(\$2,495) = \$251,995$	$\$240,500 + 500(101) = \$291,000$
102	\$2,490	$102(\$2,490) = \$253,980$	$\$240,500 + 500(102) = \$291,500$
103	\$2,485	$103(\$2,485) = \$255,955$	$\$240,500 + 500(103) = \$292,000$
$\vdots$	$\vdots$	$\vdots$	$\vdots$
$100 + x$	$\$2,500 - 5x$	$(100 + x)(\$2,500 - 5x)$	$\$240,500 + 500(100 + x)$

We want the income and expenses to be equal, so we have

$$\text{INCOME} = \text{EXPENSES}$$
$$(100 + x)(2,500 - 5x) = 240,500 + 500(100 + x)$$
$$250,000 + 2,000x - 5x^2 = 240,500 + 50,000 + 500x$$
$$-5x^2 + 1,500x - 40,500 = 0$$
$$x^2 - 300x + 8,100 = 0$$
$$(x - 30)(x - 270) = 0$$
$$x = 30, 270$$

Disregard $x = 270$ because it exceeds the capacity of the plane. Thus, for the income and expense to be the same, 30 additional people should be signed up and the cost per person is $\$2,500 - 5(30) = \$2,350$.

Page 184

57. The reading would not change. Think of fish in an aquarium. As the fish swims around in the water would the scale change? Of course not. Air is a medium with weight just as water is; the bird flying is the same as a fish swimming.

59. Let x be the number of soldiers on one side of the initial formation.

$$x^2 + 284 = (x+1)^2 - 25$$
$$x^2 + 284 = x^2 + 2x + 1 - 25$$
$$284 = 2x - 24$$
$$2x = 308$$
$$x = 154$$

We see the total number of soldiers is $154^2 + 284 = 24{,}000$.

Chapter 6 Review Questions, page 327

STOP

Studying for a chapter examination is a personal process, one which nobody else can do for you. Simply take the time to review what you have done. Here are the new terms in Chapter 6.

Addition property [6.4]

Addition property of inequality [6.5]

Algebra [6.1]

Base [6.8]

Binomial [6.1]

Binomial theorem [6.1]

Cell [6.3]

Common factor [6.2]

Comparison property [6.5]

Completely factored [6.2]

Consecutive integers [6.6]

Degree [6.1]

Difference [6.6]

Difference of squares [6.2]

Division property [6.4]

Equation [6.4]

Equation properties [6.4]

Equivalent equations [6.4]

Evaluate an expression [6.3]

Evolve [6.6]

Expand [6.1]

Factor [6.2]

FOIL [6.1]

Inequality [6.5]

Inequality symbols [6.5]

Like terms [6.1]

Linear [6.1]

Linear equation [6.4]

Monomial [6.1]

Multiplication property [6.4]

Multiplication property of inequality [6.5]

Multiplicity [6.4]

Numerical coefficient [6.1]

Percent [6.8]

Percentage [6.8]

Percent problem [6.8]

Polynomial [6.1]

Product [6.6]

Property of proportions [6.7]

Property of substitution [6.6]

Proportion [6.7]

Quadratic [6.1]

Quadratic equation [6.4]

Quadratic formula [6.4]

Quotient [6.6]

Rate [6.8]

Ratio [6.7]

Replication [6.3]

Root [6.4]

Satisfy [6.4]

Similar terms [6.1]

Simplify [6.1]

Solution [6.4]

Solve a proportion [6.7]

Solve an equation [6.4]

Solve an inequality [6.5]

Spreadsheet [6.3]

Substitution [6.6]

Subtraction property [6.4]	equality [6.4]	Trinomial [6.1]
Sum [6.6]	Term [6.1]	Variable [6.1]
Symmetric property of	Translate [6.6]	Zero-product rule [6.4]

If you can describe the term, read on to the next one; if you cannot, then look it up in the text (the section number is shown in brackets). Next, review the types of problems in Chapter 6.

TYPES OF PROBLEMS
Simplify an algebraic expression. [6.1]
Multiply binomials mentally. [6.1]
Expand a binomial using the binomial theorem. [6.1]
Factor a polynomial. [6.2]
Evaluate an algebraic expression. [6.3]
Write spreadsheet expressions in ordinary algebraic notation. [6.3]
Use a spreadsheet to evaluate (including the replicate command). [6.3]
Use binomials to answer genetic applications. [6.3]
Solve equations (both linear and quadratic). [6.4]
Use a calculator to solve an equation. [6.4]
Solve linear inequalities.[6.5]
Solve applied problems; in particular, those involving number, distance, and Pythagorean
    relationships. [6.6]
Set up and simplify ratios. [6.7]
Compare the size of fractions, decimals, and radicals. [6.7]
Solve proportion problems. [6.7]
Be able to change forms: fractions/decimals/percents. [6.8]
Estimate percents. [6.8]
Solve applied percent problems.[6.8]
Use problem-solving methods to answer uncategorized questions. [6.9]

Once again, see if you can verbalize (to yourself) how to do each of the listed types of problems.
    Work all of Chapter 6 Review Questions (whether they are assigned or not). Work through all of the problems before looking at the answers, and *then* correct each of the problems. The entire solution is shown in the answer section at the back of the text. If you worked the problem correctly, move on to the next problem, but if you did not work it correctly (or you did not know what to do), look back in the chapter to study the procedure, or ask your instructor.
    Finally, go back over the homework problems you have been assigned. If you worked a problem correctly, move on to the next problem, but if you missed it on your homework, then you should look back in the book or talk to your instructor about how to work the problem.

Page 186

If you follow these steps, you should be successful with your review of this chapter.

We give all of the answers to the Chapter Review questions (not just the odd-numbered questions).

Chapter 6 Review Questions, page 327

1. a. Algebra refers to a structure as a set of axioms that forms the basis for what is accepted and what is not.

b. The four main processes are *simplify* (carry out all operations according to the order-of-operations agreement and write the result in a prescribed form), *evaluate* (replace the variable(s) with specified numbers and then simplify), *factor* (write the expression as a product), and *solve* (find the replacement(s) for the variable(s) that make the equation true).

2. a.
$$(x-1)(x^2+2x+8) = (x-1)(x^2) + (x-1)(2x) + (x-1)(8)$$
$$= x^3 - x^2 + 2x^2 - 2x + 8x - 8$$
$$= x^3 + x^2 + 6x - 8$$

b.
$$x^2(x^2-y) - xy(x^2-1) = x^4 - x^2y - x^3y + xy$$
$$= x^4 - x^3y - x^2y + xy$$

3. a.
$$x^3 - y^3 = 2^3 - 3^3$$
$$= -19$$

b.
$$(x-y)(x^2+xy+y^2) = (2-3)(2^2+2\cdot3+3^2)$$
$$= -19$$

4. a.
$$3x^2 - 27 = 3(x^2-9)$$
$$= 3(x-3)(x+3)$$

b. $x^2 - 5x - 6 = (x-6)(x+1)$

5. a.
$$8x - 12 = 0$$
$$8x = 12$$
$$x = \frac{12}{8} = \frac{3}{2}$$

b.
$$8x - 12 = -2x^2$$
$$2x^2 + 8x - 12 = 0$$
$$x^2 + 4x - 6 = 0$$
$$x = \frac{-4 \pm \sqrt{16 - 4(1)(-6)}}{2}$$
$$= \frac{-4 \pm \sqrt{40}}{2}$$
$$= \frac{2(-2 \pm \sqrt{10})}{2}$$
$$= -2 \pm \sqrt{10}$$

Page 187

6.

	$x+3$		
x^2	x	x	x

$x+2$

x	1	1	1
x	1	1	1

$x^2 + 5x + 6 = (x+3)(x+2)$

7. a. $2x + 5 = 13$ **b.** $3x + 1 = 7x$

$\qquad\quad 2x = 8 \qquad\qquad\qquad 1 = 4x$

$\qquad\quad\;\; x = 4 \qquad\qquad\qquad\; \dfrac{1}{4} = x$

8. a. $\dfrac{2x}{3} = 6$ **b.** $2x - 7 = 5x$

$\qquad\quad 2x = 18 \qquad\qquad\quad -7 = 3x$

$\qquad\quad\;\; x = 9 \qquad\qquad\quad -\dfrac{7}{3} = x$

9. a. $\dfrac{P}{100} = \dfrac{3}{20}$ **b.** $\dfrac{25}{W} = \dfrac{80}{12}$

$\qquad\quad P = 15 \qquad\qquad\quad W = 3.75$

10. $\qquad\qquad x^2 = 4x + 5$

$\qquad x^2 - 4x - 5 = 0$

$\qquad (x - 5)(x + 1) = 0$

$\qquad\qquad\qquad x = 5, -1$

11. $\qquad 4x^2 + 1 = 6x$

$\qquad 4x^2 - 6x + 1 = 0$

$$x = \frac{6 \pm \sqrt{36 - 4(4)(1)}}{2(4)}$$

$$= \frac{6 \pm \sqrt{20}}{8}$$

$$= \frac{6 \pm 2\sqrt{5}}{8}$$

$$= \frac{3 \pm \sqrt{5}}{4}$$

12. $3 < -x$

$\quad\;\; x < -3$

13. $2 - x \geq 4$

$\quad -x \geq 2$

$\quad x \leq -2$

14. $3x + 2 \leq x + 6$

$\quad 2x \leq 4$

$\quad x \leq 2$

15. $14 > 5x - 1$

$\quad 15 > 5x$

$\quad 3 > x$

$\quad x < 3$

16 a. $<$ **b.** $=$ **c.** $<$ **d.** $<$ **e.** $>$

17. 1 is to 5 as 2.5 is to how much? $\frac{1}{5} = \frac{2.5}{x}$; this means $x = 5(2.5) = 12.5$. $12\frac{1}{2}$ cups of flour.

18. Let x, $x + 2$, $x + 4$, and $x + 6$ be the four consecutive even integers. Then

$$x + (x + 2) + (x + 4) + (x + 6) = 100$$

$$4x + 12 = 100$$

$$4x = 88$$

$$x = 22$$

The integers are 22, 24, 26, and 28.

19. $(T + t)^2 = T^2 + 2Tt + t^2$

Genotypes: 27% tall: $T^2 = (0.52)^2 = 0.2704$;

50% tall (recessive short): $2Tt = 2(0.52)(0.48) = 0.4992$;

23% short: $t^2 = (0.48)^2 = 0.2304$

Phenotypes: 77% tall: $0.2704 + 0.4992 = 0.7696$;

23% short: 0.2304

20.   (NEW YORK TO CHICAGO) + (CHICAGO TO SF) + (SF TO HONOLULU) $=$ 4,980

$\qquad\qquad\qquad\uparrow\qquad\qquad\qquad\qquad\qquad\qquad\qquad\qquad\qquad\uparrow$

(CHICAGO TO SF $-$ 1,140)   (CHICAGO TO SF $+$ 540)

Let $x =$ DISTANCE FROM CHICAGO TO SF; $(x - 1,140) + x + (x + 540) = 4,980$

$$3x - 600 = 4,980$$

$$3x = 5,580$$

$$x = 1,860$$

The distance from New York to Chicago is 720 mi, Chicago to San Francisco is 1,860 mi, and from San Francisco to Honolulu is 2,400 mi.

Page 189

CHAPTER 7 The Nature of Geometry

Chapter Overview
This is the second of the two great pillars of mathematics — algebra and geometry. As with the last chapter (algebra), this chapter (geometry) contains ideas that you may have already encountered in previous courses. It is our goal with this chapter, to present these ideas in new and fresh ways so that this does not become a "redo" with little interest.

7.1 Geometry, page 331
New Terms Introduced in this Section

Axiom	Compass	Congruent	Construct
Euclidean geometry	Euclid's postulates	Half-line	Line
Line segment	Line of symmetry	Non-Euclidean geometry	Parallel lines
Plane	Point	Postulate	Ray
Reflection	Similar figures	Similarity	Straightedge
Surface	Symmetry (line)	Theorem	Transformation
Transformational geometry	Undefined terms		

Level 1, page 337

1. Answers vary; both a young woman and an old woman.

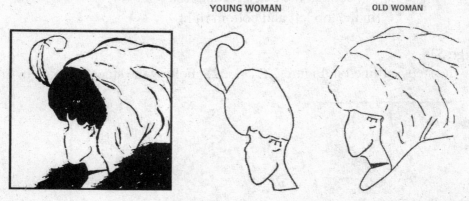

YOUNG WOMAN   OLD WOMAN

By the way, this "You See My Wife, and Mother-in-Law," also known as the "Boring Figure" was done in 1941 by the psychologist Edwund Boring was adapted freely from the 1880 anonymous German postcard shown in Figure 7.2.

3. Answers vary; you cannot give an argument based on what you think you see in a geometric figure.

5. Given a point and a radius of length $|AB|$, as shown in Figure 7.6. Set the legs of the compass on the ends of radius $\overline{AB}$; move the pointer to point O without changing the setting, as shown in the middle figure. Hold the pointer at point O and move the pencil end of the compass to draw the circle.

7. Line symmetry is best described by saying that a figure has line symmetry if the paper is folded along a line so that the left side folds onto the right side such that the figures on the left and on the right are identical.

9. The cartoon shows a reflection.

11. $\underset{P \quad Q}{\bullet\!-\!\!\bullet}$ 13. $\underset{P \quad Q}{\longleftrightarrow}$ 15. $\underset{P \quad Q}{\bullet\bullet\!-\!\!}$ 17. $\underset{P \quad Q}{\longleftrightarrow}$ 19. $\underset{P \qquad\qquad S}{\bullet\!-\!\!\bullet\!-\!\!\bullet}$

21. symmetric

23.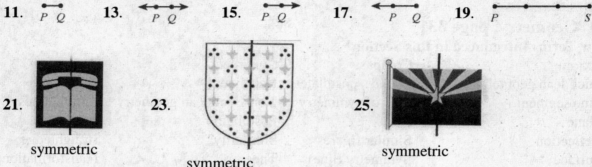

symmetric

Not symmetric; you might try to.
draw a line like the one shown,
but this is not a line of symmetry
— look at the black rectangles
in the top left and bottom right.

25. symmetric

27.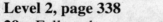

Level 2, page 338

29. Follow the steps outlined in Figure 7.5.

31. Follow the steps outlined in Figure 7.5.

33. Follow the steps outlined in Figure 7.6.   **35**. Follow the steps outlined in Figure 7.6.

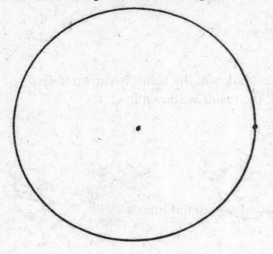
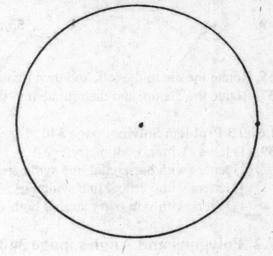

37. Follow the steps outlined in Figure 7.7.   **39**. Follow the steps outlined in Figure 7.7.

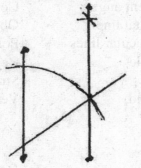
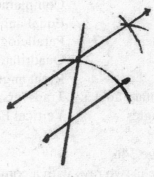

41. The butterfly is almost symmetric, but if you look at the little loop under the black patch are the top of the left wing is not there on the right wing; not symmetric.   **43**. The dodecahedron itself is symmetric, but the words in the caption are not.   **45**. The restaurant at the Los Angeles International Airport is a beautiful example of a symmetric building.
47. The Empire State Building is another great example of a symmetric building.

Level 3, page 339
Answers for Problems 48-53 may vary. If you have difficulty with this type of spatial visualization, try finding a die or other cube around your house and use post 'em on the faces.

Page 193

49. 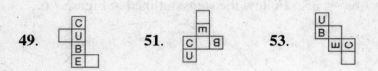   **51.**   **53.**

55. Rotate the die to the left, and then rotate the result down; the result is shown in C.

57. Rotate the die up, and then rotate it to the left; the result is shown in C.

Level 3 Problem Solving, page 340

59. (1) letters with no symmetry;

     (2) letters with horizontal line symmetry;

     (3) letters with vertical line symmetry;

     (4) letters with symmetry around both vertical and horizontal lines

7.2 Polygons and Angles, page 340

New Terms Introduced in this Section

Acute angle	Adjacent angles	Alternate exterior angles	Alternate interior angles
Angle	Complementary angles	Congruent angles	Corresponding angles
Degree	Equal angles	Horizontal line	Obtuse angle
Parallel lines	Parallelogram	Perpendicular lines	Polygon
Protractor	Quadrilateral	Rectangle	Regular polygon
Rhombus	Right angle	Square	Straight angle
Supplementary angles	Transversal	Trapezoid	Vertex (pl. vertices)
Vertical angles	Vertical line		

Level 1, page 346

1. An angle is two rays with a common endpoint.

3. A ray includes the endpoint and a half-line may or may not include the endpoint.

5. In a diagram on a printed page, any line that is parallel to the top and bottom edge of the page is considered horizontal. Lines that are perpendicular to a horizontal line are considered to be vertical.

7. Consider two intersecting lines forming four angles; adjacent angles are two angles which, when taken together, form a straight line and vertical angles are angles that are not adjacent. Now, consider two parallel lines cut by a transversal; corresponding angles are two nonadjacent angles whose interiors lie on the same side of the transversal such that one angle lies between the parallel lines and the other does not. Corresponding angles are congruent.

9. a. Count 4 sides, so it is a quadrilateral.   **b.** Count 5 sides, so it is a pentagon.
11. a. It has 3 sides, so it is a triangle.
   b. Count 6 sides, so it is a hexagon.
13. a. There are four sides, so it is a quadrilateral.
   b. There are seven sides, so it is a heptagon.
15. a. A square is a rectangle (true); note that a rectangle is not a square.
   b. A square is a parallelogram (true).
17. a. A square is a quadrilateral (true); note that a quadrilateral is not a square.
   b. A rectangle is a parallelogram, but a parallelogram is not a rectangle (false).
19. a. A trapezoid is a quadrilateral, but a quadrilateral is not a trapezoid (false).
   b. A parallelogram is not a trapezoid (false).
21. A square satisfies all parts. (All answers are "yes".)
23. A trapezoid does not satisfy any part. (All answers are "no".)

Level 2, page 347

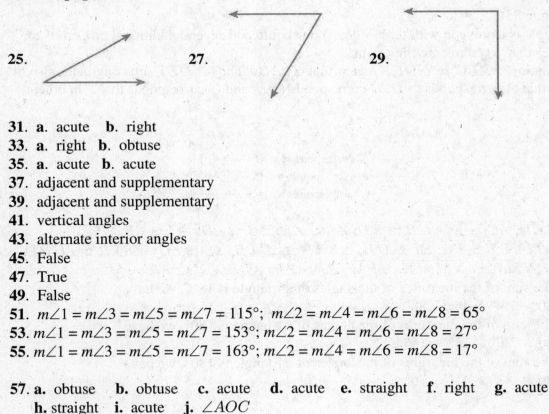

25.   **27.**   **29.**

31. a. acute   **b.** right
33. a. right   **b.** obtuse
35. a. acute   **b.** acute
37. adjacent and supplementary
39. adjacent and supplementary
41. vertical angles
43. alternate interior angles
45. False
47. True
49. False
51. $m\angle 1 = m\angle 3 = m\angle 5 = m\angle 7 = 115°$; $m\angle 2 = m\angle 4 = m\angle 6 = m\angle 8 = 65°$
53. $m\angle 1 = m\angle 3 = m\angle 5 = m\angle 7 = 153°$; $m\angle 2 = m\angle 4 = m\angle 6 = m\angle 8 = 27°$
55. $m\angle 1 = m\angle 3 = m\angle 5 = m\angle 7 = 163°$; $m\angle 2 = m\angle 4 = m\angle 6 = m\angle 8 = 17°$

57. a. obtuse   **b.** obtuse   **c.** acute   **d.** acute   **e.** straight   **f.** right   **g.** acute
   h. straight   **i.** acute   **j.** $\angle AOC$

Page 195

Level 3, Problem Solving, page 349

59.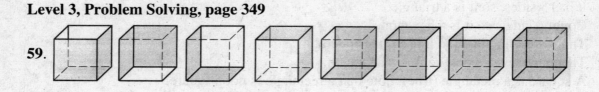

7.3 Triangles, page 349
New Terms Introduced in this Section

Acute triangle	Base angles	Base of a triangle	Congruent triangles
Corresponding parts	Equilateral triangle	Exterior angle	Exterior angle property
Isosceles triangle	Isosceles triangle property	Obtuse triangle	Right triangle
Scalene triangle	Similar triangle theorem	Sum of the measures in a triangle	
Triangle	Vertex angle		

Level 1, page 354

1. A *triangle* is a polygon with three sides. You should add several additional properties of triangles that you think are important.

3. The notation $\triangle ABC \simeq \triangle DEF$ means that $\triangle ABC$ and $\triangle DEF$ are congruent; also, it means that A corresponds to D, B corresponds to E, and C corresponds to F. In other words,

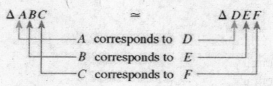

5. $\overline{AB} \simeq \overline{ED}$; $\overline{AC} \simeq \overline{EF}$; $\overline{CB} \simeq \overline{FD}$; $\angle A \simeq \angle E$; $\angle B \simeq \angle D$; $\angle C \simeq \angle F$

7. $\overline{RS} \simeq \overline{TU}$; $\overline{RT} \simeq \overline{TR}$; $\overline{ST} \simeq \overline{UR}$; $\angle SRT \simeq \angle UTR$; $\angle S \simeq \angle U$; $\angle STR \simeq \angle URT$

9. $\overline{JL} \simeq \overline{PN}$; $\overline{LK} \simeq \overline{NM}$; $\overline{JK} \simeq \overline{PM}$; $\angle J \simeq \angle P$; $\angle L \simeq \angle N$; $\angle K \simeq \angle M$

11. Since the sum of the measures of the angles of a triangle is $180°$, we have
    $x = 180° - 46° - 46° = 88°$.

13. Since the sum of the measures of the angles of a triangle is $180°$, we have
    $x = 180° - 20° - 15° = 145°$.

15. Since the sum of the measures of the angles of a triangle is $180°$, we have
    $x = 180° - 83° - 41° = 56°$.

Page 196

17. Since the measure of the exterior angles of a triangle equals the sum of the measures of the two opposite angles, we have $x = 40° + 35° = 75°$.

19. Since x and $100°$ form a straight angle, we have $x = 180° - 100° = 80°$.

21. Since the measure of the exterior angles of a triangle equals the sum of the measures of the two opposite angles, we have $x = 70° + 30° = 100°$.

Compare your constructions in Problems 23-28 with those given in the problem set.

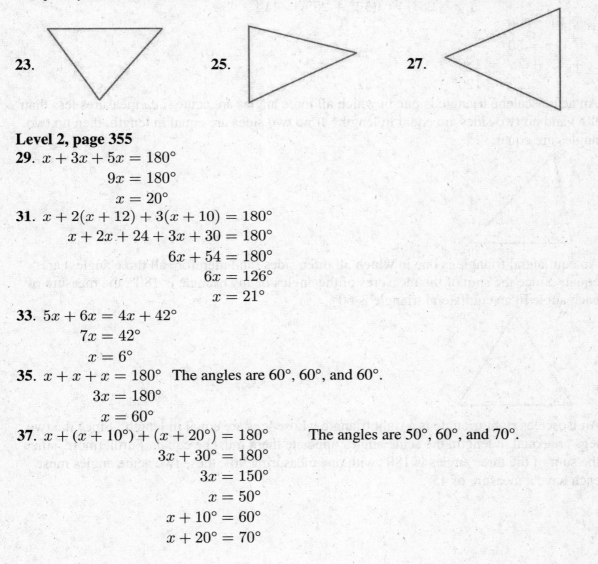

23.   **25**.   **27**.

Level 2, page 355

29. $x + 3x + 5x = 180°$
$$9x = 180°$$
$$x = 20°$$

31. $x + 2(x + 12) + 3(x + 10) = 180°$
$$x + 2x + 24 + 3x + 30 = 180°$$
$$6x + 54 = 180°$$
$$6x = 126°$$
$$x = 21°$$

33. $5x + 6x = 4x + 42°$
$$7x = 42°$$
$$x = 6°$$

35. $x + x + x = 180°$ The angles are $60°$, $60°$, and $60°$.
$$3x = 180°$$
$$x = 60°$$

37. $x + (x + 10°) + (x + 20°) = 180°$ The angles are $50°$, $60°$, and $70°$.
$$3x + 30° = 180°$$
$$3x = 150°$$
$$x = 50°$$
$$x + 10° = 60°$$
$$x + 20° = 70°$$

Page 197

39. $x + (14° + 3x) + 3(x + 25°) = 180°$ The angles are 13°, 53°, and 114°.

$$x + 14° + 3x + 3x + 75° = 180°$$
$$7x + 89° = 180°$$
$$7x = 91°$$
$$x = 13°$$
$$14° + 3x = 14° + 3(13°) = 53°$$
$$3(x + 25°) = 3(13° + 25°) = 114°$$

41. $x + 40° = 90°$
$$x = 50°$$

43. $x + 32° + 28° = 180°$
$$x = 120°$$

45. An acute scalene triangle is one in which all three angles are acute (*i.e.*, measures less than 90°) and no two sides are equal in length. If no two sides are equal in length, then no two angles are equal.

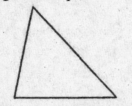

47. An equilateral triangle is one in which all three sides (and therefore all three angles) are equal. Since the sum of the measures of the angles in any triangle is 180°, the measure of each angle in an equilateral triangle is 60°.

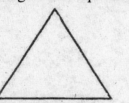

49. An isosceles right triangle is a right triangle whose legs are equal in length. Since the two legs are equal in length, the acute angles opposite them must be equal. Furthermore, since the sum of the three angles is 180° with one measuring 90°, these two acute angles must each have a measure of 45°.

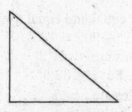

51. An obtuse scalene triangle is a triangle in which there is one obtuse angle and no sides whose lengths are equal.

Level 3, page 355

53. Construct three sides of the new triangle congruent to three sides of given triangle.

55. Construct one of the angles of the given triangle congruent to a new angle. Then, on one of the rays of that angle, construct a length congruent to the length of one of the given sides of the original triangle. Finally, construct the third angle of the new triangle congruent to the corresponding angle on the given triangle.

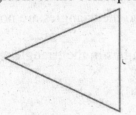

57. a. There are two triangles one on the top and one on the bottom. Looking at the one at the bottom, we note that this is an isosceles triangle, so the base angles are equal and have measure 45°. Since the top of the table and the ground are parallel, we see that

Page 199

$y = 45°$. Since the inner angles are vertical angles, they are equal and equal to 90°, we find the value of x using the exterior angle theorem: $x = 45° + 90° = 135°$.

b. Following the reasoning in part **a,** we note that the base angles are $\frac{1}{2}(180° - 85°) = 47.5°$, so $y = 47.5°$. Finally, $x = 47.5° + 85° = 132.5°$.

c. Following the reasoning in part **a,** we note that the base angles are $\frac{1}{2}(180° - 92°) = 44°$, so $y = 44°$. Finally, $x = 44° + 92° = 136°$.

Level 3 Problem Solving, page 356

59. Use reasoning similar to that outlined in Problem 58, to find a general formula, $S = 180°(n - 2)$ where S is the sum of the measures of the angles and n is the number of sides. Use this formula for $n = 5$ (pentagon):

$$\begin{aligned} S &= 180(n - 2) \\ &= 180(5 - 2) \\ &= 540° \end{aligned}$$

7.4 Similar Triangles, page 356
New Terms Introduced in this Section

Adjacent side	Corresponding angles	Corresponding sides	Hypotenuse
Opposite side	Similar triangles	Similar triangle theorem	

Level 1, page 359

1. Answers vary; congruent triangles have the same size and shape, whereas similar triangles assure only the same shape. You should add some illustrative examples.

3. Two angles of one are congruent to two angles of the other so they are similar. Contrast this with the next problem where the triangles are not similar, since you only know one pair of angles is congruent.

5. Only one angle of one is congruent to one angle of the other, so the triangles are not similar.

7. Since the sum of the angles of a triangle is 180°, we see the angles in the triangles are 60°, 30°, and 90°; thus the triangles are similar.

9. $m\angle A = m\angle A' = 25°$; $m\angle B = m\angle B' = 75°$; $m\angle C = m\angle C' = 80°$

11. $m\angle A = m\angle D = 38°$; $m\angle B = m\angle E = 68°$; $m\angle C = m\angle F = 74°$

13. $m\angle A = m\angle A' = 54°$; $m\angle B = m\angle B' = 36°$; $m\angle C = m\angle C' = 90°$

15. $|\overline{AC}| = |\overline{AB}| = 11$; $|\overline{A'C'}| = |\overline{A'B'}| = 22$; $|\overline{BC}| = 5$; $|\overline{B'C'}| = 10$

17. $|\overline{GH}| = 14$; $|\overline{HI}| = 12$; $|\overline{GI}| = 16$; $|\overline{DF}| = 20$; $\frac{|\overline{DE}|}{20} = \frac{14}{16}$, so $|\overline{DE}| = 17.5$; $\frac{|\overline{EF}|}{20} = \frac{12}{16}$, so $|\overline{EF}| = 15$

Page 200

19. $\left|\overline{AB}\right| = 10$; $\left|\overline{AC}\right| = 6$; $\left|\overline{BC}\right| = 8$; $\left|\overline{B'C'}\right| = 5$; $\frac{\left|\overline{A'B'}\right|}{5} = \frac{10}{8}$, so $\left|\overline{A'B'}\right| = 6.25$; $\frac{\left|\overline{A'C'}\right|}{5} = \frac{6}{8}$, so $\left|\overline{A'C'}\right| = 3.75$

21. $a^2 + b^2 = c^2 \qquad c = \sqrt{80} = \sqrt{16 \cdot 5} = 4\sqrt{5}$

$4^2 + 8^2 = c^2$

$\qquad 80 = c^2$

$\qquad \sqrt{80} = c$

23. $\dfrac{a}{a'} = \dfrac{b}{b'}$

$\dfrac{4}{2} = \dfrac{8}{b'}$

$b' = \dfrac{2(8)}{4}$

$\quad = 4$

25. $\dfrac{a}{c} = \dfrac{a'}{c'}$

$\dfrac{4}{6} = \dfrac{a'}{8}$

$a' = \dfrac{4(8)}{6}$

$\quad = \dfrac{16}{3}$

27. $\dfrac{b}{c} = \dfrac{b'}{c'}$

$\dfrac{b}{4} = \dfrac{8}{12}$

$b = \dfrac{4(8)}{12}$

$\quad = \dfrac{8}{3}$

29. $\dfrac{h}{30 + 25} = \dfrac{10}{25}$

$\dfrac{h}{55} = \dfrac{10}{25}$

$h = \dfrac{55(10)}{25}$

$\quad = 22$

Page 201

31. $\dfrac{x}{18} = \dfrac{10}{5 + 18}$

$\dfrac{x}{18} = \dfrac{10}{23}$

$x = \dfrac{18(10)}{23}$

≈ 7.8

33. Let h = height of the larger triangle. Then,

$$h^2 + (20 + 30)^2 = 55^2 \qquad \dfrac{h}{20 + 30} = \dfrac{y}{30}$$

$$h^2 = 525 \qquad \dfrac{\sqrt{525}}{50} = \dfrac{y}{30}$$

$$h = \sqrt{525} \qquad y \approx 13.7$$

Level 2, page 361

35. $m\angle BMA = m\angle BMC$ *Right angles because* $\overline{AC} \perp \overline{MB}$

    $m\angle A = m\angle C$ *Isosceles triangle property because* $\triangle ABC$ *is equilateral*

   $\triangle ABM \sim \triangle CBM$ *Similar triangle theorem*

37. $m\angle ACT = m\angle TCO$ *Given that they are bisectors.*

   $m\angle ATC = m\angle OTC$ *Given*

   $\triangle CAT \sim \triangle COT$ *Similar triangle theorem*

39. Let L = length of the lake;

$\dfrac{L}{210 + 140} = \dfrac{50}{140}$ The lake is 125 ft long.

$L = 125$

41. Let h = height of the house;

$\dfrac{6}{4} = \dfrac{h}{16}$ The height of the building is 24 ft.

$h = 24$

43. Let h = height of the building. Since a yardstick is 3 ft,

$\dfrac{3}{5} = \dfrac{h}{75}$ The building is 45 ft tall.

$h = 45$

45. Let h = height of the tower. Since a yardstick is 36 in.,

Page 202

$$\frac{36 \text{ in.}}{23 \text{ in.}} = \frac{h \text{ ft}}{45 \text{ ft}} \qquad \text{The bell tower is 70 ft tall.}$$
$$h \approx 70.4348$$

47. Let t = height of the tree;
$$\frac{t}{10} = \frac{5}{3} \qquad \text{The tree is 17 ft tall.}$$
$$t \approx 16.7$$

49. Let t = height of tree
$$\frac{t}{53} = \frac{69}{46} \qquad \text{Convert to inches} \qquad \text{The tree is 6 ft 8 in. tall.}$$
$$t = 79.5$$

51. $\dfrac{\text{New Orleans to Denver on map}}{\text{Chicago to New Orleans on map}} = \dfrac{\text{New Orleans to Denver}}{\text{Chicago to New Orleans}}$ Also,

$$\frac{10.8 \text{ cm}}{9.5 \text{ cm}} = \frac{\text{New Orleans to Denver}}{950 \text{ mi}}$$
$$\text{New Orleans to Denver} = \frac{(950 \text{ mi})(10.8 \text{ cm})}{9.5 \text{ cm}}$$
$$= 1{,}080 \text{ mi}$$

$$\frac{\text{Chicago to Denver on map}}{\text{Chicago to New Orleans on map}} = \frac{\text{Chicago to Denver}}{\text{Chicago to New Orleans}}$$
$$\frac{10.2 \text{cm}}{9.5 \text{ cm}} = \frac{\text{Chicago to Denver}}{950 \text{ mi}}$$
$$\text{Chicago to Denver} = \frac{(950 \text{ mi})(10.2 \text{ cm})}{9.5 \text{ cm}}$$
$$= 1{,}020 \text{ mi}$$

Denver to New Orleans is 1,080 mi and Chicago to Denver is 1,020 mi.

53. a. $\triangle HSP$ and $\triangle BTP$

  b. Yes; $m\angle HPS = m\angle TPB$ (vertical angles) and $m\angle HSP = m\angle PTB$ (right angles); thus $\triangle HSP \sim \triangle BTP$ (similar triangle theorem).

  c. $\dfrac{6}{10} = \dfrac{|\overline{BT}|}{35} \qquad \text{The footbridge is 21 ft above the base.}$
$$|\overline{BT}| = \frac{6(35)}{10}$$
$$= 21$$

Level 3, page 363

55. Given $m\angle D = m\angle E$, show $\triangle ADC \sim \triangle CEA$

*	$m\angle D = m\angle E$	*Given*
	$m\angle DBA = m\angle EBC$	*Vertical angles*
$m\angle D + m\angle DBA + m\angle BAD = 180°$		*Sum of the measures of the angles of a triangle.*
$m\angle E + m\angle EBC + m\angle BCE = 180°$		*Sum of the measures of the angles of a triangle.*
Thus,	$m\angle BAD = m\angle BCE$	*Substitution*
	$m\angle BAC = m\angle BCA$	*Isosceles triangle property*
	$m\angle ECA = m\angle BAD + m\angle BCA$	
*Therefore,	$m\angle DAC = m\angle ECA$	
so that	$\triangle ADC \sim \triangle CEA$	*Similar triangle property (two angles are equal from steps marked *).*

57. Let $\overline{CD}$ in Figure7.48 be the given segment. Draw $\overline{AB}$ so that $\left|\overline{AF}\right| = 3$ and $\left|\overline{BF}\right| = 7$. Repeat the steps outlined in Problem 56 to find E on the given segment in a 3-to-7 ratio.

59. $\angle C$ is a right angle, so $m\angle ACD + m\angle BCD = 90°$. Also, $m\angle A + m\angle ACD = 90°$ so $m\angle A = m\angle BCD$. Since $\overline{CD} \perp \overline{AB}$, $\angle ADC$ and $\angle BDC$ are right angles, so $m\angle ADC = m\angle BDC$. Thus, by the similar triangle property $\triangle ADC \sim \triangle BDC$.

7.5 Right-Triangle Trigonometry, page 363
New Terms Introduced in this Section

Angle of depression	Angle of elevation	Cosine	Inverse tangent
Inverse trigonometric ratios	Legs of a triangle	Pythagorean theorem	Sine
Tangent	Trigonometric ratios		

Level 1, page 368

1. For any right triangle with sides with lengths a and b and hypotenuse with length c, $a^2 + b^2 = c^2$. Also, if a, b, and c are lengths of sides of a triangle so that $a^2 + b^2 = c^2$, then the triangle is a right triangle.

3. In a right triangle ABC with right angle at C, $\cos A = \dfrac{\text{length of adjacent side of } A}{\text{length of hypotenuse}}$.

5. Since $5^2 + 12^2 = 13^2$, we see that Rope A would form a right triangle. On the other hand, $2^2 + 3^2 = 13 \neq 4^2$, so Rope B does not form a right triangle.

7. We see b is opposite $\angle B$.

9. Only the legs of right triangles are adjacent particular angles (not the hypotenuse). In this case, the side adjacent to $\angle B$ is the side labeled a.

11. We use the definition to say $\sin A = \frac{a}{c}$.

13. We use the definition to say $\cos A = \frac{b}{c}$.

15. We use the definition to say $\tan A = \frac{a}{b}$.

17. $\sin 56° = 0.8290$

19. $\sin 61° = 0.8746$

21. $\cos 54° = 0.5878$

23. $\cos 90° = 0$

25. $\tan 24° = 0.4452$

27. $\tan 75° = 3.7321$

29. $\sin^{-1} \frac{1}{2} = 30°$

31. $\tan^{-1} 1 = 45°$

33. $\tan^{-1} \sqrt{3} = 60°$

35. $\tan^{-1} 1.5 = 56°$

37. $\cos^{-1} 0.8 = 37°$

Level 2, page 368

39. $|\overline{AB}| = \sqrt{5^2 + 12^2} = \sqrt{25 + 144} = \sqrt{169} = 13$; this is the hypotenuse.

Look at $\angle A$ and note that $|\overline{BC}| = 12$ is the opposite side and $|\overline{AC}| = 5$ is the adjacent side. Thus, $\sin A = \frac{12}{13}$; $\cos A = \frac{5}{13}$; $\tan A = \frac{12}{5}$.

41. $|\overline{BC}| = \sqrt{6^2 - 1^2} = \sqrt{36 - 1} = \sqrt{35} \neq 5$; this is not a right triangle.

43. $|\overline{AC}| = \sqrt{6^2 - 1^2} = \sqrt{36 - 1} = \sqrt{35}$; this is the hypotenuse.

Look at $\angle A$ and note that $|\overline{BC}| = \sqrt{37}$ is the opposite side and $|\overline{AB}| = 6$ is the adjacent side. Thus, $\sin A = \frac{1}{6}$; $\cos A = \frac{\sqrt{35}}{6}$; $\tan A = \frac{1}{6}$.

45. $|\overline{AB}| = \sqrt{(\sqrt{3})^2 + 1^2} = \sqrt{3 + 1} = \sqrt{4} = 2$; this is the hypotenuse.

Look at $\angle A$ and note that $|\overline{BC}| = 1$ is the opposite side and $|\overline{AC}| = \sqrt{3}$ is the adjacent side. Thus, $\sin A = \frac{1}{2} = 0.5000$; $\cos A = \frac{\sqrt{3}}{2} \approx 0.8660$; $\tan A = \frac{1}{\sqrt{3}} \approx 0.5774$.

47. $|\overline{AB}| = \sqrt{(\sqrt{2})^2 + (\sqrt{7})^2} = \sqrt{2 + 7} = \sqrt{9} = 3 \neq 3^2$; this is not a right triangle.

49. Begin with a diagram. Let h be the height of the tower.

$$\tan 52° = \frac{h}{85}$$
$$h = 85 \tan 85°$$
$$\approx 108.795$$

The height of the tower is 109 ft.

Page 205

51. Begin with a diagram. Let x be the distance from the ship to the point directly below the observer.

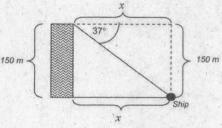

Don't forget that the angle of depression is the acute angle measured down from a horizontal line, as shown.

$$\tan 37° = \frac{150}{x}$$

$$x = \frac{150}{\tan 37°}$$
$$\approx 199.0567$$

The distance is 199 meters.

53. Begin with a diagram. Let h be the distance the ladder reaches up the wall.

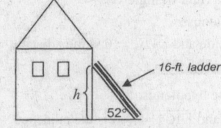

$$\sin 52° = \frac{h}{16}$$
$$h = 16 \sin 52°$$
$$\approx 12.608$$

The top of the ladder is 12 ft 7 in. from the ground.

55. Begin with a diagram. Let h be the height of the chimney stack.

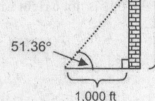

$$\tan 51.36° = \frac{h}{1,000}$$
$$h = 1,000 \tan 51.36°$$
$$\approx 1,250.8863$$

The chimney stack is 1,251 ft tall.

Level 3 Problem Solving, page 370

57. The legs are x, $\sqrt{3}\,x$, so by the Pythagorean theorem, $x^2 + 3x^2 = c^2$, where c is the

Page 206

hypotenuse. Solving (note, variables are positive):

$$x^2 + 3x^2 = c^2$$
$$4x^2 = c^2$$
$$\sqrt{4x^2} = c^2$$
$$2x = c$$

Thus, we have the following right triangle.

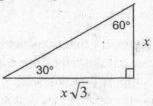

From the definition of the trigonometric ratios, we have:

$$\cos 30° = \frac{\sqrt{3}x}{2x} \qquad \sin 30° = \frac{x}{2x} \qquad \tan 30° = \frac{x}{\sqrt{3}x}$$

$$= \frac{\sqrt{3}}{2} \qquad\qquad = \frac{1}{2} \qquad\qquad = \frac{1}{\sqrt{3}} \cdot \frac{\sqrt{3}}{\sqrt{3}}$$

$$= \frac{\sqrt{3}}{3}$$

59. Plot the center of the circle and form a right triangle as shown in the following diagram:

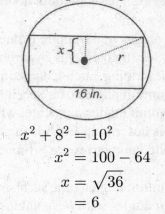

$$x^2 + 8^2 = 10^2$$
$$x^2 = 100 - 64$$
$$x = \sqrt{36}$$
$$= 6$$

The width is $2x$, so $w = 12$ in.

Page 207

7.6 Mathematics, Art, and Non-Euclidean Geometries, page 370

New Terms Introduced in this Section

Bolyai-Lobachevsky geometry	Divine proportion	Elliptic geometry
Golden ratio	Golden rectangle	Great circle
Hyperbolic geometry	Lobachevskian postulate	Non-Euclidean geometries
Projective geometry	Pseudosphere	Saccheri quadrilateral

Level 1, page 380

1. A non-Euclidean geometry is a geometry that does not assume Euclid's fifth postulate.

3. Answers vary; it is prevalent because it describes the real world around us. You should write a few paragraphs and then support your answer.

5. The divine proportion is the proportion $\frac{h}{w} = \frac{w}{h+w}$. It sometimes also refers to a rectangle whose width and height form this proportion.

7. The symbol τ (tau) is used to denote the golden ratio; it is the irrational number $\frac{1+\sqrt{5}}{2}$.

9. Answers vary, for example, start with 4 and 4.
   a. The sequence is 4, 4, 8, 12, 20, 32, $\cdots$.
   b. The ratios are $\frac{4}{4} = 1$, $\frac{8}{4} = 2$, $\frac{12}{8} = 1.5$, $\frac{20}{12} = 1.\overline{6}$, $\frac{32}{20} = 1.6$. It seems that these ratios oscillate around $\tau \approx 1.62$.
   You should do this again selecting two other numbers.

11. $\frac{97}{60} = 1.61\overline{6} \approx 1.62$ (about the same).

13. Ratio of b to s is 1.7; $\frac{1}{2}b$ to h is 1.62; these are both about the same as τ.

15. Ratio is 2.17; this is not close to τ at all.

17. The sum of the measures of the angles in a triangle is one of the best indicators of the type of geometry: 180° is Euclidean, less than 180° is hyperbolic, and more than 180° is elliptic. Thus, this problem is describing Euclidean geometry.

19. In an Euclidean geometry, the summit angles of a Saccheri quadrilateral are right angles. In a hyperbolic geometry the summit angles are acute, while in an elliptic geometry, they are obtuse. Thus, this problem is describing Euclidean geometry.

21. Lines in Euclidean and hyperbolic geometries are infinitely long, so this problem is describing an elliptic geometry.

23. In a hyperbolic geometry, the summit angles of a Saccheri quadrilateral are acute angles. Be careful not to confuse *summit* angles with *base* angles.

25. This rectangle has been rotated 90° counterclockwise from the rectangle shown in Figure 7.68, so it is a Saccheri quadrilateral.

27. This is not a Saccheri quadrilateral since the sides of equal length are not opposite sides.

Level 2, page 381

29. a. 5 in. by 3 in. *Note:* technically, 3 in. by 5 in. is not correct. The question asks for length and width, not width and length.

   b. $\frac{5}{3} = 1.\overline{6}$; this is about the same as τ.

31. a. You need to measure the length and width of your textbook. The length is $10\frac{1}{4}$in. and the width is 8 in.

   b. The ratio of length to width is $\frac{10\frac{1}{4}}{8} = 1.28125$; this is not close to τ.

33. These answers depend on your own body measurements. Use Figure 7.60, and ask someone to help you with these measurements.

35. a. There will be 2 branches in March.   **b.** There will be 8 branches in June.

   c. Complete the pattern: $1, 1, 2, 3, 5, 8, 13, 21, 34, \cdots$. Notice that each term is the sum of its two immediate predecessors. This is called a Fibonacci sequence. Therefore, after 12 months there will be 144 branches.

   d. Find the ratios: $\frac{1}{1} = 1$, $\frac{2}{1} = 2$, $\frac{3}{2} = 1.50$, $\frac{5}{3} \approx 1.67$, $\frac{8}{5} = 1.60$, $\frac{13}{8} = 1.63$, $\frac{21}{13} \approx 1.615$, $\frac{34}{21} \approx 1.619$, $\frac{55}{34} \approx 1.618$, $1\frac{144}{89} \approx 1.618$. The sequence is oscillating about the number τ.

37. In order to form the ratios, you must also work the preceding problem to find the sequence of reflections. The sequence is $1, 2, 4, 8, 13, 21, 34, \cdots$. The ratios of the number of successive paths are: $\frac{1}{1} = 1$, $\frac{2}{1} = 2$, $\frac{3}{2} = 1.50$, $\frac{5}{3} \approx 1.67$, $\frac{8}{5} = 1.60$, $\frac{13}{8} = 1.63$, $\frac{21}{13} \approx 1.615$, $\frac{34}{21} \approx 1.619$, $\frac{55}{34} \approx 1.618$, $\frac{89}{55} \approx 1.618$, $\frac{144}{89} \approx 1.618$. The sequence is oscillating about the number τ.

39. We don't know if the height is shorter than the width, so there are two possibilities. Thus, if we let h be the height of the rectangle, we have:

$$\frac{5}{h} = \frac{h}{5+h} \qquad\qquad\text{or}\qquad\qquad \frac{h}{5} = \frac{5}{h+5}$$

$$h^2 = 5(5+h) \qquad\qquad\qquad\qquad\qquad h(h+5) = 25$$

$$h^2 - 5h - 25 = 0 \qquad\qquad\qquad\qquad h^2 + 5h - 25 = 0$$

$$h = \frac{-(-5) \pm \sqrt{(-5)^2 - 4(1)(-25)}}{2(1)} \qquad\qquad h = \frac{-5 \pm \sqrt{5^2 - 4(1)(-25)}}{2(1)}$$

$$= \frac{5 \pm \sqrt{25 + 100}}{2} \qquad\qquad\qquad = \frac{-5 \pm \sqrt{25 + 100}}{2}$$

$$= \frac{5 + 5\sqrt{5}}{2} \;\text{ reject negative value} \qquad = \frac{-5 + 5\sqrt{5}}{2} \;\text{ reject negative value}$$

$$\approx 8.1 \qquad\qquad\qquad\qquad\qquad \approx 3.1$$

The window should have height either 3.1 ft or 8.1 ft.

Page 209

41. $\frac{\text{LENGTH}}{\text{WIDTH}} = \tau$, so HEIGHT $= 9\tau$

$$= 9\left(\frac{1 + \sqrt{5}}{2}\right)$$

$$\approx 14.56$$

The photograph should be 15 cm high.

43. The best answer is B. You should add a few sentences to support this response.

45. The best answer is C. You should add a few sentences to support this response.

47. Globes (spheres) model an elliptic geometry, and is such a geometry that the sum of the angles of a triangle is greater than 180°. You should actually confirm this by taking measurements on a globe.

Level 3, page 382

49. a. A great circle is a circle on a sphere with a diameter equal to the diameter of the sphere.

    b. The great circle ℓ is a line, but m is not.

    c. Yes, it does apply.

51. Lines are great circles, and the circle labeled m is not a great circle.

53. This is a true statement. Write a paragraph explaining why this is the case.

Level 3 Problem Solving, page 383

55. The length-to-width ratio will remain unchanged if

$$\frac{L}{W} = \frac{2W}{L}$$

$$\left(\frac{L}{W}\right)^2 = 2$$

$$\frac{L}{W} = \sqrt{2}$$

$$\approx 1.41421$$

57. a. $x^2 - x - 1 = 0$ **b.** They are negative reciprocals.

$$x = \frac{1 \pm \sqrt{1 - 4(-1)}}{2(1)}$$

$$= \frac{1 \pm \sqrt{5}}{2}$$

59. Just down from the North Pole there is a parallel that has a circumference of exactly one mile. If you begin anywhere on the circle that is a parallel 300 ft north of this parallel, the conditions are satisfied.

Chapter 7 Review Questions, page 384

STOP Studying for a chapter examination is a personal process, one which nobody else can do for you. Simply take the time to review what you have done. Here are the new terms in Chapter 7.

Acute angle [7.2]	Elliptic geometry [7.6]	[7.1, 7.6]
Acute triangle [7.3]	Equal angles [7.2]	Obtuse angle [7.2]
Adjacent angles [7.2]	Equilateral triangle [7.3]	Obtuse triangle [7.3]
Adjacent side [7.4]	Euclidean geometry [7.1]	Opposite side [7.4]
Alternate exterior angles [7.2]	Euclid's postulates [7.1]	Parallel lines [7.1]
Alternate interior angles [7.2]	Exterior angle [7.3]	Parallelogram [7.2]
Angle [7.2]	Exterior angle property [7.3]	Perpendicular lines [7.2]
Angle of depression [7.5]	Golden ratio [7.6]	Plane [7.1]
Angle of elevation [7.5]	Golden rectangle [7.6]	Point [7.1]
Axiom [7.1]	Great circle [7.6]	Polygon [7.2]
Base angles [7.3]	Half-line [7.1]	Postulate [7.1]
Base of a triangle [7.3]	Horizontal line [7.2]	Projective geometry [7.6]
Bolyai-Lobachevsky geometry [7.6]	Hyperbolic geometry [7.6]	Protractor [7.2]
	Hypotenuse [7.4]	Pseudosphere [7.6]
Compass [7.1]	Inverse tangent [7.5]	Pythagorean theorem [7.5]
Complementary angles [7.2]	Inverse trigonometric ratios [7.5]	Quadrilateral [7.2]
Congruent [7.1]		Ray [7.1]
Congruent angles [7.2]	Isosceles triangle [7.3]	Rectangle [7.2]
Congruent triangles [7.3]	Isosceles triangle property [7.3]	Reflection [7.1]
Construct [7.1]		Regular polygon [7.2]
Corresponding angles [7.2, 7.4]	Legs of a triangle [7.5]	Rhombus [7.2]
	Line [7.1]	Right angle [7.2]
Corresponding parts [7.3]	Line segment [7.1]	Right triangle [7.3]
Corresponding sides [7.4]	Line of symmetry [7.1]	Saccheri quadrilateral [7.6]
Cosine [7.5]	Lobachevskian postulate [7.6]	Scalene triangle [7.3]
Degree [7.2]		Similar figures [7.1]
Divine proportion [7.6]	Non-Euclidean geometries	Similar triangle theorem [7.4]

Similar triangles [7.4]	Surface [7.1]	Triangle [7.3]
Similarity [7.1]	Symmetry (line) [7.1]	Trigonometric ratios [7.5]
Sine [7.5]	Tangent [7.5]	Undefined terms [7.1]
Square [7.2]	Theorem [7.1]	Vertex (pl vertices) [7.2]
Straight angle [7.2]	Transformation [7.1]	Vertex angle [7.3]
Straightedge [7.1]	Transformational	Vertical angles [7.2]
Sum of the measures of angles	geometry [7.1]	Vertical line [7.2]
in a triangle [7.3]	Transversal [7.2]	
Supplementary angles [7.2]	Trapezoid [7.2]	

If you can describe the term, read on to the next one; if you cannot, then look it up in the text (the section number is shown in brackets). Next, study the types of problems listed at the end of Chapter 7.

TYPES OF PROBLEMS
Construct line segments. [7.1]
Construct circles, given the radius. [7.1]
Construct parallel lines. [7.1]
Find a line of symmetry for a given piece of art. [7.1]
Decide whether a given picture is symmetric. [7.1]
Visualize objects in three dimensions. [7.1]
Classify polygons with three to twelve sides. [7.2]
Construct an angle congruent to a given angle. [7.2]
Classify angles. [7.2]
Classify quadrilaterals [7.2]
Identify vertical, horizontal, intersecting, and parallel lines. [7.2]
Name the corresponding parts of congruent triangles. [7.3]
Find the measure of the third angle of a triangle. [7.3]
Find the measure of the exterior angles of a triangle. [7.3]
Construct a triangle congruent to a given triangle. [7.3]
Classify triangles and use the terminology associated with triangle classifications. [7.3]
Use algebra to find the measures of angles in a triangle. [7.3]
Decide whether a pair of given triangles is similar. [7.4]
List all six angles for a given pair of triangles. [7.4]
List all six sides of a given pair of triangles [7.4]
Given a right triangle, find the length of a missing side. [7.4]
Given similar triangles, find the length of one of the sides. [7.4]
Show that a given pair of triangles is similar. [7.4]

Solve applied problems using similar triangles. [7.4]
Evaluate a trigonometric ratio. [7.5]
Find the sine, cosine, and tangent for a given angle. [7.5]
Find $\sin^{-1}x$, $\cos^{-1}x$, and $\tan^{-1}x$. [7.5]
Solve applied problems using triangles. [7.5]
Know the terminology associated with right triangles. [7.5]
Work applied problems involving the golden ratio. [7.6]
Decide whether a figure is a Saccheri quadrilateral. [7.6]

Once again, see if you can verbalize (to yourself) how to do each of the listed types of problems.
    Work all of Chapter 7 Review Questions (whether they are assigned or not). Work through all of the problems before looking at the answers, and *then* correct each of the problems. The entire solution is shown in the answer section at the back of the text. If you worked the problem correctly, move on to the next problem, but if you did not work it correctly (or you did not know what to do), look back in the chapter to study the procedure, or ask your instructor.
    Finally, go back over the homework problems you have been assigned. If you worked a problem correctly, move on the next problem, but if you missed it on your homework, then you should look back in the book or talk to your instructor about how to work the problem.
    If you follow these steps, you should be successful with your review of this chapter.

We give all of the answers to the chapter review questions (not just the odd-numbered questions).

Chapter 7 Review Questions, page 384

1. Here is a cut-away picture of the cube:

    a. 0 **b.** 8 (corners of cube) **c.** 24 (2 such pieces on each of the 12 edges)
    d. 24 (4 on each face)   **e.** 8 (interior unpainted pieces)
2. There are two men (rotate the picture 180°).
3. It illustrates a rotation.

4. The New Yorker cartoon is almost a reflection except for the man and the word "STAR" reflecting to the word "RATS."   The photograph looks like a landscape reflected in a lake, so the illustration that best shows a reflection is the photograph.

5. If you rotate each of the illustrations, you will see the only one that has been rotated is the photograph.

6. **a**. Flag is symmetric; picture is not (if you include the flagpole).
   b. This flag is not symmetric.
   c. This logo does not have a line of symmetry.
   d. This shield is symmetric.

7. **a**. A golden rectangle is a rectangle satisfying the divine proportion.
   b. It does look like it forms a golden rectangle. (Compare with the Parthenon shown in Example 2 on page 373.)

8. If we assume the building forms a golden rectangle, the height and width must satisfy the proportion

$$\frac{h}{w} = \frac{w}{h+w}$$

for height h and width w. We are given $w = 280$ ft, so

$$\frac{h}{280} = \frac{280}{h+280}$$
$$h(h+280) = 280^2$$
$$h^2 + 280h - 280^2 = 0$$
$$h = \frac{-280 \pm \sqrt{280^2 - 4(1)(-280^2)}}{2(1)}$$
$$\approx 173, -453$$

If we ignore the negative value, we see the height of the building is 173 ft.

9. If C represents the capacity of the boat, ℓ the length of the boat, and w the width of the boat, we have $C = \frac{\ell w}{15}$.

10. **a**. Angle 1 is an obtuse angle.
    b. Angles 5 and 6 are adjacent angles and they are also supplementary angles.
    c. Angles 2 and 4 are vertical angles.
    d. Angle 2 is an acute angle.
    e. Angle 7 is an obtuse angle.
    f. Angles 1 and 3 are vertical angles.

11. **a**. Complementary angles have measures whose sum is 90°: 90° − 49° = 41°.
    b. Supplementary angles have measures whose sum is 180°: 180° − 49° = 131°.
    c. The sum of the acute angles of a right triangle is 90°, so 90° − 49° = 41°.

Page 214

12. $(3x + 20°) + (2x − 40°) + (x − 16°) = 180°$

$$6x − 36° = 180°$$
$$6x = 216°$$
$$x = 36°$$

13.
$$\frac{h}{10} = \frac{10}{h + 10}$$
$$h(h + 10) = 100$$
$$h^2 + 10h − 100 = 0$$
$$h = \frac{-10 \pm \sqrt{10^2 − 4(1)(−100)}}{2(1)}$$
$$= \frac{-10 \pm \sqrt{500}}{2}$$
$$= \frac{-10 \pm 10\sqrt{5}}{2}$$
$$= −5 \pm 5\sqrt{5} \approx 6.18, \; − 16.18 \qquad \textit{Disregard negative value.}$$

The width is about 6 in.

14. $13^2 = x^2 + 12^2$

$$169 − 144 = x^2$$
$$x = \pm 5 \qquad \textit{Disregard negative value.}$$

The other leg is 5 in.

15. If the triangle is isosceles, then by the isosceles triangle property the base angles must be the same. Also, since it is a right triangle, and since the sum of the angles of a triangle is 180°, we see the base angles must be 45°, we can use one of the trigonometric ratios to find the length of the side. If x represents the length of one of the legs of the triangle, we have

$$\sin 45° = \frac{x}{10}$$
$$x = 10 \sin 45°$$
$$= (10)\frac{\sqrt{2}}{2} \qquad \textit{Value from Example 3, page 365.}$$
$$= 5\sqrt{2} \qquad \textit{This is about 7 inches.}$$

The length of the leg is $5\sqrt{2}$ in., or about 7 in.

16. a. $\sin 59° \approx 0.8572$
   b. $\tan 0° = 0$
   c. $\cos 18° \approx 0.9511$
   d. $\tan 82° \approx 7.1154$

17. Use the definition of the tangent ratio: Let d be the distance to the lighthouse.

$$\tan 12° = \frac{160}{d}$$

$d \tan 12° = 160$ *Multiply both sides by d.*

$$d = \frac{160}{\tan 12°}$$ *By calculator: Display: 752.7408175*

$$\approx 753$$

The distance to the lighthouse is about 753 feet.

18.

19. A Saccheri quadrilateral is a quadrilateral *ABCD* with base angles *A* and *B* right angles and with sides *AC* and *BD* with equal lengths.

20. In a quadrilateral *ABCD*, if the summit angles *C* and *D* are right angles, then the result is Euclidean geometry. If they are acute, then the result is hyperbolic geometry. If they are obtuse, the result is elliptic geometry.

CHAPTER 8 The Nature of Networks and Graph Theory

Chapter Overview
This chapter is the second of three chapters dealing with geometry. The first introduces the *nature* of geometry, and this chapter presents some aspects of geometry that will be new to most students. The next chapter on measurement returns to more familiar geometric topics. In this chapter we consider such famous problems as the Königsberg bridge problem, the traveling salesperson problem, and map coloring.

8.1 Euler Circuits and Hamiltonian Cycles, page 389

New Terms Introduced in this Section

Arc	Connected network	Degree (vertex)	Edge
Euler circuit	Euler's circuit theorem	Even vertex	Graph
Hamiltonian cycle	Network	Odd vertex	Operations research
Region	Sorted-edge method	TSP	Traversable network
Vertex			

Level 1, page 397

1. In the 18th century, in the German town of Königsberg (now a Russian city), a popular pastime was to walk along the bank of the Pregel River and cross over some of the seven bridges that connected two islands, as shown in the chapter opener and in Figure 8.1. The problem was to take a walk and cross each of the seven bridges once and only once, and return to the starting location.

3. See Table 8.1. If there are no odd vertices, the network is traversable and any point may be a starting point. The point selected will also be the ending point. If there is one odd vertex, the network is not traversable. A network cannot have only one starting or ending point without the other. If there are two odd vertices, the network is traversable; one odd vertex must be a starting point and the other odd vertex must be the ending point. If there are more than two odd vertices, the network is not traversable. A network cannot have more than one starting point and one ending point.

5. The roles of vertices and paths reverse with the Euler circuits and Hamiltonian cycles. That is, with the Euler circuit problem we travel each path exactly once, and with a Hamiltonian cycle we visit each vertex exactly once.

Results shown for Problems 6-11 may vary.

7. Traversable ($A \to B \to C \to D \to A \to F \to E \to D$), but not an Euler circuit because we do not return to the starting point.

9. Not an Euler circuit; not traversable (more than 2 odd vertices).

11. Not an Euler circuit; not traversable (more than 2 odd vertices).

Routes shown for Problems 12-17 may vary.

13. Hamiltonian cycle: $A \to B \to C \to D \to A$

15. Hamiltonian cycle: $A \to B \to C \to G \to H \to F \to E \to D \to A$

17. Hamiltonian cycle: $A \to B \to C \to D \to A$

Possible paths in Problems 18-23 may vary.

19. There are five rooms (vertices), three of which are odd, so it is not traversable.

21. There are five rooms (vertices), three of which are odd, so it is not traversable.

23. There are five rooms (vertices), two of which are odd, so it is traversable. Here is one possibility:

Level 2, page 398

25. There are more odd vertices, so it is not an Euler circuit; since there are more than two odd vertices, it is not traversable.

27. There are odd vertices, so it is not an Euler circuit; since there are more than two odd vertices, it is not traversable.

29. You can confirm that it is not possible by exhausting all possible paths (trail-and-error).

31. Clearly, there is no way to visit vertex H without having visited vertex G at least twice, no matter whether H is the original starting/ending vertex or not.

33. There is more than one possibility. $1 \to 2 \to 3 \to 10 \to 11 \to 12 \to 4 \to 5$
$\to 14 \to 13 \to 19 \to 18 \to 17 \to 9 \to 8 \to 7 \to 16 \to 20 \to 15 \to 6 \to 1$

35. Transform the problem into a network; it is not traversable since there are more than two odd vertices (J and H).

37. There are two odd vertices, so it is traversable; begin at either Queens or Manhattan.

39. Yes; 2 odd vertices.

Level 3, page 400

41. A cube has eight corners (vertices), each with a degree of 3. Since any network with more than two odd vertices is not traversable, we know the edges of a cube do not form a traversable network.

43. a. NYC to Boston to Washington, D.C. to NYC; $216 + 441 + 235 = 892$ mi; forms a loop without including Cleveland.

    b. NYC to Boston to Washington, D.C. to Cleveland to NYC; $216 + 441 + 375 + 481 = 1,513$ mi

45. Brute force:

    Denver $\rightarrow$ S.L. $\rightarrow$ LA $\rightarrow$ N.O. $\rightarrow$ Denver; $879 + 1,844 + 2,009 + 1,344 = 6,076$

    Denver $\rightarrow$ S.L. $\rightarrow$ N.O. $\rightarrow$ LA $\rightarrow$ Denver; $879 + 677 + 2,009 + 1,062 = 4,627$

    Denver $\rightarrow$ LA $\rightarrow$ S.L $\rightarrow$ N.O. $\rightarrow$ Denver; $1,062 + 1,844 + 677 + 1,344 = 4,927$

    Denver $\rightarrow$ LA $\rightarrow$ N.O. $\rightarrow$ S.L. $\rightarrow$ Denver; $1,062 + 2,009 + 677 + 879 = 4,627$

    Denver $\rightarrow$ N.O. $\rightarrow$ S.L. $\rightarrow$ LA $\rightarrow$ Denver; $1,344 + 677 + 1,844 + 1,062 = 4,927$

    Denver $\rightarrow$ N.O. $\rightarrow$ LA $\rightarrow$ S.L. $\rightarrow$ Denver; $1,344 + 2,009 + 1,844 + 879 = 6,076$

    There is one route (along with its reverse route) with the mileage 4,627 mi. Note this is the same as the sorted-edge solution from Problem 44b.

Level 3 Problem Solving, page 400

47. Use the sorted edge method:

    NYC $\rightarrow$ B $\rightarrow$ DC $\rightarrow$ C $\rightarrow$ A $\rightarrow$ NYC; $216 + 441 + 375 + 780 + 887 = 2,699$ mi

49. A tetrahedron has four triangular faces; the sum of the measures of the angles on each face is $180°$, so the sum of the measures of the face angles of a tetrahedron is $720°$.

51.

Luke in this room:	1	2	3	4	5	6	7	8	9	10	11	12	13
Number of routes:	1	2	3	5	8	13	21	34	55	89	144	233	377

    a. Room 1: 1 path; room 2: 2 paths; room 3: 3 paths; room 4: 5 paths

    b. From the table, we see it is 89.

    c. From the table, we see it is 377.

53. a. Start at the top (or at the right) at one of the two odd vertices.

    b. The chance that you will choose one of the two odd vertices at random is small.

55. The result is a twisted band twice as long and half as wide as the band with which you began.

57. Two interlocking pieces, one twice as long as the other.

59. One edge; one side; after cutting, you will have a single loop with a knot.

8.2 Trees and Minimum Spanning Trees, page 402

New Terms Introduced in this Section

Connected graph	Kruskal's algorithm	Minimum spanning tree	Number-of-edges theorem
Spanning tree	Tree	Weight	Weighted graph

Level 1, page 409

1. A *tree* is a graph which is connected and has no circuits.

3. To construct the minimum spanning tree from a weighted graph:
    1. Select any edge with minimum weight.
    2. Select the next edge with minimum weight among those not yet selected.
    3. Continue to choose edges of minimum weight from those not yet selected, but make sure not to select any edge that forms a circuit.
    4. Repeat this process until the tree connects all of the vertices of the original graph.
5. There is a circuit (see the diamond in the middle), so it is not a tree.
7. This is not a tree because it has a circuit (see the quadrilateral in the center).
9. This is not a tree because it has a circuit and is not connected.
11. This is a tree because it is connected and it has no circuits.

Answers to Problems 12-19 vary.

13.   15.

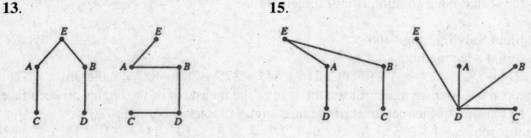

17.   19.

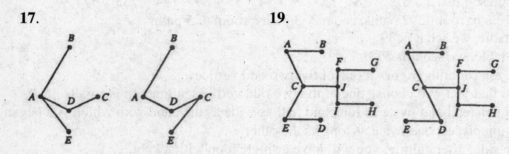

21.

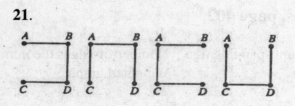

Page 220

22.

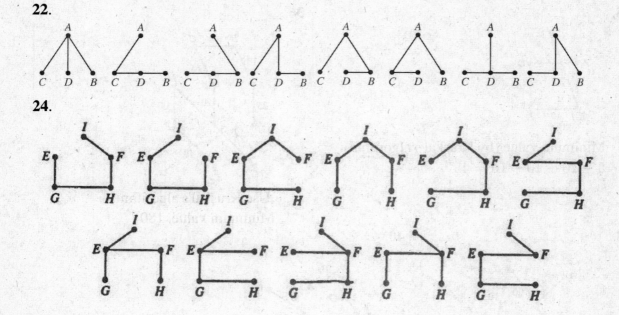

24.

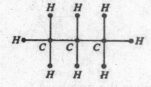

26. The propane molecule is a tree.   **28.** The isobutane molecule is a tree.

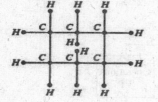

30. Not a tree; because there is a circuit.   **32.**

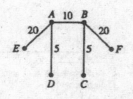

Minimum value (by Kruskal's algorithm):
$20 + 10 + 5 + 5 + 20 = 60$

Page 221

34.

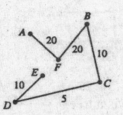

Minimum value (by Kruskal's algorithm):
$5 + 20 + 10 + 10 = 45$

36.

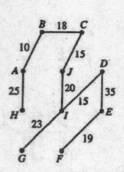

Use Kruskal's algorithm;
Minimum value, 180

38.

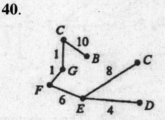

Use Kruskal's algorithm;
Minimum value; 65

40.

Use Kruskal's algorithm:
Minimum value, 30

Level 2, page 410

42. From the number-of-edges-and-vertices-in-a-tree theorem, and since the number of vertices is 65, the number of edges (or bounds) is $65 - 1 = 64$.

44. No, because some sites lead you back to previously visited sites, which completes a circuit.

46.

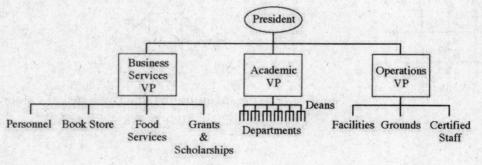

Page 222

48. From the number-of-vertices-and-edges-in-a-tree theorem, and since the number of edges is 48, the number of vertices is $48 + 1 = 49$.

50. Use Kruskal's algorithm; there are six buildings using the walkways of length 10 ft, 15 ft, 25 ft, 30 ft, and 30 ft. The total is 110 ft. Since the cost is \$350/ft, we estimate the minimum cost for the project to be (\$350/ft)(100 ft) = \$35,000. The actual amount is (\$350/ft)(110 ft) = \$38,500.

52. The minimum spanning tree connects San Francisco to Oakland (smallest distance, 13 miles); Oakland to San Jose (next smallest distance, 46 miles); Manteca to Merced (54 miles); Merced to Fresno (56 miles); Santa Rosa to San Francisco (58 miles); Oakland to Manteca (66 miles), and then complete the graph with Merced to Yosemite Village (82 miles). The minimum spanning tree is 375 miles.

Level 3, page 412

54. **a.**   **b.**   **c.** The minimum cost is
$1 + 1 + 1 + 1 = 4$;
that is, \$4,000,000.

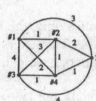

56. **a.**   **b.**   **c.** The minimum distance is
$172 + 160 + 105 + 237$
$\qquad\qquad = 674$ miles.
The minimum cost is
\$85(674) = \$57,290.

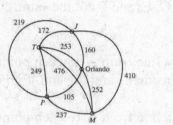   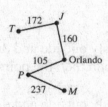

Level 3, Problem Solving, page 412

58. **a.** If a graph is a tree with n vertices, then the number of edges is $n - 1$.

    b. There are no edges; yes, the property holds because $1 - 1 = 0$.

    c. Can only have one edge to have a tree. If there are two vertices and two or more edges, then there is a circuit, which is not permitted.

    d. If there are three vertices, then it takes two edges to form a tree. If there are three or more edges, then there is a circuit.

8.3 Topology and Fractals, page 413
New Terms Introduced in this Section

Closed curve	Four-color problem	Fractal geometry	Genus
Jordan curve	Planar curve	Simple curve	Tessellation
Topologically equivalent	Topology		

Level 1, page 418

1. *Topology* is that branch of mathematics that is concerned with discovering and analyzing the essential similarities and differences between sets and figures. An important idea of topology is that of elastic motion.

3. Note that A and F each have an inside and outside, so these are topologically equivalent. B and G have two interior regions and one exterior region, so these are also topologically equivalent. Finally, C, D and H are topologically equivalent, with E in a class by itself.

5. Figure A and F are simple closed curves.

7. You need to use the font shown in the problem. If we consider the holes only, the letters A, D, O, P, Q, and R have genus one, so these are all in the same class. The letters B has genus two, so this letter is in a class by itself. Finally, all the other letters have genus 0 (no holes): C, E, F, G, H, I, J, K, L, M, N, S, T, U, V, W, X, Y, and Z

9. A drinking straw and a sewing needle both have one hole (genus 1), so B and D are of the same class. A funnel with a handle has two holes (genus), so G is in its own class. Finally, all the other objects have genus 0 (no holes): A, C, E, and F are the same class.

Level 2, page 419

11. Take some point X which is obviously outside and draw a line from X to each point in term. If the number of crossings is odd then the point is inside (A and C) and if it is even then the point (B) is outside.

13. Take some point X which is obviously outside and draw a line from X to each point in term. If the number of crossings is even then the point is outside (C) and if it is odd, then the point (A and B) is inside.

15. This map can be colored with two colors.

17. Answers vary; first region is one color, second another; then the third region (that lies on each of those) needs a third color.

19.

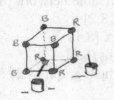

21. The letters shown here conform to common usage and are used for easy identification.

Level 3, page 420

23. Answers vary; construct a tessellation.

25. There are a total of 364 different possibilities. We show one possibility here.

Level 3 Problem Solving, page 420

27. a. No, it will now fit the left hand.

   b. Yes, they are topologically equivalent.

Chapter 8 Review Questions, page 422

Studying for a chapter examination is a personal process, one which nobody else can do for you. Simply take the time to review what you have done. Here are the new terms in Chapter 8.

Arc [8.1]	Edge [8.1]	Fractal geometry [8.3]
Closed curve [8.3]	Euler circuit [8.1]	Genus [8.3]
Connected graph [8.1]	Euler's circuit theorem [8.1]	Graph [8.1]
Connected network [8.1]	Even vertex [8.1]	Hamiltonian cycle [8.1]
Degree (vertex) [8.1]	Four-color problem [8.3]	Jordan curve [8.3]

Page 225

Kruskal's algorithm [8.2]	Simple curve [8.3]	Traversable network [8.1]
Minimum spanning tree [8.2]	Sorted-edge method [8.1]	Tree [8.2]
Network [8.1]	Spanning tree [8.2]	Vertex [8.1]
Number-of-edges theorem [8.2]	Tessellation [8.3]	Weight [8.2]
Odd vertex [8.1]	Topologically equivalent [8.3]	Weighted graph [8.2]
Operations research [Overview]	Topology [8.3]	
Planar curve [8.3]	Traveling salesperson	
Region [8.1]	problem (TSP) [8.1]	

If you can describe the term, read on to the next one; if you cannot, then look it up in the text (the section number is shown in brackets). Next, study the types of problems listed at the end of Chapter 8.

TYPES OF PROBLEMS
Decide whether a network is an Euler circuit. [8.1]
Work floor-plan problems. [8.1]
Solve applied problems involving traversable networks. [8.1]
Determine whether a given graph is a tree. [8.2]
Find spanning trees for a given graph. [8.2]
Given a weighted graph, find the minimum spanning tree. [8.2]
Sort figures into topologically equivalent classes. [8.3]
Decide whether a given point is an interior or exterior point. [8.3]
Solve applied problems involving the four color-theorem. [8.3]
Design mosaics (tessellations). [8.3]

Once again, see if you can verbalize (to yourself) how to do each of the listed types of problems.

Work all of Chapter 8 Review Questions (whether they are assigned or not). Work through all of the problems before looking at the answers, and *then* correct each of the problems. The entire solution is shown in the answer section at the back of the text. If you worked the problem correctly, move on to the next problem, but if you did not work it correctly (or you did not know what to do), look back in the chapter to study the procedure, or ask your instructor.

Finally, go back over the homework problems you have been assigned. If you worked a problem correctly, move on to the next problem, but if you missed it on your homework, then you should look back in the book or talk to your instructor about how to work the problem.

If you follow these steps, you should be successful with your review of this chapter.

We give all of the answers to the chapter review questions (not just the odd-numbered questions).

Chapter 8 Review Questions, page 422

1. Classify the vertices; they are all odd (degree 3), so there are more than 2 odd vertices. Thus it is not traversable.

2. Classify the vertices; the outside corners and the corner at the bottom of the inner rectangle are even; the intersecting lines at the top show an even vertex, and the two vertices at the top of the inner rectangle are odd (degree 5). Start at either of these odd vertices, and end at the other to show that the network is traversable.

3. Classify the vertices; all of them are even vertices so there is an Euler circuit and the network is traversable.

4. There are 7 rooms, including the exterior. There are two rooms with an odd number of doors, so begin in either of those rooms and end in the other — it is possible.

5. **a.** Because the vertex 3 is odd, this cannot be an Euler circuit.

   b. Yes, there is at least one possibility: $3 \to 2 \to 1 \to 5 \to 4 \to 7 \to 8 \to 9 \to 10 \to 11 \to 12 \to 13 \to 14 \to 15 \to 6$.

6. Answers vary.

   a.   **b.**   **c.**

   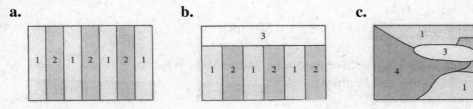

   d. No such map exists.

7. Using the formula from Example 8 in Section 8.1, we see there are 8 cities, so there are
   $$\frac{8 \cdot 7 \cdot 6 \cdot 5 \cdot 4 \cdot 3 \cdot 2 \cdot 1}{2} = 20{,}160 \text{ different routes.}$$

8. SF $\to$ LA $\to$ SD $\to$ P $\to$ SF; $369 + 133 + 353 + 755 = 1{,}610$
   SF $\to$ LA $\to$ P $\to$ SD $\to$ SF; $369 + 372 + 353 + 502 = 1{,}596$
   SF $\to$ SD $\to$ LA $\to$ P $\to$ SF; $502 + 133 + 372 + 755 = 1{,}762$
   SF $\to$ SD $\to$ P $\to$ LA $\to$ SF; $502 + 353 + 372 + 369 = 1{,}596$
   SF $\to$ P $\to$ SD $\to$ LA $\to$ SF; $755 + 353 + 133 + 369 = 1{,}610$
   SF $\to$ P $\to$ LA $\to$ SD $\to$ SF; $755 + 372 + 133 + 502 = 1{,}762$

   It looks like either of the routes shown in boldface is the best brute-force route.

9. SF $\to$ LA $\to$ SD $\to$ LA; $369 + 133 + 133 = 635$, but this does not include Phoenix, so this method does not give a best route.

10. The weighted graph is shown:

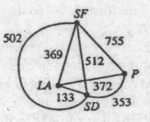

The minimum spanning tree is:

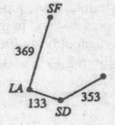

The minimum distance is $369 + 133 + 353 = 855$ miles.

11. Yes, there are many possibilities, for example, Minnesota, Wisconsin, Iowa, Illinois, Indiana, Michigan, Ohio, and Pennsylvania.

12. Actually no, but if you use the ASC ferry system from Wisconsin to Pennsylvania, then the answer is yes; Iowa, Minnesota, Wisconsin, Pennsylvania, Ohio, Michigan, Indiana, Illinois, and Iowa. (There are other possibilities if you use the ferry system.)

13. $A \xrightarrow{2} B \xrightarrow{1} G \xrightarrow{4} L \xrightarrow{3} K \xrightarrow{2} F \xrightarrow{8} A$; the cost of this trip is $2 + 1 + 4 + 3 + 2 + 8 = 20$.

14. $M \xrightarrow{2} H \xrightarrow{1} I \xrightarrow{1} N \xrightarrow{3} O \xrightarrow{5} J \xrightarrow{3} E \xrightarrow{6} D \xrightarrow{3} C \xrightarrow{5} B \xrightarrow{1} G \xrightarrow{4} L \xrightarrow{3} K \xrightarrow{2} F \xrightarrow{8} A$; the cost of this trip is 47.

15. There are 15 vertices, so the number of possibilities is

$$\frac{14 \cdot 13 \cdots 3 \cdot 2 \cdot 1}{2} = 4.3589 \times 10^{10}$$

Page 228

16. The cost of this trip is 39.

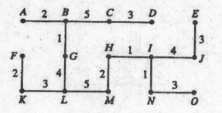

17. No, because there are more than two odd vertices.

18. There are many possibilities; $A \rightarrow B \rightarrow C \rightarrow D \rightarrow I \rightarrow E \rightarrow F \rightarrow G \rightarrow H \rightarrow A$

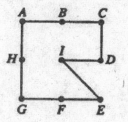

19. Answers vary; here is one:

20. There are two edges and two sides; the result after cutting is two interlocking loops.

CHAPTER 9 The Nature of Measurement

Chapter Overview
This chapter is dealing with the idea of measurement, in particular the U.S. measurement system and the SI (metric) measurement system. In particular we discuss one, two, and three dimensions. This is a chapter in the book which will be used many times (outside this text) in your life, since measurements are part of our daily lives.

9.1 Perimeter, page 427

New Terms Introduced in this Section

Accuracy	Center of a circle	Centi-	Circle
Circumference	Deci-	Deka-	Diameter
Equilateral triangle	Foot	Hecto-	Inch
Kilo	Length	Measure	Meter
Metric system	Mile	Milli-	Perimeter
Pi (π)	Precision	Radius	Rectangle
Semicircle	SI system	Square	United States System
Yard			

Level 1, page 433

1. Precision refers to the unit of measurement used, whereas the accuracy refers to the answers. In this book, the accuracy of an answer should not exceed the precision of the measurement. For example, a measurement of the length of the side of a square to the nearest tenth, say 2.1 in. might be used in calculating the area. On a calculator, $2.1^2 = 4.41$. The precision of the measurement is 2.1 (one decimal place) and the accuracy 4.41 is two decimal places. Since the accuracy should not exceed the precision of the measurement (one decimal places), we say the area of the square is 4.4 in.^{2} Make up some examples on your own and include those.

3. Perimeter of a square: $P = 4s$;
   Perimeter of a rectangle: $P = 2(\ell + w)$;
   Perimeter of an equilateral triangle: $P = 3s$;
   Perimeter of a regular pentagon: $P = 5s$

5. The number π is the ratio of the circumference to the diameter of a circle. It is an irrational number without an exact decimal or fractional representation.

7. The segments show here are the proper size, but if you use a ruler then you have missed the intent of this question. The idea here is that you *estimate* these lengths.

    a. ———————————————   **b.** ——   **c.** ———

9. This is an *estimation* problem — do not measure your book! Answer B (10 cm) is way too small, and answer C (1 ft) is too large, so we guess that 10 in. (A) is the correct size.

11. This is an *estimation* problem — do not measure a dollar bill. Answer A (3 in. is closer to the width than the length) and 9 inches is way too big, so we guess that 6 in. (B) is the correct size.

13. This is an *estimation* problem — do not measure a dollar bill. Answer B (6 in.) is about the length of the dollar bill and answer C (46 in.) is way too big, so we guess that 18 in. (A) is the correct answer.

15. This is an *estimation* problem — do not measure a pencil. A pencil which is 4 in. long is stubby, and certainly not new whereas 18 in. is way too long, so we see the correct answer must be 7 in. (C).

17. This is an *estimation* problem — do not measure a tire. The most reasonable answer is 60 in., which is choice B.

19. This is an *estimation* problem — do not measure a credit card. The most reasonable answer is 30 cm, which is choice B.

21. This is an *estimation* problem — do not measure a television screen. The most reasonable answer is 100 in., which is choice C.

23. This is an *estimation* problem; the most reasonable answer is 1 km, which is choice C.

25. This is an *estimation* problem; the most reasonable answer is 170 cm, which is choice C.

27. This is an *estimation* problem; the most reasonable answer is 2.5 m, which is choice A.

29. This is an *estimation* problem. Since a yard is a bit shorter than a meter, 100 yards are a bit shorter than 100 meters, so the correct answer is C.

31. This is an *estimation* problem. Since a meter is a bit longer than a yard, 100 yards is a bit shorter than 100 meters, so it should take a bit shorter time, so the correct answer is A.

33. The correct answer is A. You should learn (memorize) the meanings of the metric prefixes.

35. This is not an estimation problem, and you should use a ruler to measure the segment and then round your answer to the requested length. **a.** 3 cm **b.** 3.4 cm **c.** 1 in. **d.** $1\frac{3}{8}$ in.

Level 2, page 434

37. $P = 2(\ell + w)$
    $= 2(4 \text{ in.} + 5 \text{ in.})$
    $= 18 \text{ in.}$

39. $P = 3s$
    $= 3(3 \text{ dm})$
    $= 9 \text{ dm}$

41. $C = 2\pi r$
    $= 2\pi(2.40 \text{ m})$
    $\approx 15.08 \text{ m}$

43. The ends form a circle with diameter 50.00 ft and the top and bottom each have length
120.00 ft. $P = \pi(50 \text{ ft}) + 2(120 \text{ ft})$
$\approx 397.08 \text{ ft}$

45. $P = 2(\ell + w)$
$= 2(14 \text{ ft} + 9 \text{ ft})$
$= 46 \text{ ft}$

47. This is not as difficult as it looks. A little reflection shows that the perimeter is the same as the
perimeter of a rectangle with width 1.7 cm and height $1.4 \text{ cm} + 1.0 \text{ cm} + 3.5 \text{ cm} = 5.9 \text{ cm}$.
$P = 2(\ell + w)$
$= 2(1.7 \text{ cm} + 5.9 \text{ cm})$
$= 15.2 \text{ cm}$

49. We estimate the curved portion to be one-quarter of the circumference of a circle with
radius 10.0 in. We round the answer to the nearest tenth of an inch.
$P = \dfrac{1}{4}\pi(2)(10.0 \text{ in.}) + 10.0 \text{ in.} + 10.0 \text{ in.}$
$= 35.7 \text{ in.}$

51. Notice that the perimeter of this figure is the same as the perimeter of the previous figure.
Round your answer to the nearest hundredth centimeter.
$2.00 \text{ cm} + 3.00 \text{ cm} + 3.00 \text{ cm} + \frac{1}{2}\pi(2.00 \text{ cm}) \approx 11.14 \text{ cm}$

53. $P = 2(\ell + w)$ *Perimeter formula*   The length is 300 m.
$750 \text{ m} = 2(\ell + 75 \text{ m})$ *Substitute known values.*
$375 \text{ m} = \ell + 75 \text{ m}$ *Perimeter formula*
$300 \text{ m} = \ell$

Level 3, page 435

55. $P = a + b + c$ The sides are $2x = 26$ in., $3x = 39$ in., and $3x = 52$ in.
$117 \text{ in} = 2x + 3x + 4x$
$117 \text{ in.} = 9x$
$13 \text{ in.} = x$

57. Answers vary; 48 cm on my body is smaller than the 52.5 cm in l'Louvre. How long is
your body cubit?

Level 3 Problem Solving, page 435

59. From Problem 57, 1 cubit ≈ 0.525 m; thus, 300 cubits ≈ 157.50 m, 50 cubits ≈ 26.25 m,
and 30 cubits ≈ 15.75 m. The approximate measurements of the ark are 158 m long, 26
m wide, and 16 m high.

9.2 Area, page 435
New Terms Introduced in this Section

Acre	Area	Base	Estimate
Parallelogram	Square unit	Trapezoid	

Level 1, page 442

1. The number representing the number of square units that are contained within a planar region.

3. **a.** Count the squares to estimate 3 cm^2. Use the formula, $A = \frac{1}{2}bh = \frac{1}{2}(2)(3) = 3$.
    b. Count the squares to estimate $4(\frac{3}{4}\text{ cm}^2) = 3$ cm^2. Use the formula,
    $A = \pi r^2 = \pi(1)^2 = \pi$ which is 3 when rounded to the nearest square centimeter.

5. Count the squares: 6 complete squares plus about 4 half-squares which is 8 cm^2. By formula, $A = bh = 4(2) = 8$.

7. Count the squares; 8 complete squares plus about 5 half-squares which is 10 or 11 cm^2.
    By formula, there is one rectangle and two triangles:
    rectangle: $A = \ell w = 4(2) = 8$
    top triangle: $A = \frac{1}{2}bh = \frac{1}{2}3(1) = \frac{3}{2}$
    side triangle: $A = \frac{1}{2}bh = \frac{1}{2}3(1) = \frac{3}{2}$
    Total area: $8 + 2(\frac{3}{2}) = 11$ cm^2

9. Notice that choice A is impossible because areas are measured in square units. By remembering the size of a dollar bill we know that the area is closer to 18 in.^{2} than to 6 in.^{2}, so we see the correct answer is C.

11. Notice that choice C is impossible because areas are measured in square units. By remembering the size of your textbook we know it is closer to 70 in.^{2} than 70 cm^2, so we see the correct answer is A.

13. This is an estimation problem, and we see that 90 cm^2 could be a 9 cm by 10 cm rectangle, 10 in.^{2} could be a 1 in. by 10 in. rectangle, and 600 cm^2 could be 20 cm by 30 cm. A sheet of notebook is closer in size to the later size, so we see the answer is C.

15. We have seen advertisements for TVs and know that a 40 in. screen means the diagonal is 40 in. Since most plasma sets are fairly large (48 in. = 4 ft by 24 in. = 2 ft) we estimate the area to be about 8 ft^2. The closest choice is 10 ft^2, or choice A.

17. Choice A might be 1 meter by 1 meter.... imagine feet like that! Choice B might be 20 in. by 20 in. — wow! We see the only reasonable answer is C.

Level 2, page 443

19. $A = \ell w$
$\quad = (5 \text{ in.})(3 \text{ in.})$
$\quad = 15 \text{ in.}^2$

21. $A = \ell w$
$\quad = (23 \text{ m})(52 \text{ m})$
$\quad = 1{,}196 \text{ m}^2$

23. $A = s^2$
$\quad = (10 \text{ mm})^2$
$\quad = 100 \text{ mm}^2$

25. $A = bh$
$\quad = (120 \text{ ft})(63 \text{ ft})$
$\quad = 7{,}560 \text{ ft}^2$

27. $A = \frac{1}{2}bh$
$\quad = \frac{1}{2}(21 \text{ dm})(13 \text{ dm})$
$\quad = 136.5 \text{ dm}^2$

29. $A = \frac{1}{2}h(B + b)$
$\quad = \frac{1}{2}(30 \text{ cm})(210 \text{ cm} + 160 \text{ cm})$
$\quad = (15 \text{ cm})(370 \text{ cm})$
$\quad = 5{,}550 \text{ cm}^2$

31. $A = \frac{1}{2}h(B + b)$
$\quad = \frac{1}{2}(4 \text{ in.})(9 \text{ in.} + 5 \text{ in.})$
$\quad = (2 \text{ in.})(14 \text{ in.})$
$\quad = 28 \text{ in.}^2$

33. Since $d = 20 \text{ in.}, r = 10 \text{ in.}$
$\quad A = \pi r^2$
$\quad = \pi(10 \text{ in.})^2$
$\quad \approx 314.2 \text{ in.}^2$

Page 235

35. Since $d = 10$ in., $r = 5$ in.

$$A = \pi r^2$$
$$= \pi(5\,\text{in.})^2$$
$$\approx 78.5\,\text{in.}^2$$

37. Since $d = 28$ in., $r = 14$ in. We also note that we are finding the area of a half-circle.

$$\frac{1}{2}A = \frac{1}{2}\pi r^2$$
$$= \frac{1}{2}\pi(14\,\text{in.})^2$$
$$\approx 307.9\,\text{in.}^2$$

39. The shaded area, A, is the area of a rectangle plus the area of a half-circle of radius 1 cm.

$$A = (2\,\text{cm})(3\,\text{cm}) + \frac{1}{2}\pi(1\,\text{cm})^2$$
$$= 6\,\text{cm}^2 + 0.5\pi\,\text{cm}^2$$
$$\approx 7.6\,\text{cm}^2$$

41. The area of lot A: $\frac{1}{2}(135\,\text{ft})(70\,\text{ft}) = 4{,}725\,\text{ft}^2$; Cost per ft^2: $\frac{\$13{,}500}{4{,}725\,\text{ft}^2} \approx \$2.86/\text{ft}^2$

the area of lot B: $(50\,\text{ft})(160\,\text{ft}) = 8{,}000\,\text{ft}^2$; Cost per ft^2: $\frac{\$25{,}500}{8{,}000\,\text{ft}^2} \approx \$3.19/\text{ft}^2$

Lot A costs less per square foot.

43. $\dfrac{(750\,\text{ft})(1{,}290\,\text{ft})}{43{,}560\,\text{ft}^2} \approx 22.2$ acres

45. $A = 12\,\text{in.} \times 18\,\text{in.} = 216\,\text{in.}^2$

47. $A = 8.5\,\text{in.} \times 11\,\text{in.} = 93.5\,\text{in.}^2$

49. Estimate area (do not use a calculator): $A = 100\,\text{ft} \times 30\,\text{ft} = 3{,}000\,\text{ft}^2$

Now, since 1 pound of seed covers 150 ft^2 we can find $\frac{3{,}000\,\text{ft}^2}{150\,\text{ft}^2} = 20$ pounds.

Finally, the cost is about $\$6.00 \times 20 = \$120$.

Calculate area: $A = 100\,\text{ft} \times 30\,\text{ft} = 3{,}000\,\text{ft}^2$

Calculate amount of seed: $\dfrac{3{,}000\,\text{ft}^2}{150\,\text{ft}^2/\text{lb}} = 20\,\text{lb}$

Calculate the cost: $\$5.85 \times 20 = \$117$.

The seed will cost $\$117$. The estimate is consistent with the calculated amount.

Level 3, page 444

51. First find $s = 0.5(a + b + c) = 0.5(5\,\text{ft} + 8\,\text{ft} + 10\,\text{ft}) = 11.5\,\text{ft}$

$$A = \sqrt{s(s-a)(s-b)(s-c)}$$
$$= \sqrt{11.5(11.5-5)(11.5-8)(11.5-10)}$$
$$\approx 19.81 \quad \text{by calculator}$$

The area of the triangle is about 20 ft^2.

53. The area we are looking for is the area, A, of a rectangle plus the area of a triangle.

Area of the rectangle: $(240 \text{ ft})(180 \text{ ft}) = 43{,}200 \text{ ft}^2$

Area of the triangle: $s = 0.5(a+b+c) = 0.5(180 \text{ ft} + 130 \text{ ft} + 150 \text{ ft}) = 230 \text{ ft}$

$$\text{AREA OF TRIANGLE} = \sqrt{s(s-a)(s-b)(s-c)}$$
$$= \sqrt{230(230-180)(230-130)(230-150)}$$
$$\approx 9{,}591.66 \quad \text{by calculator}$$

$A = 43{,}200 \text{ ft}^2 + 9{,}592 \text{ ft}^2 = 52{,}792 \text{ ft}^2$.

55. Since a pentagon has five sides, there are 10 right triangles with $b = 5$ and $h \approx 6.88$.

$$A = 10\left(\frac{1}{2} bh\right)$$
$$\approx 5(5)(6.88)$$
$$= 172$$

The area of the pentagon is 172 in.^{2}.

57. Since an octagon has eight sides, there are 16 right triangles with $b = 5$ and $h \approx 12.07$.

$$A = 16\left(\frac{1}{2} bh\right)$$
$$\approx 8(5)(12.07)$$
$$= 482.8$$

The area of the octagon is 483 in.^{2}.

Level 3, Problem Solving, page 445

59. Notice that the shaded portion is a circle that has been "turned inside out."

$\text{AREA OF CIRCLE} = 5^2\pi \approx 79 \text{ in.}^2$.

9.3 Surface Area, Volume, and Capacity, page 445

New Terms Introduced in this Section

Capacity	Cube	Cubic unit	Cup
Gallon	Liter	Ounce	Parallelepiped
Quart	Rectangular parallelepiped	Surface area	Volume

Level 1, page 453

1. Length is one dimensional (in., ft, cm, etc.), area is two dimensional (in.^{2}, ft^2, cm^2, etc.), and volume is three dimensional (in.^{3}, ft^3, cm^3, etc.). You should show some examples of each.

3. Volume is the number of cubic units a solid object contains, whereas capacity is the amount of liquid a container will hold.

5. A liter is a bit larger than a quart.

7. Bottom layer is 5 cm by 3 cm, for 15 cm^2. There are four such layers for a total of 60 cm^3.

9. $V = s^3$
   $$= (5\text{ ft})^3$$
   $$= 125\text{ ft}^3$$

11. $V = s^3$
    $$= (20\text{ cm})^3$$
    $$= 8,000\text{ cm}^3$$

13. $V = \ell w h$
    $$= (3\text{ ft})(2\text{ ft})(4\text{ ft})$$
    $$= 24\text{ ft}^3$$

15. $V = \ell w h$
    $$= (30\text{ cm})(40\text{ cm})(80\text{ cm})$$
    $$= 96,000\text{ cm}^3$$

17. **a.** 2 c   **b.** 16 oz

19. **a.** 13 oz   **b.** 380 mL

21. **a.** 25 mL   **b.** 75 mL   **c.** 75 mL

Level 2, page 454

23. Choice A is way too small (not much more than a spoonful); choice C would be a large paint bucket, so it seem that choice B is the most appropriate.

25. Choice B (200 mL) is more than half a can of Coke (too much for a cough medicine) and 2 L is the size of container of milk, so we see A is the appropriate choice.

27. Choice A (15 L) is about $7\frac{1}{2}$ large bottles of Coke (plastic container) and choice B (200 mL) is a bit more than half a can of Coke, so choice C (70 L) is the most appropriate choice.

29. 50 kL is 50,000 L; way too much for a bath, but certainly not enough for the drinking water for a large city, so choice B is the most appropriate answer.

31. The correct answer is B; milli means one-thousandth.

Page 238

33. $V = \ell wh$

    $= (7 \text{ m})(8 \text{ m})(2 \text{ m})$

    $= 112 \text{ m}^3$

    $= 112 \text{ kL}$ One kL is the same as 1 m^3.

The swimming pool contains 112 kL.

35. There are 5 faces (since the box is open), each with an area of $(25 \text{ cm})^2$

    $S = 5(25 \text{ cm})^2$

      $= 3{,}125 \text{ cm}^2$

37. There are six faces (since the box is closed):

Front/back:	$(10 \text{ cm})(4 \text{ cm}) = 40 \text{ cm}^2$
left/right:	$(25 \text{ cm})(4 \text{ cm}) = 100 \text{ cm}^2$
top/bottom:	$(25 \text{ cm})(10 \text{ cm}) = 250 \text{ cm}^2$

TOTAL: $= 2(40 \text{ cm}^2 + 100 \text{ cm}^2 + 250 \text{ cm}^2) = 780 \text{ cm}^2$

39. There are 5 faces (since the box is open):

Front/back:	$(50 \text{ cm})(50 \text{ cm}) = 2{,}500 \text{ cm}^2$
left/right:	$(200 \text{ cm})(50 \text{ cm}) = 10{,}000 \text{ cm}^2$
bottom:	$(200 \text{ cm})(50 \text{ cm}) = 10{,}000 \text{ cm}^2$

TOTAL: $= 2(2{,}500 \text{ cm}^2 + 10{,}000 \text{ cm}^2) + 10{,}000 \text{ cm}^2$

                $= 35{,}000 \text{ cm}^2$

41. This problem is similar to Example 2, p. 446. This can has a bottom, but not a top. The surface area is the side plus the bottom. The circumference of the bottom is

$\pi d = 2\pi r = 2\pi(2 \text{ cm}) = 4\pi \text{ cm}$

$(4\pi \text{ cm})(8 \text{ cm}) + \pi(2 \text{ cm})^2 = 32\pi \text{ cm}^2 + 4\pi \text{ cm}^2$

                                  $= 36\pi \text{ cm}^2$

                                  $\approx 113.1 \text{ cm}^2$

43. There are six faces (since the box is closed):

Front rectangular side:	$(2 \text{ in.})(1 \text{ in.}) = 2 \text{ in.}^2$
Left rectangular side:	$(1.95 \text{ in.})(1 \text{ in.}) = 1.95 \text{ in.}^2$
right rectangular side:	$(1.35 \text{ in.})(1 \text{ in.}) = 1.35 \text{ in.}^2$
Top/bottom sides	$\frac{1}{2}(2 \text{ in.})(1.25 \text{ in.}) = 1.25 \text{ in.}^2$

TOTAL: $= 2 \text{ in.}^2 + 1.95 \text{ in}^2 + 1.35 \text{ in.}^2 + 2(1.25 \text{ in.}^2)$

                $= 7.8 \text{ in.}^2$

45. There are five faces: Bottom: $(125 \text{ m})(125 \text{ m}) = 15{,}625 \text{ m}^2$

one side: $\dfrac{1}{2}(125 \text{ m.})(55 \text{ m}) = 3{,}437.5 \text{ m}^2$

TOTAL: $15{,}625 \text{ m}^2 + 4(3{,}437.5 \text{ m}^2) = 29{,}375 \text{ m}^2$

47. First find the volume, then the capacity.

$V = s^3$ Since 231 in.^{3} = 1 gallon, we divide by 231: Capacity $= \dfrac{1{,}000}{231}$

$\quad = (10 \text{ in.})^3$ $= 4.32$

$\quad = 1{,}000 \text{ in.}^3$

We round this answer (per directions) to the nearest tenth of a gallon: 4.3 gal.

49. First find the volume, then the capacity. $V = \ell w h$

$= (16 \text{ cm})(10 \text{ cm})(25 \text{ cm})$

$= 4{,}000 \text{ cm}^3$

$= 4 \text{ L}$ 1,000 cm³ = 1 L

The orange juice container holds 4 L.

51. First find the volume, then the capacity. $V = \ell w h$

$= (20 \text{ cm})(15 \text{ cm})(30 \text{ cm})$

$= 9{,}000 \text{ cm}^3$

$= 9 \text{ L}$ 1,000 cm³ = 1 L

The gasoline container holds 9 L.

53. First find the volume, then the capacity. $V = \ell w h$

$= (50 \text{ cm})(200 \text{ cm})(50 \text{ cm})$

$= 500{,}000 \text{ cm}^3$

$= 500 \text{ L}$ 1,000 cm³ = 1 L

The container holds 500 L.

Level 3, page 455

55. **a**. $V = \ell w h$ **b**. $45.375 \text{ ft}^3 - 19 \text{ ft}^3 = 26.375 \text{ ft}^3$

$= (36 \text{ in.})(33 \text{ in.})(66 \text{ in.})$

$= 78{,}408 \text{ in.}^3$

$= 78{,}408 \text{ in.}^3 \cdot \dfrac{1 \text{ ft}^3}{12^3 \text{ in.}^3}$

$= 45.375 \text{ ft}^3$

Page 240

57. $V = \ell w h$

$\quad = (50 \text{ ft})(4 \text{ ft})(4 \text{ in.})$

$\quad = \left(\dfrac{50}{3}\text{yd}\right)\left(\dfrac{4}{3}\text{ yd}\right)\left(\dfrac{4}{36}\text{ yd}\right)$

$\quad = \dfrac{50 \cdot 4 \cdot 4}{3 \cdot 3 \cdot 36}\text{ yd}^3$

$\quad \approx 2.469135$

Rounding to the nearest half cubic yard, we see that 2.5 yd^3 is needed.

Level 3 Problem Solving, page 455

59. a. $\dfrac{2.8 \times 10^7 \text{ ft}^2}{1 \text{ mi}^2} \times \dfrac{1 \text{ person}}{50 \text{ ft}^2} = 560{,}000 \text{ people/mi}^2$

$\quad$ **b.** $\dfrac{1 \text{ mi}^2}{560{,}000 \text{ people}} \times (6.2 \times 10^9 \text{ people}) \approx 11{,}071 \text{ mi}^2.$

$\quad\quad$ This is about the size of the state of Maryland.

$\quad$ **c.** $(5.2 \times 10^7) \times 640$ acres; divide this by the population 6.2×10^9 and the result is about 5.4 acres per person!

9.4 Miscellaneous Measurements, page 456

New Terms Introduced in this Section

Celsius	Cone	Cylinder	Decimal
Fahrenheit	Gram	Mass	Ounce
Pound	Prism	Pyramid	Sphere
Temperature	Ton	Weight	

Level 1, page 464

1. You might wish to re-read the first few pages of this chapter before formulating your answer. You should write a few paragraphs, giving support for your comments.

3. Convert from one metric unit to another by moving the decimal according the memory aid, "Karl Has Developed My Decimal Craving for Metrics" to the set of the names of the metric columns: "Kilometer, Hectometer, Dekameter, Meter, Decimeter, Centimeter, Millimeter." We may substitute liter or gram for the word meter. It might be a good idea to include a few examples.

5. a. centimeter **b.** meter

7. a. liter **b.** kiloliter

9. **a**. Celsius **b**. Celsius

11. Compare with some known weight; a liter of water (or milk) weighs about one kilogram, so the most reasonable answer is 0.4 kg, or C.

13. Compare with some known weight; a liter of water (or milk) weighs about one kilogram, so the most reasonable answer is 12 kg, namely answer C.

15. While 32°F is cold (freezing), 32°C is hot; the correct answer is B.

17. Look at the thermometers in Figure 9.27 (oven temperatures) to see that the most reasonable answer is C. Another way (without using the thermometer) would be to reason that 100°C is boiling, 120° is barely enough for cooking a steak, whereas 500° is much too hot, leaving us with the reasonable choice of 290° which is roughly triple the boiling point.

19. The correct response is A.

21. It is desirable that you memorize the metric prefixes. The correct answer is B.

23. Kilogram is a measurement of weight, so the answer is C.

25. Liter is a measure of capacity, so the answer is B.

27. Milliliter is a measure of capacity, so the answer is B.

29. Centimeter is a measure of length, so the answer is A.

31. Find the answer by moving the decimal point: 0.000063 kiloliter; 0.00063 hectoliter; 0.0063 dekaliter; 0.063 liter; 0.63 deciliter; 6.3 centiliter; **63 milliliter**

33. Find the answer by moving the decimal point: 0.0035 kiloliter; 0.035 hectoliter; 0.35 dekaliter; **3.5 liter**; 35 deciliter; 350 centiliter; 3,500 milliliter

35. Find the answer by moving the decimal point: 0.08 kiloliter; 0.8 hectoliter; **8 dekaliter**; 80 liter; 800 deciliter; 8,000 centiliter; 80,000 milliliter

37. Find the answer by moving the decimal point: 0.31 kiloliter; **3.1 hectoliter**; 31 dekaliter; 310 liter; 3,100 deciliter; 31,000 centiliter; 310,000 milliliter

Level 2, page 465

39. **a**. From mm to m is three places (smaller to larger, so move to the left): 0.001

   b. From mm to km is six places (smaller to larger, so move to the left): 0.000001

   c. From m to mm is three paces (larger to smaller, so move to the right): 1,000

   d. From m to cm is two places (larger to smaller, so move to the right): 100

41. **a**. From L to mL is three places (larger to smaller, so move to the right): 1,000

   b. From L to dL is one place (larger to smaller, so move to the right): 10

   c. From L to kL is three places (smaller to larger, so move to the left): 0.001

   d. From kL to mL is six places (larger to smaller, so move to the right): 1,000,000

43. There are 5 faces (since the box is open):

Front/back: $(3.5 \text{ cm})(4.2 \text{ cm}) = 14.7 \text{ cm}^2$

left/right: $(3.5 \text{ cm})(4.2 \text{ cm}) = 14.7 \text{ cm}^2$

bottom: $(3.5 \text{ cm})^2 = 12.25 \text{ cm}^2$

$S = (3.5 \text{ cm})^2 + 4(3.5 \text{ cm})(4.2 \text{ cm}) = 71.05 \text{ cm}^2$; round to the nearest unit for a surface area of 71 cm².

45. $S = 4\pi r^2$

$\quad = 4\pi(15 \text{ cm})^2$

$\quad = 900\pi \text{ cm}^2$

$\quad \approx 2{,}827 \text{ cm}^2$

47. $V = (5 \text{ ft})(3 \text{ ft})(2 \text{ ft}) = 30 \text{ ft}^3$

49. The base of this figure is the triangle, so $B = \frac{1}{2}bh$ where $b = 10$ cm and $h = 6$ cm.

$V = Bh$

$\quad = \left[\frac{1}{2}(10 \text{ cm})(6 \text{ cm}) \right](20 \text{ cm}) \qquad \text{where } h = 20 \text{ cm}$

$\quad = 600 \text{ cm}^3$

51. Note this is a sphere with $r = 6$ in.

$S = 4\pi r^2 \qquad \text{and } V = \frac{4}{3}\pi r^3$

$\quad = 4\pi(6 \text{ in.})^2 \qquad\qquad = \frac{4}{3}\pi(6 \text{ in.})^3$

$\quad = 144\pi \text{ in.}^2 \qquad\qquad = 288\pi \text{ in.}^3$

$\quad \approx 452 \text{ in.}^2 \qquad\qquad \approx 905 \text{ in.}^3$

53. $A = \ell w = (2\ell)(3w) = 6\,\ell w$; The area is increased six-fold.

55. $S = 4\pi r^2 = 4\pi(3r)^2 = 36\pi r^2$; The area is increased nine-fold.

Level 3, page 466

57. $A = \pi r^2$, or (in terms of the diameter),

$$A = \pi \left(\frac{d}{2} \right)^2 = \frac{\pi d^2}{4}$$

Now, if we double the diameter, we have

$$A = \frac{\pi(2d)^2}{4} = \frac{4\pi d^2}{4}$$

Page 243

so we see that the area is increased four-fold.

Level 3 Problem Solving, page 466

59. Here is a spreadsheet which will work:

Here is an example of the output:

	A	B
1	**Temperature Conversions**	
2		
3	Celsius	Fahrenheit
4	0	+(9/5)*A4+32
5	+A4+5	replicate
6	replicate	
7		
8		
9		
10		
11		
12		

	A	B
1	**Temperature Conversions**	
2		
3	Celsius	Fahrenheit
4	0	32
5	5	41
6	10	50
7	15	59
8	20	68
9	25	77
10	30	86
11	35	95
12	40	104

Chapter 9 Review Questions, page 468

Studying for a chapter examination is a personal process, one which nobody else can do for you. Simply take the time to review what you have done. Here are the new terms in Chapter 9.

Accuracy [9.1]
Acre [9.2]
Area [9.2]
Base [9.2]
Capacity [9.3]
Celsius [9.4]
Center of a circle [9.1]
Centi- [9.1]
Circle [9.1]
Circumference [9.1]
Cone [9.4]
Cube [9.3]
Cup [9.3]
Cubic unit [9.3]

Cylinder [9.4]
Decimal [9.4]
Deci- [9.1]
Deka- [9.1]
Diameter [9.1]
Equilateral triangle [9.1]
Fahrenheit [9.4]
Foot [9.1]
Gallon [9.3]
Gram [9.4]
Hecto- [9.1]
Inch [9.1]
Kilo- [9.1]
Length [9.1]

Liter [9.3]
Mass [9.4]
Measure [9.1]
Meter [9.1]
Metric System [9.1]
Mile [9.1]
Milli- [9.1]
Ounce [9.3, 9.4]
Parallelepiped [9.3]
Parallelogram [9.2]
Perimeter [9.1]
Pi (π) [9.1]
Pound [9.4]
Precision [9.1]

Page 244

Prism [9.4]	Semicircle [9.1]	Ton [9.4]
Pyramid [9.4]	SI system [9.1]	Trapezoid [9.2]
Quart [9.3]	Sphere [9.4]	United States system [9.1]
Radius [9.1]	Square [9.1]	Volume [9.3]
Rectangle [9.1]	Square unit [9.3]	Weight [9.4]
Rectangular	Surface area [9.3]	Yard [9.1]
parallelepiped [9.3]	Temperature [9.4]	

If you can describe the term, read on to the next one; if you cannot, then look it up in the text (the section number is shown in brackets). Next, study the types of problems listed at the end of Chapter 9.

TYPES OF PROBLEMS

Distinguish between the concepts of precision and accuracy. [9.1]

Be able to measure length in both the United States and metric measurement systems. [9.1]

Estimate lengths; choose an appropriate unit for measuring a given length. [9.1]

Find the perimeter, circumference, or distance around a given figure. [9.1]

Solve applied problems involving length. [9.1]

Be able to measure area in both the United States and metric measurement systems. [9.2]

Estimate areas; choose an appropriate unit for measuring a given area. [9.2]

Find the area of a given figure. [9.2]

Solve applied problems involving area. [9.2]

Find the surface area of an object. [9.3]

Estimate volumes; choose an appropriate unit for measuring a given volume. [9.3, 9.4]

Find the volume of a given solid. [9.3, 9.4]

Solve applied problems involving volume. [9.3, 9.4]

Estimate capacities. [9.3]

Find the capacity of a given container. [9.3]

Solve applied problems involving capacity. [9.3]

Measure the amount of a liquid. [9.3]

Be able to measure mass (weight) in both the United States and metric measurement
    systems. [9.4]

Estimate weights; choose an appropriate unit for measuring a given mass. [9.4]

Be able to measure temperature in both the United States and metric measurement
    systems. [9.4]

Be able to measure volume in both the United States and metric measurement systems.
    [9.3, 9.4]

Estimate temperatures. [9.4]

Change units within the metric system. [9.4]

Change units within the United States system.[9.4]

Change units between the metric and United States systems. [9.5]

Once again, see if you can verbalize (to yourself) how to do each of the listed types of problems.

Work all of Chapter 9 Review Questions (whether they are assigned or not). Work through all of the problems before looking at the answers, and *then* correct each of the problems. The entire solution is shown in the answer section at the back of the text. If you worked the problem correctly, move on to the next problem, but if you did not work it correctly (or you did not know what to do), look back in the chapter to study the procedure, or ask your instructor.

Finally, go back over the homework problems you have been assigned. If you worked a problem correctly, move on to the next problem, but if you missed it on your homework, then you should look back in the book or talk to your instructor about how to work the problem.

If you follow these steps, you should be successful with your review of this chapter.

We give all of the answers to the chapter review questions (not just the odd-numbered questions).

Chapter 9 Review Questions, page 468

1. **a.** ————————————————————

    b. 1.9 cm

2. **a.** feet and meters

    b. Answers vary; 40°C; 100°F

    c. Answers vary; $\frac{1}{2}$ in.; 1.5 cm

3. **a.** one-thousandth

    b. capacity

    c. 10 km = 1,000,000 cm (Move the decimal point 5 places to the right.)

4. First find the distance around the semicircle: $C = \dfrac{1}{2}\pi d$

    $$= \frac{1}{2}\pi(8\text{ in.})$$

    $$= 4\pi$$

    The distance around is $10 + 7 + 4\pi + 7 + 6$. To the nearest inch, the perimeter is 43 in.

5. First find the area of the semicircle: $A = \dfrac{1}{2}\pi r^2$

    $$= \frac{1}{2}\pi(4)^2$$

    $$= 8\pi$$

    The area of the trapezoid is $A = \frac{1}{2}(8)(7 + 13) = 80$. Thus, the area of the entire figure is

Page 246

$$8\pi + 80 \approx 105.1327412$$

To the nearest square inch, the area is 105 in.^{2}.

6. Since the area of the base is 105 in.^{2} and since 144 in.^{2} is one square foot, we see that the volume must be less than 1 ft^3 because 105 in.^{2} < 144 in.^{2}.

7. Do not use the rounded answer from Problem 5, but use the original measurements.

$\begin{aligned} V &= Bh \\ &= \left[(80 + 8\pi)\text{in.}^2\right](12\,\text{in.}) \\ &\approx 1{,}262\ \text{in.}^3 \end{aligned}$ The volume of the prism is 1,260 in.^{3}.

8. a. $\begin{aligned} S &= 4\pi r^2 \\ &= 4\pi (1\,\text{ft})^2 \\ &= 4\pi\ \text{ft}^2 \end{aligned}$ **b.** $4\pi\ \text{ft}^2 \approx 12.6\ \text{ft}^2$

9. a. $\begin{aligned} V &= \frac{4}{3}\pi r^3 \\ &= \frac{4}{3}\pi (1\,\text{ft})^3 \\ &= \frac{4}{3}\pi\ \text{ft}^3 \end{aligned}$ **b.** $\frac{4}{3}\pi \approx 4.188790205$; the volume is 4.2 ft^3.

10. a. $\begin{aligned} V &= \ell w h \\ &= (3\,\text{dm})(2\,\text{dm})(5\,\text{dm}) \\ &= 30\ \text{dm} \end{aligned}$ The volume is 30 dm^3.

  b. Since 1 L = 1 dm^3, we see from part **a** that the box holds 30 liters.

11. BASE + 2(SIDE) + 2(FRONT) = (2 dm)(3 dm) + 2(2 dm)(5 dm) + 2(3 dm)(5 dm)

$$= 56\ \text{dm}^2$$

12. a. Since 1 dm = 10 cm, we have 30 cm × 20 cm × 50 cm.

  b. First, find the volume: $V = 30\ \text{in.} \times 20\ \text{in.} \times 50\ \text{in.} = 30{,}000\ \text{in.}^3$
  Since 1 gal occupies 231 in.^{3} we see there are
  $$\frac{30{,}000}{231}\ \text{gal} \approx 129.87\ \text{gal}$$
  If we round to the nearest gallon, we find the capacity is 130 gallons.

13. The surface area (including the top and the bottom) is found by finding the surface area of those parts plus the surface area of the side of the can. If we unroll the can, we see it is a rectangle of length equal to the circumference of the circle and height equal to the height of the cylinder.

Area of the top (and bottom) $\pi r^2 = \pi (1.3\,\text{cm})^2 = 1.69\pi\ \text{cm}^2$

Area of the side   Circumference: $\pi d = \pi (2.6\,\text{cm}) = 2.6\,\pi\ \text{cm}$

                 Thus, the area of the side is $(2.6\pi\ \text{cm})(8\,\text{cm}) = 20.8\pi\ \text{cm}^2$

The total surface area: $2(1.69\pi\ \text{cm}^2) + 20.8\pi\ \text{cm}^2 = 24.18\pi\ \text{cm}^2 \approx 76\ \text{cm}^2$.

Page 247

14. a. The 6-in. pizza is about $0.13/in.^{2}; the 10-in. and 12-in. pizzas are about $0.10/in.^{2}, and the 14-in. pizza is about $0.09/in.^{2}. The best value is the large size.

    b. Answers vary; the size is 201 in.^{2}; price for a large should be about $0.09 per square inch; I would price it at $17.95 ($16.95 to $18.25 is acceptable; calculate $201 \times 0.09 \approx 18.09$).

15. $S = (80 \text{ ft})(30 \text{ ft}) = 2,400 \text{ ft}^2$

16. $V = (80 \text{ ft})(30 \text{ ft})(3 \text{ ft}) = 7,200 \text{ ft}^3$

17. Since $1 \text{ ft}^2 \approx 7.48$ gal, and the volume is $7,200 \text{ ft}^3$, the total capacity of the swimming pool is $(7.48 \text{ gal})(7,200) \approx 53,856$ gal

18. $11 \text{ ft} \times 16 \text{ ft} = 3\frac{2}{3} \text{ yd} \times 5\frac{1}{3} \text{ yd}$ You need to purchase 20 square yards.

$$= \frac{11}{3} \times \frac{16}{3} \text{ yd}^2$$

$$= \frac{176}{9} \text{ yd}^2$$

$$\approx 19.6 \text{ yd}^2$$

19. a.

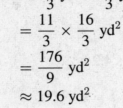

    b. $25^2 + 312^2 = 313^2$, so the answer is yes.

$$5^2 + 12^2 = 25 + 144 = 169 = 13^2$$

20. The diagram gives a geometric justification of the distributive law.

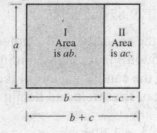

Area of rectangle I: ab

Area of rectangle II: ac

Area of I + II: $ab + ac$

Area of large rectangle is:

$$ab + ac$$

Page 248

CHAPTER 10 The Nature of Growth

Chapter Overview
This chapter introduces one of the most important ideas not only in mathematics, but also in understanding the world around us — growth and decay.   The growth of a population or organism, or money as well as the decay of radioactive materials, income, or resources are all important applications that require the knowledge of the material in this chapter.

10.1 Exponential Equations, page 473
New Terms Introduced in this Section

Argument	Change of base theorem	common logarithm	Exact solution
Exponential	Exponential equation	Logarithm	Micrometer
Natural logarithm			

Level 1, page 480

1. Short answer: a *logarithm* is an exponent. Complete answer: $\log_b x$ is the exponent on a base b that gives the answer x.

3. A *natural logarithm*, written $\ln N$, is the exponent on a base e that gives N.

5. **a.** $\log N = \log_{10} N$ **b.** $\ln N = \log_e N$ **c.** $\log_b N$ is the exponent on a base b that gives N.

7. An *exponential equation* is an equation in which a variable appears in an exponent.

9. **a.** Six is the exponent (log) on a base 2 which gives 64: $6 = \log_2 64$
   b. Two is the exponent (log) on a base 10 which gives 2: $2 = \log 100$
   c. p is the exponent (log) on a base n which gives m: $p = \log_n m$

11. a. Negative one is the exponent (log) on a base 10 which gives $\frac{1}{10}$: $-1 = \log \frac{1}{10}$
   b. Two is the exponent (log) on a base 6 which gives 36: $2 = \log_6 36$
   c. n is the exponent (log) on a base t which gives s: $n = \log_t s$

13. a. 1 is the exponent on a base 10 which gives 10.
   b. 3 is the exponent on a base 10 which gives 1,000.
   c. -5 is the exponent on a base 10 which gives 10^{-5}.

15. a. -3 is the exponent on a base 5 which gives 5^{-3}.
   b. $\log 0.1 = \log 10^{-1}$; -1 is the exponent on a base 10 which gives $\frac{1}{10}$.
   c. 1 is the exponent on a base e which gives e.

17. a. 1 is the exponent on a base b which gives b.
   b. 3 is the exponent on a base b which gives b^3.
   c. -6 is the exponent on a base b which gives b^{-6}.

19. a. $4^x = \dfrac{1}{16}$ **b.** $x = \log_5 8$ **c.** $x = \log_6 4.5$ **d.** $x = \log_4 5$

$2^{2x} = 2^{-4}$

$2x = -4$

$x = -2$

21. a. $216^x = 36$ **b.** $x = \log 2.5$ **c.** $x = \log 45$ **d.** $x = \log 15$

$\left(6^3\right)^x = 6^2$

$6^{3x} = 6^2$

$3x = 2$

$x = \dfrac{2}{3}$

23. a. $\log 4.27 \approx 0.630427875025 \approx 0.63$

    b. $\log_b b^2 = 2$ by definition of logarithm; to two decimal places, we write 2.00.

    c. $\log_t t^3 = 3$ by definition of logarithm; to two decimal places, we write 3.00.

    d. $\ln 10 \approx 2.30258509299 \approx 2.30$

25. a. $\log 71{,}600 \approx 4.85491302231 \approx 4.85$

    b. $\log_3 9 = 2$ by definition of logarithm; to two decimal places, we write 2.00.

    c. $\log_{19} 1 = 0$ by definition of logarithm; to two decimal places, we write 0.00.

    d. $\ln 1{,}000 \approx 6.90775527898 \approx 6.91$

27. a. $\ln 2.27 \approx 0.819779821493 \approx 0.82$

    b. $\ln 16.77 \approx 2.81959157584 \approx 2.82$

    c. $\ln 7.3 \approx 1.98787434815 \approx 1.99$

    d. $\ln 0.321 \approx -1.13631415585 \approx -1.14$

29. a. $\log_3 45 = \dfrac{\log 45}{\log 3} \approx 3.4650$

    b. $\log_5 91 = \dfrac{\log 91}{\log 5} \approx 2.8028$

31. a. $\log_7 13 = \dfrac{\log 13}{\log 7} \approx 1.3181$

    b. $\log_2 556 = \dfrac{\log 556}{\log 2} \approx 9.1189$

33. a. $\log_2 0.0056 = \dfrac{\ln 0.0056}{\ln 2} \approx -7.4804$

    b. $\log_{8.3} 105 = \dfrac{\ln 105}{\ln 8.3} \approx 2.1991$

35. a. $\log_7 56 = \dfrac{\ln 56}{\ln 7} \approx 2.0686$

    b. $\log_2 10.85 = \dfrac{\ln 10.85}{\ln 2} \approx 3.4396$

Level 2, page 481

37. a. $x = \log_8 3 \approx 0.528320834$ **b.** $x = \log_{64} 5 \approx 0.386988016$

Page 250

39. a. $x = \log 42 \approx 1.623249290$ **b.** $\log 0.0234 \approx -1.630784143$

41. a. $x = -\ln 8 \approx -2.07944154168$ **b.** $x = -\log 125 \approx -2.09691001301$

43. a. $5 - 3x = \log 0.041$

$-3x = -5 + \log 0.041$

$x = \frac{1}{3}(5 - \log 0.041)$

≈ 2.129072048

b. $2x - 1 = \log 515$

$2x = 1 + \log 515$

$x = \frac{1}{2}(1 + \log 515)$

≈ 1.855903615

45. $a = \ln x - 0.14$
$= \ln 25 - 0.14$
≈ 3.07

47. $a = \ln x - 0.14$
$= \ln 10 - 0.14$
≈ 2.16

49. $t = 50 \ln\left(\dfrac{d}{7.25}\right)$
$= 50 \ln\left(\dfrac{8.25}{7.25}\right)$
≈ 6.5

The year is 2015

51. $t = 50 \ln\left(\dfrac{d}{8.55}\right)$
$= 50 \ln\left(\dfrac{10}{8.55}\right)$
≈ 7.83

The year is 2016

Level 3, page 481

53. $2 \cdot 3^x + 7 = 61$

$2 \cdot 3^x = 54$

$3^x = 27$

$x = 3$

55. $\left(1 + \dfrac{0.08}{360}\right)^{360x} = 2$

$360x = \log_b 2$ where $b = 1 + \dfrac{0.08}{360}$

$x = \dfrac{\log_b 2}{360}$

$= \dfrac{1}{360}\left[\dfrac{\log 2}{\log\left(1 + \frac{0.08}{360}\right)}\right]$

≈ 8.665302427

57. $P = P_0 e^{rt}$

  $\dfrac{P}{P_0} = e^{rt}$

  $rt = \ln \dfrac{P}{P_0}$

  $t = \dfrac{1}{r} \ln \dfrac{P}{P_0}$

59. $V = \frac{4}{3}\pi r^3 = \frac{4}{3}\pi(3{,}963 \text{ mi})^3 \approx 2.60711883 \times 10^{11} \text{ mi}^3$

  Change this to in.^{3} using $(1 \text{ mi})^3 = (5{,}280 \text{ ft})^3 = [5{,}280(12 \text{ in.})]^3$ and $(1 \text{ ft})^3 = (12 \text{ in.})^3$;
  $V \approx 6.631416905 \times 10^{25} \text{ in.}^3$. Since 1 gallon = 231 in.^{3}, $V = 2.870743249 \times 10^{23} \text{ gal}$
  Solve the exponential equation

  $$2^x = V$$
  $$x = \log_2 V$$
  $$= \frac{\log V}{\log 2}$$
  $$\approx 77.92577049$$

  It would take about 78 doublings, the machine would reach the required capacity in the
  next minute; that is in about 1 hr 18 minutes.

10.2 Logarithmic Equations, page 482

New Terms Introduced in this Section

Addition law of logarithms   Grant's tomb properties   Laws of logarithms
Log of both sides theorem   Logarithmic equations   Multiplicative law of logarithms
Subtractive law of logarithms

Level 1, page 488

1. A *logarithmic equation* is an equation in which there is a logarithm on one or both sides.
3. Solve for the logarithm; this will lead to one of the forms: (1) $\log_b M = \log_b N$ which
   implies $M = N$; (2) $\log_b M = N$ which implies $b^N = M$.
5. False; a common logarithm is a logarithm in which the base is 10.
7. False; the exponent is $\log_b N$.
9. True; note that $81^2 = 6{,}561$.
11. False; x is the exponent on 1.5.
13. False; try $A = 10$ and $B = 20$ for a counterexample.
15. False; try $A = 10$ and $B = 20$ for a counterexample.
17. False; try $A = 10$ and $B = 20$ for a counterexample.

Page 252

19. False; $\log_b N$ is not defined when N is negative.

21. a. From the Grant's tomb property, $\ln 23$ is the exponent on a base e which gives 23.

  b. From the Grant's tomb property, $\log 3.4$ is the exponent on a base 10 which gives 3.4.

  c. From the Grant's tomb property, $\log_4 x$ is the exponent on a base 4 which gives x.

  d. From the Grant's tomb property, x is the exponent on a base b which gives b^x.

23. a. 2 is the exponent on a base 5 with gives 25.

  b. 7 is the exponent on a base 2 which gives 128.

  c. 4 is the exponent on a base 3 which gives 81.

  d. 3 is the exponent on a base 4 which gives 64.

25. a. 5 is the exponent on a base 10 which gives x, so $x = 10^5$.

  b. 1 is the exponent on a base x which gives e, so $x = e$.

  c. $x = e^2$ by the definition of logarithm.

  d. $x = e^3$ by the definition of logarithm.

27. a. $\log 2 + \log 3 + \log 4 = \log(2 \cdot 3 \cdot 4)$
$$= \log 24$$

  b. $\log 40 - \log 10 - \log 2 = \log(40/10) - \log 2$
$$= \log 4 - \log 2$$
$$= \log(4/2)$$
$$= \log 2$$

  c. $2 \ln x + 3 \ln y - 4 \ln z = \ln x^2 + \ln y^3 - \ln z^4$
$$= \ln\left(\frac{x^2 y^3}{z^4}\right)$$

29. a. $\ln 3 - 2 \ln 4 + \ln 8 = \ln 3 - \ln(2^2)^2 + \ln(2^3)$
$$= \ln\left(\frac{3 \cdot 2^3}{2^4}\right)$$
$$= \ln\left(\frac{3}{2}\right)$$

  b. $\ln 3 - 2 \ln(4 + 8) = \ln 3 - \ln(12)^2$
$$= \ln\left(\frac{3}{12^2}\right)$$
$$= \ln\left(\frac{1}{4 \cdot 12}\right)$$
$$= \ln \frac{1}{48}$$

Page 253

$$\textbf{c.}\quad \ln 3 - 2(\ln 4 + \ln 8) = \ln 3 - 2[\ln(4 \cdot 8)]$$
$$= \ln 3 - \ln\left(2^5\right)^2$$
$$= \ln\left(\frac{3}{2^{10}}\right)$$
$$= \ln\frac{3}{1{,}024}$$

31. $pH = -\log[H^+]$
$$= -\log[2.86 \times 10^{-4}]$$
$$\approx 3.5$$

33. $pH = -\log[H^+]$
$$= -\log[6.31 \times 10^{-7}]$$
$$\approx 6.2$$

35. $N = 1{,}500 + 300\ln a$
$$= 1{,}500 + 300\ln 1{,}000$$
$$= 3{,}572 \ \text{ (rounded to the nearest unit)}$$

Level 2, page 489

*The **a** and **b** parts of Problems 37-55 are designed to show the students that logarithmic equations are essentially the same as first degree equations.*

37. a. $\dfrac{1}{2}x - 2 = 2$ **b.** $\dfrac{1}{2}\log x - \log 100 = 2$

$$\frac{1}{2}x = 4 \qquad\qquad \frac{1}{2}\log x = 4 \qquad \text{\textit{Note: log 100}} = 2$$
$$x = 8 \qquad\qquad \log x = 8$$
$$x = 10^8 \qquad \text{Definition of logarithm}$$

39. a. $\dfrac{1}{2}x = 3 - x$ **b.** $\dfrac{1}{2}\log_b x = 3\log_b 5 - \log_b x$

$$\frac{3}{2}x = 3 \qquad\qquad \frac{3}{2}\log_b x = 3\log_b 5 \qquad \text{Add } log_b x \text{ to both sides.}$$
$$x = 2 \qquad\qquad \log_b x = 2\log_b 5$$
$$\log_b x = \log_b 25 \qquad \text{\textit{Note: 2 log}}_b 5 = log_b 5^2$$
$$x = 25 \qquad \text{log of both sides theorem}$$

Page 254

41. a. $1 = x - 1$ **b.** $\ln e = \ln \dfrac{\sqrt{2}}{x} - \ln e$
$\quad\quad 2 = x$

$$1 = \ln \frac{\sqrt{2}}{x} - 1 \quad \textit{Note: ln e = 1}$$

$$e^2 = \frac{\sqrt{2}}{x}$$

$$x = \frac{\sqrt{2}}{e^2}$$

43. a. $3 - x = 1$ **b.** $\ln e^3 - \ln x = 1$
$\quad\quad 3 = x + 1$ $\quad\quad\quad 3 - \ln x = 1 \quad \textit{Note: ln e}^3 \textit{= 3}$
$\quad\quad 2 = x$ $\quad\quad\quad\quad\quad 2 = \ln x$
$\quad\quad\quad\quad\quad\quad\quad\quad\quad x = e^2 \quad \textsf{Definition of logarithm}$

45. $\log(\log x) = 1$ **47.** $x^2 5^x = 5^x$
$\quad\quad \log x = 10^1$ $\quad\quad\quad\quad\quad\quad x^2 = 1$
$\quad\quad\quad x = 10^{10}$ $\quad\quad\quad\quad\quad x^2 - 1 = 0$
$\quad\quad\quad\quad\quad\quad\quad\quad (x-1)(x+1) = 0$
$\quad\quad\quad\quad\quad\quad\quad\quad\quad\quad x = 1, -1$

49. $\quad\quad\quad \log x = 1.8 + \log 4.8$
$\quad \log x - \log 4.8 = 1.8$
$\quad\quad\quad \log \dfrac{x}{4.8} = 1.8$
$\quad\quad\quad\quad\quad x = 4.8 \cdot 10^{18}$

51. $\ln x - \ln 8 = 12$
$\quad\quad \ln \dfrac{x}{8} = 12$
$\quad\quad\quad \dfrac{x}{8} = e^{12}$
$\quad\quad\quad\quad x = 8e^{12}$

53. $\log 2 = \dfrac{1}{4} \log 16 - x$
$\quad\quad x = \log(2^4)^{1/4} - \log 2$
$\quad\quad\quad = \log 2 - \log 2$
$\quad\quad\quad = 0$

Page 255

55. $\log x + \log(x - 3) = 2$

$\qquad \log x(x - 3) = 2$

$\qquad\qquad x(x - 3) = 10^2$

$x^2 - 3x - 100 = 0$

$$x = \frac{3 \pm \sqrt{9 - 4(1)(-100)}}{2}$$

$$= \frac{3 \pm \sqrt{409}}{2}$$

Level 3, page 489

57. a. $\quad 10 = 80 - 27 \ln t \quad$ After about 13 seconds. $\qquad$ **b.** $\qquad R = 80 - 27 \ln t$

$27 \ln t = 70 \qquad\qquad\qquad\qquad\qquad\qquad\qquad 27 \ln t = 80 - R$

$\ln t = \dfrac{70}{27} \qquad\qquad\qquad\qquad\qquad\qquad\qquad \ln t = \dfrac{80 - R}{27}$

$t = e^{70/27} \qquad\qquad\qquad\qquad\qquad\qquad\qquad t = e^{(80-R)/27}$

≈ 13.364

Level 3, Problem Solving, page 489

59. a. $M = \dfrac{\log 15^{15} - 11.8}{1.5} \approx 3.89$

$\quad$ **b.** $M = \dfrac{\log 10^{25} - 11.8}{1.5} = 8.8$

$\quad$ **c.** $\quad 8.0 = \dfrac{\log E - 11.8}{1.5} \quad$ The energy is about $10^{23.8}$ ergs.

$\qquad 12 = \log E - 11.8$

$\qquad 23.8 = \log E$

$\qquad E = 10^{23.8}$

$\quad$ **d.** $\qquad\qquad M = \dfrac{\log E - 11.8}{1.5}$

$\qquad\qquad 1.5M = \log E - 11.8$

$\qquad 1.5M + 11.8 = \log E$

$\qquad\qquad\quad E = 10^{1.5M+11.8}$

Page 256

10.3 Applications of Growth and Decay, page 490
New Terms Introduced in this Section

Decay formula	Decibel	Decibel formula	Growth formula
Half-life	Logarithmic scale	Richter number	Richter scale

Level 1, page 497

1. The *growth/decay* formula is $A = A_0 e^{rt}$ where A_0 is the initial amount, A is the future amount, t is the time (in years), and r is the annual growth rate (if it is positive) and the annual decay rate (if it is negative).

3. A *logarithmic scale* is a scale on a graph in which logarithms are used to make data more manageable by expanding small variations and compressing large ones. (See Problem 58, for example.)

5. $D = 10 \log(10^{-13}/10^{-16}) = 30$ dB

7. $D = 10 \log(5.23 \times 10^{-6}/10^{-16}) \approx 107$ dB

9. $D = 10 \log(2.53 \times 10^{-5}/10^{-16}) \approx 114$ dB

Level 2, page 497

11. $r \approx 0.1614549977$ (from Example 3), so $A \approx 209{,}693 e^{9r} \approx 896{,}716$

13. $r = \frac{1}{30} \ln 0.5 \approx -0.023104906$

15. $t = \ln 0.5/(-0.0246) \approx 28.18$; The half-life is about 28 years.

17. $M = \log(5.1 \times 10^2) + 4.0 \approx 6.7$

19. $6 - 4 = \log \frac{A_1}{A_2}$, so $10^2 = \frac{A_1}{A_2}$ or $A_1 = 100 A_2$. This means that the earthquake with magnitude 6 is 100 times stronger than an earthquake of magnitude 4.

21. $\log E = 11.8 + 1.5(8.3)$, so $E = 10^{24.25}$ ergs

23. $\quad A = A_0 e^{rt}$

$\quad \dfrac{A}{A_0} = e^{rt}$

$\quad 0.28 = e^{rt} \qquad$ where $r \approx -1.209680943\text{E} - 4$ (from Example 5)

$\quad rt = \ln 0.28$

$\quad t = \dfrac{\ln 0.28}{r}$

$\quad\quad \approx 10{,}523$

The artifact is about 10,500 years old.

25.

$$A = A_0 e^{rt}$$
$$\frac{A}{A_0} = e^{rt}$$
$$0.5 = e^{rt} \quad \text{where } t = 2.6$$
$$r(2.6) = \ln 0.5$$
$$r = \frac{\ln 0.5}{2.6}$$
$$\approx -0.2665950694$$

We have,
$$A = A_0 e^{rt}$$
$$15.5 = 100 e^{rt}$$
$$\frac{15.5}{100} = e^{rt}$$
$$rt = \ln\left(\frac{15.5}{100}\right)$$
$$t = \frac{1}{r}\ln\left(\frac{15.5}{100}\right)$$
$$\approx 6.993115686$$

The elapsed time is about 7 years.

27. From Example 9,

$$\frac{E_1}{E_2} = \frac{10^{1.5(8.3)+11.8}}{10^{1.5(9.5)+11.8}} = 10^{-1.8}$$

so $10^{1.8}E_1 = E_2$. The earthquake with a magnitude of 9.5 releases $10^{1.8} \approx 63$ times more energy than an earthquake with magnitude 8.3.

29.
$$A = A_0 e^{rt}$$
$$\frac{A}{A_0} = e^{rt}$$
$$0.85 = e^{rt} \quad \text{where } r = -1.20968094\text{E} - 4$$
$$rt = \ln 0.85$$
$$t = \frac{\ln 0.85}{r}$$
$$\approx 1{,}343$$

The artifact is about 1,300 years old.

31. From Example 9,

$$\frac{E_1}{E_2} = \frac{10^{1.5(7.0)+11.8}}{10^{1.5(7.1)+11.8}} = 10^{-0.15}$$

so $10^{0.15}E_1 = E_2$. The earthquake with a magnitude of 7.1 released $10^{0.015} \approx 1.41$ times more energy than an earthquake with magnitude 7.0.

33. From Example 9,

$$\frac{E_1}{E_2} = \frac{10^{1.5(8)+11.8}}{10^{1.5(7)+11.8}} = 10^{1.5}$$

so $10^{1.5}E_2 = E_1$. The earthquake with a magnitude of 8 releases $10^{1.5} \approx 31.6$ times more energy than an earthquake with magnitude 7.

35. $T = A + (B - A)10^{-kt}$ The cookie is about 128°F.

$= 74 + (375 - 74)10^{-0.075(10)}$

≈ 127.53

37. $k = \dfrac{1}{t} \log \dfrac{B - A}{T - A}$

$= \dfrac{1}{30} \log \dfrac{100 - 22}{45 - 22}$

≈ 0.0176788922

We use this value of k to find the temperature:

$T = A + (B - A)10^{-kt}$

$= 22 + (100 - 22)10^{-k(40)}$

≈ 37.31

The temperature in 40 minutes is about 37°C.

39. $P = 14.7e^{-0.21a}$

$11.9 = 14.7e^{-0.21a}$

$-0.21a = \ln\left(\dfrac{11.9}{14.7}\right)$

$a = \dfrac{1}{-0.21}\ln\left(\dfrac{11.9}{14.7}\right)$

≈ 1.006233779

Since this altitude is in miles, we convert to feet: $5,280a \approx 5,313$ ft.

41. $P = 50e^{-t/250}$

$e^{-t/250} = \dfrac{P}{50}$

$-\dfrac{t}{250} = \ln \dfrac{P}{50}$

$t = -250 \ln \dfrac{P}{50}$

If $P = 30$ W, then $t = -250 \ln \frac{30}{50} \approx 127.7$; The satellite will operate for about 128 days.

Page 259

Level 3, page 498

For Problems 43-50,

a. *The growth rate is* $r = 0.1\ln\left(\frac{P}{P_0}\right)$ *where P is the population in 2000 and P_0 is the population in 1990.*

b. *Use the number r from part **a** (not the rounded value) with $P = P_0 e^{4r}$, where P_0 is the population in 1990.*

c. *Answers vary.*

d. *Use r from part **a** (not the rounded value) with $P = P_0 e^{6r}$, where P_0 is the population in 2000.*

43. a. $r = 0.1\ln\left(\frac{P}{P_0}\right)$ Rounding r to the nearest tenth percent, we see $r = 1.5\%$

$\quad = 0.1\ln\left(\frac{735,616}{636,070}\right)$

$\quad \approx 0.014539962327$

b. $P = P_0 e^{4r}$

$\quad = 636,070 e^{4r}$ Use r from part a.

$\quad \approx 674,161$ Round to the nearest unit.

c. Overestimate; actual rate is lower than r from part a.

d. $P = P_0 e^{6r}$

$\quad = 735,616 e^{6r}$ Use r from part a.

$\quad \approx 802,673$ Round to the nearest unit.

45. a. $r = 0.1\ln\left(\frac{P}{P_0}\right)$ Rounding r to the nearest tenth percent, we see $r = 1.1\%$

$\quad = 0.1\ln\left(\frac{545,524}{488,374}\right)$

$\quad \approx 0.011066529487$

b. $P = P_0 e^{4r}$

$\quad = 488,374 e^{4r}$ Use r from part a.

$\quad \approx 510,478$ Round to the nearest unit.

c. Overestimate; actual rate is lower than r from part a.

d. $P = P_0 e^{6r}$

$\quad = 545,524 e^{6r}$ Use r from part a.

$\quad \approx 582,976$ Round to the nearest unit.

Page 260

47. a. $r = 0.1 \ln\left(\dfrac{P}{P_0}\right)$ Rounding r to the nearest tenth percent, we see $r = 8.8\%$

$\quad\quad = 0.1 \ln\left(\dfrac{876{,}593}{365{,}272}\right)$

$\quad\quad \approx 0.087540052102$

b. $P = P_0 e^{4r}$

$\quad\quad = 365{,}272 e^{4r}$ *Use r from part a.*

$\quad\quad \approx 518{,}429$ *Round to the nearest unit.*

c. Overestimate; actual rate is lower than r from part a.

d. $P = P_0 e^{6r}$

$\quad\quad = 876{,}593 e^{6r}$ *Use r from part a.*

$\quad\quad \approx 1{,}482{,}201$ *Round to the nearest unit.*

49. a. $r = 0.1 \ln\left(\dfrac{P}{P_0}\right)$ Rounding r to the nearest tenth percent, we see $r = 3.7\%$

$\quad\quad = 0.1 \ln\left(\dfrac{287{,}151}{198{,}518}\right)$

$\quad\quad \approx 0.03691284$

b. $P = P_0 e^{4r}$

$\quad\quad = 198{,}518 e^{4r}$ *Use r from part a.*

$\quad\quad \approx 230{,}104$ *Round to the nearest unit.*

c. Underestimate; actual rate is lower than r from part a.

d. $P = P_0 e^{6r}$

$\quad\quad = 287{,}151 e^{6r}$ *Use r from part a.*

$\quad\quad \approx 358{,}341$ *Round to the nearest unit.*

51. a. $\quad\quad\quad A = A_0 e^{rt}$

$\dfrac{3{,}485{,}557}{2{,}968{,}528} = e^{r(10)}$

$\quad\quad 10r = \ln\left(\dfrac{3{,}485{,}557}{2{,}968{,}528}\right)$

$\quad\quad\quad r = \dfrac{1}{10}\ln\left(\dfrac{3{,}485{,}557}{2{,}968{,}528}\right)$

$\quad\quad\quad\quad \approx 0.016056165222$

Use this value of r to find the 2004 population, A, when $t = 10$.

$$A = A_0 e^{rt}$$
$$= 3,448,613 e^{10r}$$
$$\approx 4,049,258$$

b.
$$A = A_0 e^{rt}$$
$$\frac{3,448,613}{3,485,557} = e^{r(4)}$$
$$4r = \ln\left(\frac{3,448,613}{3,485,557}\right)$$
$$r = \frac{1}{4}\ln\left(\frac{3,448,613}{3,485,557}\right)$$
$$\approx -0.0026639345$$

Use this value of r to find the 2004 population, A, when $t = 10$.
$$A = A_0 e^{rt}$$
$$= 3,448,613 e^{10r}$$
$$\approx 3,357,957$$

c.
$$A = A_0 e^{rt}$$
$$\frac{3,448,613}{2,968,528} = e^{r(14)}$$
$$14r = \ln\left(\frac{3,448,613}{2,968,528}\right)$$
$$r = \frac{1}{14}\ln\left(\frac{3,448,613}{2,968,528}\right)$$
$$\approx 0.010707565298$$

Use this value of r to find the 2004 population, A, when $t = 10$.
$$A = A_0 e^{rt}$$
$$= 3,448,613 e^{10r}$$
$$\approx 3,838,370$$

53. a.
$$A = A_0 e^{rt}$$
$$6 = 5.3 e^{0.0098t}$$
$$0.0098t = \ln \frac{6}{5.3}$$
$$t \approx 12.65843354$$

The projected date is 12 years, 8 months, or in other words February, 2003.

b. Since the world population actually reached 6 billion before the predicted date, the

Page 262

conclusion is that the actual growth rate is greater than the assumed growth rate.

c. The actual date for 5.3 billion to reach 6 billion is July 1990 to October 1999 or 9 years 3 months or $t \approx 9\frac{3}{12} = 9.25$. Use this value to find r in the growth formula:

$$A = A_0 e^{rt}$$
$$6 = 5.3 e^{r(9.25)}$$
$$9.25r = \ln \frac{6}{5.3}$$
$$t \approx 0.013411097154$$

The projected date for 7 billion is found by again using the growth formula:

$$A = A_0 e^{rt}$$
$$7 = 6 e^{rt}$$
$$rt = \ln \frac{7}{6}$$
$$t \approx 11.4942631511$$

The projected date is 11 years 6 months, or in other words April, 2011.

55. $0.73 = e^{r_1 t_1}$ *$r_1 \approx -1.209680943E - 4$ from Example 5.*

$$r_1 t_1 = \ln 0.73$$ *Definition of logarithm*

$$t_1 = \frac{\ln 0.73}{r_1}$$

$$t_1 \approx 2601.601245$$

For radium, we use this value for t_1: $0.32 = e^{r_2 t_1}$

$$r_2 t_1 = \ln 0.32$$

$$r_2 = \frac{\ln 0.32}{t_1}$$

$$r_2 \approx -4.379742227E - 4$$

Finally, the half life is found using this value for r_2 : $0.50 = e^{r_2 h}$

$$r_2 h = \ln 0.50$$

$$h = \frac{\ln 0.50}{r_2}$$

$$h \approx 1582.62095$$

The half-life of radium is about 1,583 yr.

Level 3 Problem Solving, page 499

57. From Example 5, $r \approx -1.209680943E - 4$, and from $P = P_0 e^{rt}$:

$$\frac{P}{P_0} = e^{rt}$$

$$rt = \ln 0.923$$

$$t \approx \frac{\ln 0.923}{-1.209680943\text{E} - 4} \approx 662.37337$$

Since the sample was measured in 1988, this dates the artifact at $1988 - t \approx 1326$. This means the shroud was probably new when d'Arcis wrote the memo to the Pope, and that it is probably not the burial robe of Christ.

59. $y = \dfrac{6e^{5.085-0.1156x}}{1 + e^{5.085-0.1156x}}$

$$= \frac{6e^{5.085-0.1156(75)}}{1 + e^{5.085-0.1156(75)}}$$

$$\approx 0.1619$$

If we round y to the nearest unit, we see that at $75°$, we would expect no O-rings failures.

$$y = \frac{6e^{5.085-0.1156x}}{1 + e^{5.085-0.1156x}}$$

$$= \frac{6e^{5.085-0.1156(32)}}{1 + e^{5.085-0.1156(32)}}$$

$$\approx 4.7995$$

If we round y to the nearest unit, we see that at $32°$, we would expect 5 O-rings failures.

Chapter 10 Review Questions, page 500

STOP

Studying for a chapter examination is a personal process, one which nobody else can do for you. Simply take the time to review what you have done. Here are the new terms in Chapter 10.

Addition law of logarithms [10.2]
Argument [10.1]
Change of base theorem [10.1]
Common logarithm [10.1]
Decay formula [10.3]
Decibel [10.3]
Evaluate [10.1]
Exact solution [10.1]
Exponential [10.1]

Exponential equation [10.1]
Grant's tomb properties [10.2]
Growth formula [10.3]
Half-life [10.3]
Laws of logarithms [10.2]
Log of both sides theorem [10.2]
Logarithm [10.1]
Logarithmic equation [10.2]
Logarithmic scale [10.3]

Micrometer [10.1]
Multiplicative law of
  logarithms [10.2]
Natural logarithm [10.1]
Richter number [10.3]
Richter scale [10.3]
Subtractive law of logarithms
  [10.2]

If you can describe the term, read on to the next one; if you cannot, then look it up in the text (the section number is shown in brackets). Next, study the types of problems listed at the end of Chapter 10.

TYPES OF PROBLEMS
Know the definition of a logarithm. [10.1]
Evaluate logarithms. [10.1]
Use the Grant's tomb properties to simplify logarithmic expressions. [10.1, 10.2]
Solve exponential equations. [10.1]
Solve logarithmic equations. [10.2]
Solve applied problems of growth and decay. [10.3]

Once again, see if you can verbalize (to yourself) how to do each of the listed types of problems.

Work all of Chapter 10 Review Questions (whether they are assigned or not). Work through all of the problems before looking at the answers, and *then* correct each of the problems. The entire solution is shown in the answer section at the back of the text. If you worked the problem correctly, move on to the next problem, but if you did not work it correctly (or you did not know what to do), look back in the chapter to study the procedure, or ask your instructor.

Finally, go back over the homework problems you have been assigned. If you worked a problem correctly, move on to the next problem, but if you missed it on your homework, then you should look back in the book or talk to your instructor about how to work the problem.

If you follow these steps, you should be successful with your review of this chapter.

We give all of the answers to the chapter review questions (not just the odd-numbered questions).

Chapter 10 Review Questions, page 500

1. $\log 100 + \log \sqrt{10} = 2 + \dfrac{1}{2} = \dfrac{5}{2}$ **2.** $\ln e + \ln 1 + \ln e^{542} = 1 + 0 + 542$
$$= 543$$

3. $\log_8 4 + \log_8 16 + \log_8 8^{2.3} = \log_8 64 + 2.3$ **4.** $10^{\log 0.5} = 0.5$
$$= 2 + 2.3$$
$$= 4.3$$

5. $\ln e^{\log 1,000} = \ln e^3 = 3$ **6. a.** $\log 8.43 \approx 0.93$ **b.** $\log 9,760 \approx 3.99$

7. a. $\ln 2 \approx 0.69$ **b.** $\ln 0.125 \approx -2.08$ **8. a.** $\log_2 10 \approx 3.32$ **b.** $\log_\pi \frac{1}{\pi} = \log_\pi (\pi)^{-1} = -1$

9. $10^x = 85$ **10.** $e^x = 500$ **11.** $435^x = 890$
    $= \log 85$ $x = \ln 500$ $x = \log_{435} 890$
    ≈ 1.929418926 ≈ 6.214608098 ≈ 1.117832865

Page 265

12. $e^{3x+1} = 45$

$\quad 3x + 1 = \ln 45$

$\quad\quad\quad = \dfrac{\ln 45 - 1}{3}$

$\quad\quad\quad \approx 0.935554163$

13. $\log_6 x = 4$

$\quad\quad x = 6^4$

$\quad\quad\quad = 1{,}296$

14. $2^{3x-1} = 6$

$\quad 3x - 1 = \log_2 6$

$\quad\quad\quad x = \dfrac{\log_2 6 + 1}{3}$

15. $10^{2x} = 5$

$\quad 2x = \log 5$

$\quad\; x = \dfrac{\log 5}{2}$

16. $\quad\quad\quad\quad \log(x+1) = 2 + \log(x-1)$

$\quad\quad \log(x+1) - \log(x-1) = 2$

$\quad\quad\quad\quad\quad\quad \log \dfrac{x+1}{x-1} = 2$

$\quad\quad\quad\quad\quad\quad\quad \dfrac{x+1}{x-1} = 10^2$

$\quad\quad\quad\quad\quad x = 1 = 100x - 100$

$\quad\quad\quad\quad\quad\quad\quad\quad x = \dfrac{101}{99}$

17. $\quad 3 \ln \dfrac{e}{\sqrt[3]{5}} = 3 - \ln x$

$\quad\quad \ln \dfrac{e^3}{5} + \ln x = 3$

$\quad\quad\quad\quad \ln \dfrac{e^3 x}{5} = 3$

$\quad\quad\quad\quad\; \dfrac{e^3 x}{5} = e^3$

$\quad\quad\quad\quad\quad \dfrac{x}{5} = 1$

$\quad\quad\quad\quad\quad\; x = 5$

18. $A = P(1+i)^x$

$\quad \dfrac{A}{P} = (1+i)^x$

$\quad\; x = \log_{(1+i)} \dfrac{A}{P}$

19. $\quad A = A_0 e^{-0.1t}$

$\quad\; \dfrac{A}{A_0} = e^{-0.1t}$

$\quad -0.1t = \ln \dfrac{A}{A_0}$

$\quad\quad\; t = -10 \ln \dfrac{A}{A_0}$

$\quad\quad\quad = -10 \ln 0.5$

$\quad\quad\quad \approx 6.93$

It would take about 7 days.

20. $\quad A = A_0 e^{r(14/12)}$

$\quad \dfrac{A}{A_0} = e^{r(14/12)}$

$\quad\quad r = \dfrac{12}{14} \ln 2$

$\quad\quad\quad \approx 0.59$

Equation is $A = A_0 e^{0.59t}$, where A_0 is the number of teens infected and t is the number of years after 1992.

CHAPTER 11 The Nature of Financial Management

Chapter Overview
This is one of the most useful chapters in the book. The goal of this chapter is to put real dollars in your pocket because of what you learn in this chapter. This chapter discusses both simple and compound interest, installment buying, car loans, home loans, savings, and retirement planning.

11.1 Interest, page 503
New Terms Introduced in this Section

Annual compounding	Compound interest	Compound interest formula
Continuous compounding	Daily compounding	e
Euler's number	Exact interest	Future value
Future value formula	Inflation	Interest
Interest rate	Monthly compounding	Natural base
Ordinary interest	Present value	Present value formula
Principal	Quarterly compounding	Semiannual compounding
Simple interest formula	Time	

Level 1, page 514

1. *Interest* is a rental fee paid for the use of another's money.

3. *Simple interest* is calculated according to the formula $I = Prt$, whereas *compound interest* pays interest on the accrued interest after a given length of time (called the compounding period).

5. It is about opening a savings account for a child to be used at the child's retirement. It shows how much can be accumulated if you plan ahead. Today, 12% is an unrealistic rate of return, but nevertheless the example shows that a *one-time* deposit of just over $400 at fixed rate of 12% would result in a million dollars at retirement. If you want $10 million you need a one-time deposit of $4,102.70.

7. C; interest rate is not stated, but you should still recognize a reasonable answer.

9. B is the most reasonable.

11. $P \approx \$50,000$, $r \approx 10\%$, $t = 2$, $I \approx \$50,000(2)\left(\frac{1}{10}\right) = \$10,000$; the best choice is A.

13. The decimal portion (0.52895) is about half a year, so the number closest to half of 365 is 200; the best choice is D.

15. Since one year is 360 days (ordinary interest) we see that half is 180 days; choice is B.

17. $I = Prt$

$= \$5,000(0.10)(3)$

$= \$1,500$

19. $I = Prt$

$= \$1,000(0.14)(30)$

$= \$4,200$

21. $A = P\left(1 + \dfrac{r}{n}\right)^{nt}$

$= \$835\left(1 + \dfrac{0.035}{2}\right)^{2(6)}$

$= \$1,028.25$

23. $A = P\left(1 + \dfrac{r}{n}\right)^{nt}$

$= \$9,730.50\left(1 + \dfrac{0.076}{12}\right)^{7(12)}$

$= \$16,536.79$

25. $A = Pe^{rt}$

$= \$119,400e^{(0.075)(30)}$

$= \$1,132,835.66$

27. $P = A\left(1 + \dfrac{r}{n}\right)^{-nt}$

$= \$2,500\left(1 + \dfrac{0.082}{12}\right)^{-5(12)}$

$= \$1,661.44$

29.
$$A = Pe^{rt}$$
$$\$1,000,000 = Pe^{(0.06)(30)}$$
$$\frac{\$1,000,000}{e^{(0.06)(30)}} = P$$
$$\$165,298.89 = P$$

Page 268

31. a. $n = 1$

$$A = P\left(1 + \frac{r}{n}\right)^{nt}$$

$$= \$12,000\left(1 + \frac{0.045}{1}\right)^{(1)(20)}$$

$$= \$28,940.57$$

b. $n = 2$

$$A = P\left(1 + \frac{r}{n}\right)^{nt}$$

$$= \$12,000\left(1 + \frac{0.045}{2}\right)^{(2)(20)}$$

$$= \$29,222.27$$

c. $n = 4$

$$A = P\left(1 + \frac{r}{n}\right)^{nt}$$

$$= \$12,000\left(1 + \frac{0.045}{4}\right)^{(4)(20)}$$

$$= \$29,367.30$$

d. $n = 12$

$$A = P\left(1 + \frac{r}{n}\right)^{nt}$$

$$= \$12,000\left(1 + \frac{0.045}{12}\right)^{(12)(20)}$$

$$= \$29,465.60$$

e. $n = 360$

$$A = P\left(1 + \frac{r}{n}\right)^{nt}$$

$$= \$12,000\left(1 + \frac{0.045}{360}\right)^{(360)(20)}$$

$$= \$29,513.58$$

Note: if you found $29,513.60, you used a 365-day year instead of the agreed number of 360 days.

f. $n = 525,600$

$$A = P\left(1 + \frac{r}{n}\right)^{nt}$$

$$= \$12,000\left(1 + \frac{0.045}{525,600}\right)^{(525,600)(20)}$$

$$= \$29,515.24$$

Note: many calculators will give $29,515.22, which is an acceptable answer.

g. $A = Pe^{rt}$

$$= \$12,000e^{(0.045)(20)}$$

$$= \$29,515.24$$

h. $A = P(1 + rt)$

$$= \$12,000(1 + 0.045 \cdot 20)$$

$$= \$22,800.00$$

Level 2, page 515

33. $A = P(1 + rt)$

$$= \$1,500\left(1 + 0.21 \cdot \frac{55}{360}\right)$$

$$= \$1,548.13$$

35. a. $A = Pe^{rt}$

$$= \$1.75e^{0.09(5)}$$

$$= \$2.74$$

b. $A = Pe^{rt}$

$$= \$1.25e^{0.09(5)}$$

$$= \$1.96$$

c. $A = Pe^{rt}$

$$= \$1.95e^{0.09(5)}$$

$$= \$3.06$$

d. $A = Pe^{rt}$
$\quad = \$2.95e^{0.09(5)}$
$\quad = \$4.63$

e. $A = Pe^{rt}$
$\quad = \$1,600e^{0.09(5)}$
$\quad = \$2,509.30$

f. $A = Pe^{rt}$
$\quad = \$19,000e^{0.09(5)}$
$\quad = \$29,797.93$

g. $A = Pe^{rt}$
$\quad = \$28,000e^{0.09(5)}$
$\quad = \$43,912.74$

h. $A = Pe^{rt}$
$\quad = \$16,000e^{0.09(5)}$
$\quad = \$25,092.99$

37. $t = 5; P = \$1,000; r = 0.04; n = 4$

$$A = P\left(1 + \frac{r}{n}\right)^{nt}$$

$$= \$1,000\left(1 + \frac{0.04}{4}\right)^{5(4)}$$

$$\approx \$1,220.19$$

39. $t = 15; P = \$1,000; r = 0.16$

$A = Pe^{rt}$
$\quad = \$1,000e^{(0.16)(15)}$
$\quad \approx \$11,023.18$

41. $P = \$14,500; m = \$410.83; t = 4; n = 12$

$I = A - P$
$\quad = \$410.83 \cdot 12 \cdot 4 - \$14,500$
$\quad = \$5,219.84$

43. $P = \$170,000; m = \$1,247.40; t = 30; n = 12$

$I = A - P$
$\quad = \$1,247.40 \cdot 12 \cdot 30 - \$170,000$
$\quad = \$279,064$

45. $t = 30; P = \$850; r = 0.10$

$A = Pe^{rt}$
$\quad = \$850e^{(0.10)(30)}$
$\quad \approx \$17,072.71$

47. $I = \$5,075; t = 1; r = 0.05$

$I = Prt$
$\$5075 = P(0.05)(1)$
$\dfrac{\$5,075}{0.05} = P$
$\quad P = \$101,500$

Page 270

In Problems 49-51, the amount needed in the retirement account is simple interest because the interest is removed from the account each month and consequently does not accrue any interest.

49. $I = \$10,000; t = \dfrac{1}{12}; r = 0.15$

$$I = Prt$$

$$\$10,000 = P(0.15)\left(\dfrac{1}{12}\right)$$

$$\dfrac{\$10,000(12)}{0.15} = P$$

$$P = \$800,000$$

51. $I = \$1,000; t = \dfrac{1}{12}; r = 0.06$

$$I = Prt$$

$$\$1,000 = P(0.06)\left(\dfrac{1}{12}\right)$$

$$\dfrac{\$1,000(12)}{0.06} = P$$

$$P = \$200,000$$

Level 3, page 516

53. $1,000 = 750(1 + 0.08t)$

$$\dfrac{1,000}{750} = 1 + 0.08t$$

$$0.08t = \dfrac{1}{3}$$

$$t \approx 4.1\overline{6}$$

4 years, 61 days

55. $5,000 = 3,500\left(1 + \dfrac{0.05}{365}\right)^{365t}$

$$\dfrac{5,000}{3,500} = \left(1 + \dfrac{0.05}{365}\right)^{365t}$$

$$365t = \log_{(1+0.05/365)} \dfrac{5,000}{3,500}$$

$$t \approx 7.133987462$$

7 years, 49 days

Page 271

57. a.
$$A = P\left(1 + \frac{r}{n}\right)^{nt}$$

$$\$1{,}250 = \$1{,}000\left(1 + \frac{0.07}{12}\right)^{12t}$$

$$\frac{1{,}250}{1{,}000} = \left(1 + \frac{0.07}{12}\right)^{12t}$$

$$12t = \log_{(1+0.07/12)}\frac{1{,}250}{1{,}000}$$

$$t = \frac{1}{12}\left[\log_{(1+0.07/12)}\frac{1{,}250}{1{,}000}\right] \approx 3.197053654$$

The necessary time to wait is 3 years, 3 months (or 3 years, 2.4 months).

b.
$$A = P\left(1 + \frac{r}{n}\right)^{nt}$$

$$\$2{,}000 = \$1{,}000\left(1 + \frac{0.07}{12}\right)^{12t}$$

$$2 = \left(1 + \frac{0.07}{12}\right)^{12t}$$

$$12t = \log_{(1+0.07/12)} 2$$

$$t = \frac{1}{12}\left[\log_{(1+0.07/12)}2\right] \approx 9.930955715$$

The necessary time to wait is 10 years (or 9 years, 11.2 months).

c.
$$A = P\left(1 + \frac{r}{n}\right)^{nt}$$

$$\$2{,}000 = \$1{,}000\left(1 + \frac{r}{12}\right)^{12(5)}$$

$$2 = \left(1 + \frac{r}{12}\right)^{60}$$

$$2^{1/60} = 1 + \frac{r}{12}$$

$$\frac{r}{12} = 2^{1/60} - 1$$

$$r = 12(2^{1/60} - 1) \approx 0.1394332836$$

The interest rate is about 14%.

Page 272

Level 3 Problem Solving, page 516

59. We want to find Y so that the compounded amount is equal to the amount with the simple interest formula; that is:

$$P\left(1 + \frac{r}{n}\right)^{nt} = P(1 + Yt) \quad \textit{Given}$$

$$P\left(1 + \frac{r}{n}\right)^{n} = P(1 + Y) \quad \textit{t = 1 (one year)}$$

$$\left(1 + \frac{r}{n}\right)^{n} = 1 + Y \quad \textit{Divide both sides by P.}$$

$$\left(1 + \frac{r}{n}\right)^{n} - 1 = Y \quad \textit{Subtract 1 from both sides}$$

11.2 Installment Buying, page 517

New Terms Introduced in this Section

Add-on interest	Adjusted balance method	Annual percentage rate
APR	Average daily balance method	Close-end loan
Credit card	Dealer's cost	Finance charge
Five-percent offer	Grace period	Installment loan
Line of credit	Open-end loan	Previous balance method
Revolving credit	Sticker price	

Level 1, page 523

1. *Add-on interest* is a method of calculating interest which uses the simple interest formula to calculate the amount of interest and then adds this total onto the original principal in order to find the amount that needs to be repaid.

3. *Open-end credit* authorizes a pre-approved line of credit; it usually takes the form of a credit card loan. *Closed-end credit* refers to a fixed amount that is borrowed with a promise to repay the loan in a specified manner.

5. Answers vary; answer the questions. If you have several bank cards, do it for each card you have.

7. Without interest, the amount is $\frac{\$2,400}{24} = \$100$. The interest would be added to this amount, so we see the installments would be more than $100; the correct response is B.

9. The annual fee is usually $0 or small compared to the interest changes. Likewise, the grace period is important in avoiding finance charges if the balance is paid off each month. Thus, the correct response is the APR, or choice B.

11. The adjusted balance method is most advantageous to the consumer, so the correct answer is B.

13. Estimate $1,295 \approx \$1,200$ and $9.8\% \approx 10\%$, so the interest is

$I = Prt = \$1,200(0.1)(4) = \$120(4) = \$480.$ We see the best answer is B.

15. In today's market, choice B is the most reasonable because 1% is much too small, and choice C is exorbitant.

17. You certainly would not offer more than the sticker price, and if you paid the sticker price, that would not be considered an offer, so the only reasonable choice is B.

19. Since August has 31 days, the correct response is C.

21. 12% APR would be a monthly rate of 1% and 1% of $650 is $6.50, so the best choice is A.

23. It should be the least expensive, so the best choice is B.

25. $1\frac{1}{4}\% \times 12 = 15\%$

27. $1\frac{1}{2} \times 12 = 18\%$

29. $0.02192\% \times 365 = 8.0008$; the annual rate is 8%.

31. a. $I = Prt$ **b.** $I = Prt$

$$= \$300(0.18)\left(\frac{1}{12}\right) \qquad = \$250(0.18)\left(\frac{1}{12}\right)$$

$$= \$4.50 \qquad\qquad\qquad\qquad = \$3.75$$

    c. $I = Prt$

$$= \left(\frac{\$300(10) + \$250(20)}{30}\right)(0.18)\left(\frac{30}{365}\right)$$

$$= \$3.95$$

33. a. $I = Prt$ **b.** $I = Prt$

$$= \$3,000(0.15)\left(\frac{1}{12}\right) \qquad = \$2,950(0.15)\left(\frac{1}{12}\right)$$

$$= \$37.50 \qquad\qquad\qquad\qquad = \$36.88$$

    c. $I = Prt$

$$= \left(\frac{\$3,000(10) + \$2,950(20)}{30}\right)(0.15)\left(\frac{30}{365}\right)$$

$$= \$36.58$$

35. $\$13,378(1 + 0.06) \approx \$14,181$

37. $\$17,250(1 + 0.10) = \$18,975$

Level 2, page 524

39. $\text{APR} = \dfrac{2rn}{n+1}$ **41.** $\text{APR} = \dfrac{2rn}{n+1}$

$\approx \dfrac{2(0.12)(36)}{37}$ $\approx \dfrac{2(0.11)(24)}{25}$

$\approx 23.4\%$ $\approx 21.1\%$

43. a. For 0% financing, we find the monthly payment: $m = \frac{\$20,650}{60} = \$344.17$

    b. $P = \$20,650 - \$2,000 = \$18,650;\ I = Prt$

$= \$18,650(0.025)(5)$

$= \$2,331.25$

      Thus, $m = \dfrac{\$18,650 + \$2,331.25}{60}$

$= \$349.69$

    c. $\text{APR} = \dfrac{2rn}{n+1}$ **d.** The 0% financing is better ($\$344.17$ vs $\$349.69$).

$\approx \dfrac{2(0.025)(60)}{61}$

$\approx 4.9\%$

45. a. For 0% financing, we find the monthly payment: $m = \frac{\$42,700}{60} = \$711.67$

    b. $P = \$42,700 - \$5,100 = \$37,600;\ I = Prt$

$= \$37,600(0.025)(5)$

$= \$4,700.00$

      Thus, $m = \dfrac{\$37,600 + \$4,700.00}{60}$

$= \$705.00$

    c. $\text{APR} = \dfrac{2rn}{n+1}$ **d.** The 2.5% financing is better ($\$711.67$ vs $\$705.00$).

$\approx \dfrac{2(0.025)(60)}{61}$

$\approx 4.9\%$

Level 3, page 525

47. a. $A = \$317.50(36)$ **b.** $I = A - P$ **c.** $I = Prt$

$= \$11,430$ $= \$11,430 - \$9,000$ $2,430 = 9,000r(3)$

$= \$2,430$ $r = 0.09$

Page 275

d. $\text{APR} = \dfrac{2r(36)}{37}$

$\approx \dfrac{2(0.09)(36)}{37}$

$\approx 17.5\%$

49. a. $A = \$488.40(48)$ 　　　**b.** $I = A - P$

$= \$23,443.20$ 　　　　　　　$= \$23,443.20 - \$14,350$

$= \$9,093.20$

c. 　　　$I = Prt$ 　　　　　　**d.** $\text{APR} = \dfrac{2r(48)}{49}$

$9,093.20 = 14,350r(4)$

$r = 0.1584181185$ 　　　　　$\approx \dfrac{2(0.1584181185)(48)}{49}$

$\approx 31.0\%$

51. a. $A = \$168.51(48) + \$798$ 　　**b.** $I = A - P$

$= \$8,886.48$ 　　　　　　　　　$= \$8,886.48 - \$6,798$

$= \$2,088.48$

c. 　　　$I = Prt$ 　　　　　　**d.** $\text{APR} = \dfrac{2r(48)}{49}$

$2,088.4 = 6,000r(4)$

$r = 0.08702$ 　　　　　　　　$\approx \dfrac{2(0.08702)(48)}{49}$

$\approx 17.0\%$

53. 8% add-on rate; $\text{APR} = \dfrac{2r(48)}{49}$

$\approx \dfrac{2(0.08)48}{49}$

$\approx 15.7\%$

55. 11% add-on rate; $\text{APR} = \dfrac{2r(48)}{49}$

$\approx \dfrac{2(0.11)48}{49}$

$\approx 21.6\%$

Page 276

Level 3, Problem Solving, page 525

57. Previous balance method: $I = Prt$

$$= \$3,000(0.18)\left(\frac{1}{12}\right)$$

$$= \$45.00$$

Adjusted balance method: $I = Prt$

$$= \$2,700(0.18)\left(\frac{1}{12}\right)$$

$$= \$40.50$$

Average daily balance method: $I = Prt$

$$= \left(\frac{\$3,000(14) + \$2,700(17)}{31}\right)(0.18)\left(\frac{31}{365}\right)$$

$$= \$43.35$$

59. $P = \$1,598(1.05)$ Next, we find $I = Prt$ and

$\quad = \$1,677.90$ $= \$1,677.90(0.15)(3)$

$\qquad\qquad\qquad\qquad\qquad\qquad\qquad = \$755.06$

$\quad A = P + I$ Finally, $m = \dfrac{A}{n}$

$\quad\ = \$1,677.90 + \$755.06$

$\quad\ = \$2,432.96$ $= \dfrac{\$2,432.96}{36}$

$\qquad\qquad\qquad\qquad\qquad\qquad = \$67.58$

11.3 Sequences, page 526

New Terms Introduced in this Section

Arithmetic sequence	Common difference	Common ratio	Fibonacci sequence
Fibonacci-type sequence	Geometric sequence	Sequence	Term of a sequence

Level 1, page 536

1. A *sequence* is a list of numbers having a first term, a second term, a third term, and so on.

3. An *arithmetic sequence* is a sequence whose consecutive terms differ by the same real number, called the common difference.

5. A *Fibonacci-type sequence* is a sequence for which a first and second number are given, and then consecutive numbers are found by adding the two previous terms. **The** Fibonacci sequence is 1, 1, 2, 3, 5, 8, 13, 21, $\cdots$

Page 277

7. $s_1 = 2$, $s_2 = 4$, $s_3 = 8$, $s_4 = 16$

    a. The sequence is geometric because all of its successive ratios are equal.

    b. $r = \frac{s_2}{s_1} = \frac{s_3}{s_2} = \frac{s_4}{s_3} = 2$

    c. $s_5 = rs_4 = 2(16) = 32$

9. $s_1 = 5$, $s_2 = 15$, $s_3 = 25$

    a. The sequence is arithmetic because all of its successive differences are equal.

    b. $d = s_2 - s_1 = s_3 - s_2 = 10$

    c. $s_4 = s_3 + d = 25 + 10 = 35$

11. $s_1 = 5$, $s_2 = 15$, $s_3 = 20$

    a. The sequence does not have a common difference or a common ratio, so we check to see if it is Fibonacci-type. Since $5 + 15 = 20$, we see that it is Fibonacci-type.

    b. $s_1 = 5$, $s_2 = 15$ with $s_1 + s_2 = s_3 = 20$.

    c. $s_4 = s_3 + s_2 = 20 + 15 = 35$

13. $s_1 = 25$, $s_2 = 5$, $s_3 = 1$

    a. The sequence is geometric because all of its successive ratios are equal.

    b. $\frac{s_2}{s_1} = \frac{5}{25} = \frac{1}{5}$ and $\frac{s_3}{s_2} = \frac{1}{5}$ so $r = \frac{1}{5}$.

    c. $s_4 = rs_3 = \left(\frac{1}{5}\right)(1) = \frac{1}{5}$

15. $s_1 = 1$, $s_2 = 3$, $s_3 = 9$

    a. The sequence is geometric because all of its successive ratios are equal.

    b. $\frac{s_2}{s_1} = \frac{3}{1} = 3$ and $\frac{s_3}{s_2} = \frac{9}{3} = 3$ so $r = 3$.

    c. $s_4 = rs_3 = (3)(9) = 27$

17. $s_1 = 8$, $s_2 = 6$, $s_3 = 7$, $s_4 = 5$, $s_5 = 6$, $s_6 = 4$

    a. The sequence does not have a common difference or a common ratio, so we check to see if it is Fibonacci-type. It is not.

    b. We look for a pattern. Starting with the first term, we subtract 2, then add 1, then subtract 2, then add 1 again, and so on.

    c. We carry out the pattern to see that $s_7 = s_6 + 1 = 4 + 1 = 5$.

19. $s_1 = 3$, $s_2 = 6$, $s_3 = 12$, $s_4 = 24$, $s_5 = 48$

    a. The sequence is geometric because all of its successive ratios are equal.

    b. $\frac{s_2}{s_1} = \frac{6}{3} = 2$, $\frac{s_3}{s_2} = \frac{12}{6} = 2$, $\frac{s_4}{s_3} = \frac{24}{12} = 2$, and $\frac{s_5}{s_4} = \frac{48}{24} = 2$, so $r = 2$.

    c. $s_6 = rs_5 = 2(48) = 96$

21. $s_1 = 10$, $s_2 = 10$, $s_3 = 10$

    a. The sequence is arithmetic because there is a common difference, and it is also geometric because there is a common ratio.

    b. $d = s_3 - s_2 = s_2 - s_1 = 0$, $r = \frac{s_2}{s_1} = \frac{s_3}{s_2} = 1$.

    c. $s_4 = 10$

23. $s_1 = 3$, $s_2 = 6$, $s_3 = 9$, $s_4 = 15$

    a. The sequence does not have a common difference or a common ratio, so we check to see if it is Fibonacci-type. Since $3 + 6 = 9$, and $6 + 9 = 15$, we see it is Fibonacci-type.

    b. $s_1 = 3$, $s_2 = 6$ with $s_1 + s_2 = s_3 = 9$ and $s_2 + s_3 = 6 + 9 = 15 = s_4$.

    c. $s_5 = s_4 + s_3 = 15 + 9 = 24$

25. $s_1 = 8$, $s_2 = 12$, $s_3 = 18$, $s_4 = 27$

    a. The sequence is geometric because all of its successive ratios are equal.

    b. $\frac{s_2}{s_1} = \frac{12}{8} = \frac{3}{2}$; $\frac{s_3}{s_2} = \frac{18}{12} = \frac{3}{2}$; $\frac{s_4}{s_3} = \frac{27}{18} = \frac{3}{2}$, so $r = \frac{3}{2}$

    c. $s_5 = rs_4 = \frac{3}{2}(27) = \frac{81}{2}$

27. $s_1 = 4^5$, $s_2 = 4^4$, $s_3 = 4^3$, $s_4 = 4^2$

    a. The sequence is geometric because all of its successive ratios are equal.

    b. $\frac{s_2}{s_1} = \frac{4^4}{4^5} = \frac{1}{4}$; $\frac{s_3}{s_2} = \frac{4^3}{4^2} = \frac{1}{4}$, so $r = \frac{1}{4}$

    c. $s_5 = rs_4 = \frac{1}{4}(4^2) = 4$

29. $s_1 = \frac{1}{10}$, $s_2 = \frac{1}{5}$, $s_3 = \frac{3}{10}$, $s_4 = \frac{2}{5}$, $s_5 = \frac{1}{2}$

    a. The sequence is arithmetic because all of its successive differences are equal.

    b. $s_2 - s_1 = \frac{1}{5} - \frac{1}{10} = \frac{2}{10} - \frac{1}{10} = \frac{1}{10}$;

        $s_3 - s_2 = \frac{3}{10} - \frac{1}{5} = \frac{3}{10} - \frac{2}{10} = \frac{1}{10}$;

        $s_4 - s_3 = \frac{2}{5} - \frac{3}{10} = \frac{4}{10} - \frac{3}{10} = \frac{1}{10}$;

        $s_5 - s_4 = \frac{1}{2} - \frac{2}{5} = \frac{5}{10} - \frac{4}{10} = \frac{1}{10}$

    We see that $d = \frac{1}{10}$.

    c. $s_6 = s_5 + d = \frac{1}{2} + \frac{1}{10} = \frac{5}{10} + \frac{1}{10} = \frac{6}{10} = \frac{3}{5}$

31. $s_1 = \frac{7}{12}$, $s_2 = \frac{2}{3}$, $s_3 = \frac{3}{4}$, $s_4 = \frac{5}{6}$

    a. The sequence is arithmetic because all of its successive differences are equal.

    b. $s_2 - s_1 = \frac{2}{3} - \frac{7}{12} = \frac{8}{12} - \frac{7}{12} = \frac{1}{12}$;

        $s_3 - s_2 = \frac{3}{4} - \frac{2}{3} = \frac{9}{12} - \frac{8}{12} = \frac{1}{12}$;

        $s_4 - s_3 = \frac{5}{6} - \frac{3}{4} = \frac{10}{12} - \frac{9}{12} = \frac{1}{12}$

    We see that $d = \frac{1}{12}$.

    c. $s_5 = s_4 + d = \frac{5}{6} + \frac{1}{12} = \frac{10}{12} + \frac{1}{12} = \frac{11}{12}$

Level 2, page 537

33. a. $s_n = -3 + 3n$ is given so

    $s_1 = -3 + 3(1) = 0$

    $s_2 = -3 + 3(2) = 3$

    $s_3 = -3 + 3(3) = 6$

Page 279

   b. The sequence 0, 3, 6 is arithmetic with common difference $d = 3$.

35. a. $s_n = 2 - n$ is given so
$$s_1 = 2 - 1 = 1$$
$$s_2 = 2 - 2 = 0$$
$$s_3 = 2 - 3 = -1$$

   b. The sequence 1, 0, -1 is arithmetic with common difference $d = -1$.

37. a. $s_n = 10 - 10n$ is given so
$$s_1 = 10 - 10(1) = 0$$
$$s_2 = 10 - 10(2) = -10$$
$$s_3 = 10 - 10(3) = -20$$

   b. The sequence 0, -10, -20 is arithmetic with common difference $d = -10$.

39. a. $s_n = 1 - \frac{1}{n}$ is given so
$$s_1 = 1 - \frac{1}{1} = 0$$
$$s_2 = 1 - \frac{1}{2} = \frac{1}{2}$$
$$s_3 = 1 - \frac{1}{3} = \frac{2}{3}$$

   b. The sequence 0, $\frac{1}{2}$, $\frac{2}{3}$ is not arithmetic (no common difference) nor is it geometric (no common ratio), so we would say it is neither.

41. a. $s_n = \frac{1}{2}n(n + 1)$ is given so
$$s_1 = \frac{1}{2}(1)(1 + 1) = 1$$
$$s_2 = \frac{1}{2}(2)(2 + 1) = 3$$
$$s_3 = \frac{1}{2}(3)(3 + 1) = 3(2) = 6$$

   b. The sequence 1, 3, 6 is not arithmetic (no common difference) nor is it geometric (no common ratio), so we would say it is neither.

43. a. $s_n = (-1)^n$ is given so
$$s_1 = (-1)^1 = -1$$
$$s_2 = (-1)^2 = 1$$
$$s_3 = (-1)^3 = -1$$

   b. The sequence -1, 1, -1 is geometric with common ratio $r = -1$.

45. a. $s_n = \frac{2}{3}$ is given so $s_1 = s_2 = s_3 = \frac{2}{3}$

   b. The sequence $\frac{2}{3}$, $\frac{2}{3}$, $\frac{2}{3}$ is arithmetic with common difference $d = 0$ and it is also geometric with common ratio $r = 1$.

47. a. $s_n = (-1)^n(n + 1)$ is given so
$$s_1 = (-1)^1(1 + 1) = -2$$
$$s_2 = (-1)^2(2 + 1) = 3$$
$$s_3 = (-1)^3(3 + 1) = -4$$

   b. The sequence -2, 3, -4 is not arithmetic (no common difference) nor is it geometric

Page 280

(no common ratio), so we would say it is neither.

49. Since $s_n = 7 - 3n$, we have $s_{69} = 7 - 3(69) = -200$

51. Since $s_n = (-1)^{n+1}5^{n+1}$, we have $s_3 = (-1)^{3+1}5^{3+1} = (-1)^4 5^4 = 625$

53. $s_1 = 3$ (given);

$s_2 = \frac{1}{3}s_{2-1} = \frac{1}{3}(3) = 1$

$s_3 = \frac{1}{3}s_{3-1} = \frac{1}{3}s_2 = \frac{1}{3}(1) = \frac{1}{3}$

$s_4 = \frac{1}{3}s_{4-1} = \frac{1}{3}s_3 = \frac{1}{3}\left(\frac{1}{3}\right) = \frac{1}{9}$

$s_5 = \frac{1}{3}s_{5-1} = \frac{1}{3}s_4 = \frac{1}{3}\left(\frac{1}{9}\right) = \frac{1}{27}$

55. $s_1 = 1$ (given)

$s_2 = 2$ (given);

$s_3 = s_{3-1} + s_{3-2} = s_2 + s_1 = 2 + 1 = 3$

$s_4 = s_{4-1} + s_{4-2} = s_3 + s_2 = 3 + 2 = 5$

$s_5 = s_{5-1} + s_{5-2} = s_4 + s_3 = 5 + 3 = 8$

Level 3 Problem Solving, page 537

57. $n = 1$: $\frac{1}{\sqrt{5}}\left(\frac{1+\sqrt{5}}{2}\right)^1 \approx 0.7236 \approx 1$; $a_1 = 1$

$n = 2$: $\frac{1}{\sqrt{5}}\left(\frac{1+\sqrt{5}}{2}\right)^2 \approx 1.1708 \approx 1$; $a_2 = 1$

$n = 3$: $\frac{1}{\sqrt{5}}\left(\frac{1+\sqrt{5}}{2}\right)^3 \approx 1.8944 \approx 2$; $a_3 = 2$

$n = 4$: $\frac{1}{\sqrt{5}}\left(\frac{1+\sqrt{5}}{2}\right)^4 \approx 3.0652 \approx 3$; $a_4 = 3$

$n = 5$: $\frac{1}{\sqrt{5}}\left(\frac{1+\sqrt{5}}{2}\right)^5 \approx 4.9597 \approx 5$; $a_1 = 5$

$n = 6$: $\frac{1}{\sqrt{5}}\left(\frac{1+\sqrt{5}}{2}\right)^6 \approx 8.02492 \approx 8$; $a_1 = 8$

$n = 7$: $\frac{1}{\sqrt{5}}\left(\frac{1+\sqrt{5}}{2}\right)^7 \approx 12.984597 \approx 13$; $a_1 = 13$

It is Fibonacci. You might wish to explore this interesting formula further. The pattern follows a form known as Binet's formula:

$$f_n = \frac{1}{\sqrt{5}}\left[\left(\frac{1+\sqrt{5}}{2}\right)^n - \left(\frac{1-\sqrt{5}}{2}\right)^n\right]$$

There is a connection between this formula, the Fibonacci sequence, and the golden ratio (discussed in Chapter 7).

59. Suppose x and y represent the two numbers selected. Then the 10 numbers are found:

Page 281

1st number　　　x

2nd number:　　y

3rd number:　　$x + y$

4th number:　　$x + 2y$

5th number:　　$2x + 3y$

6th number:　　$3x + 5y$

7th number:　　$5x + 8y$

8th number:　　$8x + 13y$

9th number:　　$13x + 21y$

10th number:　$21x + 34y$

SUM:　　　　　$55x + 88y = 11(5x + 8y)$

Notice that the sum is eleven times the 7th number. This shows that no matter what two numbers are chosen, the sum of the ten numbers is always eleven times the 7th number. Since the coefficients increase in accordance with the Fibonacci sequence, it makes sense to call this the "Fibonacci magic trick."

11.4 Series, page 538

New Terms Introduced in this Section

Alternating series	Arithmetic series	Evaluate a summation	Expand a summation
Finite series	Geometric series	Infinite series	Partial sum
Series	Sigma notation	Summation notation	

Level 1, page 546

1. A *sequence* is a list of numbers having a first term, a second term, and so on. A *series* is the indicated sum of the terms of a sequence.

3. A *partial sum* of an infinite series is $G_1 = g_1$, $G_2 = g_1 + g_2$; $G_3 = g_1 + g_2 + g_3$; $\cdots$

5. Arithmetic sequence; common difference $d = -3$, $s_1 = 12$, $s_5 = 0$; $S_5 = A_5 = 5\left(\frac{12+0}{2}\right) = 30$

7. $s_1 = 10$; this is a geometric sequence with $r = 2$, so $S_4 = G_4 = \frac{10(1 - 2^4)}{1 - 2} = 150$

9. $s_1 = -1$; this is a geometric sequence with $r = -1$, so $S_7 = G_7 = \frac{-1[1 - (-1)^7]}{1 - (-1)} = -1$

11. $\sum\limits_{k=3}^{5} k = 3 + 4 + 5 = 12$

13. $\sum\limits_{k=2}^{6} k^2 = 2^2 + 3^2 + 4^2 + 5^2 + 6^2 = 90$

15. $\displaystyle\sum_{k=1}^{10}[1^k+(-1)^k]=0+2+0+2+0+2+0+2+0+2$

$$=10$$

17. $\displaystyle\sum_{k=0}^{4}3(-2)^k=3(-2)^0+3(-2)^1+3(-2)^2+3(-2)^3+3(-2)^4$

$$=3[1-2+4-8+16]$$
$$=33$$

Level 2, page 547

19. Since this is geometric, we see $g_1=1$ and $r=\frac{1/2}{1}=\frac{1}{2}$.

$$G=\frac{g_1}{1-r}$$
$$=\frac{1}{1-\frac{1}{2}}$$
$$=2$$

21. Since this is geometric, we see $g_1=1$ and $r=\frac{1/3}{1}=\frac{1}{3}$.

$$G=\frac{g_1}{1-r}$$
$$=\frac{1}{1-\frac{1}{3}}$$
$$=\frac{3}{2}$$

23. Since this is geometric, we see $g_1=-20$ and $r=\frac{10}{-20}=-\frac{1}{2}$.

$$G=\frac{g_1}{1-r}$$
$$=\frac{-20}{1-(-\frac{1}{2})}$$
$$=\frac{-20}{\frac{3}{2}}$$
$$=-\frac{40}{3}$$

25. This is an arithmetic series with $d = 2$, $a_1 = 1$, $n = 5$, and $a_5 = 9$:

$$A_5 = n\left(\frac{a_1 + a_n}{2}\right)$$
$$= 5\left(\frac{1 + 9}{2}\right)$$
$$= 25$$

27. This is an arithmetic series with $d = 1$, $a_1 = 1$, $n = 5$, and $a_5 = 5$:

$$A_5 = n\left(\frac{a_1 + a_n}{2}\right)$$
$$= 5\left(\frac{1 + 5}{2}\right)$$
$$= 15$$

29. This is an arithmetic series with $d = 2$, $a_1 = 2$, and $n = 10$:

$$A_{10} = \frac{n}{2}[2a_1 + (n - 1)d]$$
$$= \frac{10}{2}[2(2) + (10 - 1)2]$$
$$= 5[4 + 18]$$
$$= 110$$

31. This is an arithmetic series with $d = 2$, $a_1 = 1$, and $n = 100$:

$$A_{100} = \frac{n}{2}[2a_1 + (n - 1)d]$$
$$= \frac{100}{2}[2(1) + (100 - 1)2]$$
$$= 50[2 + 198]$$
$$= 10{,}000$$

33. This is an arithmetic series with $d = 1$, $a_1 = 1$, and $n = 100$:

$$A_{100} = \frac{n}{2}[2a_1 + (n - 1)d]$$
$$= \frac{100}{2}[2(1) + (100 - 1)1]$$
$$= 50[2 + 99]$$
$$= 5{,}050$$

35. This is an arithmetic series with $d = 2$, $a_1 = 2$:

$$A_n = \frac{n}{2}[2a_1 + (n-1)d]$$
$$= \frac{n}{2}[2(2) + (n-1)2]$$
$$= \frac{n}{2}[4 + 2n - 2]$$
$$= \frac{n}{2}[2n + 2]$$
$$= n(n+1)$$

37. This is an arithmetic series with $d = 50$, $a_1 = 100$, and $n = 20$:

$$A_{20} = \frac{n}{2}[2a_1 + (n-1)d]$$
$$= \frac{20}{2}[2(100) + (20-1)50]$$
$$= 10[200 + 950]$$
$$= 11,500$$

39. This is an arithmetic series with $d = 2$, $a_1 = 42$ and $a_n = 98$, and we need to find n, so we use the formula for the nth term of an arithmetic sequence:

$$a_n = a_1 + (n-1)d$$
$$98 = 42 + (n-1)2$$
$$56 = (n-1)2$$
$$28 = n - 1$$
$$29 = n$$

Now that we know n, we can find the sum:

$$A_{29} = n\left(\frac{a_1 + a_n}{2}\right)$$
$$= 29\left(\frac{42 + 98}{2}\right)$$
$$= 29(70)$$
$$= 2,030$$

41. To "run the table" means to sink all the balls, which means we need to find the sum of a arithmetic sequence where $d = 1$, $n = 15$, $a_1 = 1$, and $a_{15} = 15$.

Page 285

$$A_{15} = n\left(\frac{a_1 + a_n}{2}\right)$$
$$= 15\left(\frac{1+15}{2}\right)$$
$$= 15(8)$$
$$= 120$$

43. We use an arithmetic series for both parts of this question.
Missy sinks the even-numbered balls, so her score is $2+4+6+8+10+12+14$, so
$a_1 = 2$, $a_n = 14$, and $n = 7$: $A_7 = n\left(\frac{a_1 + a_n}{2}\right)$
$$= 7\left(\frac{2+14}{2}\right)$$
$$= 7(8)$$
$$= 56$$
Shannon sinks the odd-numbered balls, so his score is $1+3+5+7+9+11+13+15$,
so $a_1 = 1$, $a_n = 15$, and $n = 8$: $A_8 = n\left(\frac{a_1 + a_n}{2}\right)$
$$= 8\left(\frac{1+15}{2}\right)$$
$$= 8(8)$$
$$= 64$$

45. The number of letters sent in each mailing is a finite geometric series with $r = 6$:
$$6 + 26 + 216 + \cdots = 6^1 + 6^2 + 6^3 + 6^4 + 6^5$$
We seek G_5 with $g_1 = 6, n = 5$, and $r = 6$:
$$G_5 = \frac{g_1(1 - r^n)}{1 - r}$$
$$= \frac{6(1 - 6^5)}{1 - 6}$$
$$= 9,330$$

47. Look at the figure to discern the pattern: $1 + 2 + 3 + \cdots + 87$, so $n = 87$, $a_1 = 1$,
$a_{87} = 87$, so

$$A_{87} = n\left(\frac{a_1 + a_n}{2}\right)$$

$$= 87\left(\frac{1 + 87}{2}\right)$$

$$= 3,828$$

The structure would need 3,828 blocks.

49. This is an infinite geometric series with $r = 0.90$ and $g_1 = 20$. $\quad G = \dfrac{20}{1 - 0.90} = 200$

The total distance the tip of the pendulum travels is 200 cm.

51. This is an infinite geometric series with $r = \frac{3}{4} = 0.75$ and $g_1 = 375$.

$$G = \frac{375}{1 - 0.75} = 1,500 \qquad \text{The flywheel will complete 1,500 revolutions.}$$

53. After the initial drop of 10 ft, notice that the ball travels a distance of 9 ft since it goes up and then down. This is true for each trip after the initial drop. Thus,

$$10 + 2 \cdot 9 + 2 \cdot 8.1 + \cdots = 10 + 2 \cdot 10(0.9) + 2 \cdot 10(0.9)^2 + \cdots$$

$$= 10 + 2 \cdot 10(0.9)[1 + 0.9 + 0.9^2 + \cdots]$$

$$= 10 + 18\left(\frac{1}{1 - 0.9}\right)$$

$$= 10 + 18(10)$$

$$= 190$$

The distance is 190 ft.

Level 3, page 548

55. Since the culture of 1 million bacteria increases by 100% every 24 hours means that the culture doubles in size every day. This means that on day 1 the original culture of 1 million bacteria has grown to 2 million bacteria, $g_1 = 2$. The sequence is 2, 4, 8, $\cdots$ so it is geometric with $r = 2$ and $d = 10$:

$$g_n = g_1 r^{n-1}$$

$$= 2 \cdot 2^{10-1}$$

$$= 2^{10}$$

$$= 1,024$$

Since the units are in millions, there are 1,024,000,000 bacteria present after 10 days.

57. In a two-team elimination tournament with 32 teams, there are 16 games in the first round, 8 in the next, and so on, so we consider $1 + 2 + 4 + 8 + 16$ as a geometric series with

$r = \frac{1}{2}$: $\quad G_6 = \dfrac{16(1 - 0.5^5)}{1 - 0.5} = 31$ games.

Page 287

59. For this tournament, $\frac{729}{3} = 243$ which is the number playing the first round. Since this is 3^5, we see that the sequence is $243 + 81 + 27 + 9 + 3 + 1$; $r = \frac{1}{3}$ and $n = 6$:

$$G_6 = \frac{243(1 - (\frac{1}{3})^6)}{1 - \frac{1}{3}} = 364 \text{ games.}$$

11.5 Annuities, page 548
New Terms Introduced in this Section

Annuity   Lump-sum problem   Monthly payment   Periodic payment problem
Sinking fund

Level 1, page 554

1. A *lump sum problem* is a financial problem which deals with a single amount of money.
3. An *annuity* is a financial problem which seeks the future value of periodic payments to an interest bearing account.
5. An annuity seeks the future value, given a monthly payment, whereas a sinking fund seeks the monthly payment, given the future value.

It would be very beneficial to have a programmable calculator for this chapter. For Problems 7-22, use the notation shown in the computational window on page 553 for an ordinary annuity. Enter m, r, t, and $n = 12$ (compounded monthly).

7. $m = \$50, r = 0.05, t = 3$; by calculator, $A = \$1,937.67$
9. $m = \$50, r = 0.08, t = 3$; by calculator, $A = \$2,026.78$
11. $m = \$50, r = 0.05, t = 30$; by calculator, $A = \$41,612.93$
13. $m = \$50, r = 0.08, t = 30$; by calculator, $A = \$74,517.97$
15. $m = \$100, r = 0.05, t = 10$; by calculator, $A = \$15,528.23$
17. $m = \$100, r = 0.08, t = 10$; by calculator, $A = \$18,294.60$
19. $m = \$150, r = 0.05, t = 35$; by calculator, $A = \$170,413.86$
21. $m = \$650, r = 0.05, t = 30$; by calculator, $A = \$540,968.11$

For Problems 23-34, use the notation shown in the computational window on page 553 for an ordinary annuity. Enter m, r, t, and n.

23. $m = \$500, n = 1; r = 0.08, t = 30$; by calculator, $A = \$56,641.61$
25. $m = \$250, n = 2; r = 0.08, t = 30$; by calculator, $A = \$59,497.67$
27. $m = \$300, n = 4; r = 0.06, t = 30$; by calculator, $A = \$99,386.46$
29. $m = \$200, n = 4; r = 0.08, t = 20$; by calculator, $A = \$38,754.39$
31. $m = \$30, n = 12; r = 0.08, t = 5$; by calculator, $A = \$2,204.31$
33. $m = \$2,500, n = 2; r = 0.085, t = 20$; by calculator, $A = \$252,057.07$

For Problems 35-46, use the notation shown in the computational window on page 553 for a sinking fund. Enter A, n, r, and t.

35. $A = \$7,000$, $n = 1$; $r = 0.08$, $t = 5$; by calculator, $m = \$1,193.20$

37. $A = \$25,000$, $n = 2$; $r = 0.12$, $t = 5$; by calculator, $m = \$1,896.70$

39. $A = \$165,000$, $n = 2$; $r = 0.02$; $t = 10$; by calculator, $m = \$7,493.53$

41. $A = \$500,000$, $n = 4$; $r = 0.08$, $t = 10$; by calculator, $m = \$8,277.87$

43. $A = \$100,000$, $n = 4$; $r = 0.08$, $t = 8$; by calculator, $m = \$2,261.06$

45. $A = \$45,000$, $n = 12$; $r = 0.07$, $t = 30$; by calculator, $m = \$36.89$

Level 2, page 554

47. annuity; $m = \$20,000$, $n = 1$, $t = 20$, $r = 0.08$; $A = \$915,239.29$

49. annuity; $m = \$1,000$, $n = 4$, $r = 0.08$, $t = 5.5$; $A = \$27,298.98$

51. sinking fund; $A = \$70,000$, $t = 5$, $n = 4$, $r = 0.08$; $m = \$2,880.97$

Level 3, page 555

53. annuity; $m = \$20,000$, $t = 29$, $n = 1$, $r = 0.05$; $A = \$1,246,454.24$

TOTAL = ANNUITY + FINAL PAYMENT = $\$1,666,454.24$

55. annuity; $m = \$6,885$, $t = 40$, $n = 1$, $r = 0.05$; $A = \$831,706$ (nearest dollar)

57. Answers depend on your age. Even though you will calculate only your age, it is interesting to see the accumulated value of using the amount of your social security deposit for your own benefit.

Here is a spreadsheet program for printing this table of retirement amounts:

	A	B	C	D	E
1	**Age**	**Rate**	**Period**	**Deposit**	**Retirement Amount**
2	t	r	n	6885	+D2*(((1+B2/C2)^(C2*(65-A2)-1)/(B2/C2))
3	Replicate Row 2 for Rows 3 to 44				
4					

Here is an output for ages 18-60.

	A	B	C	D	E
1	Age	Rate	Period	Deposit	Retirement Amount
2	18	0.05	1	$6,885.00	$1,226,352.22
3	19	0.05	1	$6,885.00	$1,161,397.35
4	20	0.05	1	$6,885.00	$1,099,535.57
5	21	0.05	1	$6,885.00	$1,040,619.59
6	22	0.05	1	$6,885.00	$984,509.14
7	23	0.05	1	$6,885.00	$931,070.61
8	24	0.05	1	$6,885.00	$880,176.77
9	25	0.05	1	$6,885.00	$831,706.45
10	26	0.05	1	$6,885.00	$785,544.23
11	27	0.05	1	$6,885.00	$741,580.22
12	28	0.05	1	$6,885.00	$699,709.74
13	29	0.05	1	$6,885.00	$659,833.08
14	30	0.05	1	$6,885.00	$621,855.32
15	31	0.05	1	$6,885.00	$585,686.02
16	32	0.05	1	$6,885.00	$551,239.06
17	33	0.05	1	$6,885.00	$518,432.44
18	34	0.05	1	$6,885.00	$487,188.04
19	35	0.05	1	$6,885.00	$457,431.47
20	36	0.05	1	$6,885.00	$429,091.87
21	37	0.05	1	$6,885.00	$402,101.78
22	38	0.05	1	$6,885.00	$376,396.94
23	39	0.05	1	$6,885.00	$351,916.13
24	40	0.05	1	$6,885.00	$328,601.08
25	41	0.05	1	$6,885.00	$306,396.26
26	42	0.05	1	$6,885.00	$285,248.82
27	43	0.05	1	$6,885.00	$265,108.40
28	44	0.05	1	$6,885.00	$245,927.05
29	45	0.05	1	$6,885.00	$227,659.09
30	46	0.05	1	$6,885.00	$210,261.04
31	47	0.05	1	$6,885.00	$193,691.47
32	48	0.05	1	$6,885.00	$177,910.92
33	49	0.05	1	$6,885.00	$162,881.83
34	50	0.05	1	$6,885.00	$148,568.41
35	51	0.05	1	$6,885.00	$134,936.58
36	52	0.05	1	$6,885.00	$121,953.89
37	53	0.05	1	$6,885.00	$109,589.42
38	54	0.05	1	$6,885.00	$97,813.73
39	55	0.05	1	$6,885.00	$86,598.79
40	56	0.05	1	$6,885.00	$75,917.90
41	57	0.05	1	$6,885.00	$65,745.61
42	58	0.05	1	$6,885.00	$56,057.73
43	59	0.05	1	$6,885.00	$46,831.17
44	60	0.05	1	$6,885.00	$38,043.97

Level 3 Problem Solving, page 555

59. sinking fund (principal); $A = \$4,000,000$, $t = 30$, $n = 1$, $r = 0.08$; $m = \$35,309.73$;

interest: $I = Prt$

$\qquad\quad = \$4,000,000(0.055)(1)$

$\qquad\quad = \$220,000$

TOTAL = PRINCIPAL + INTEREST = $\$255,309.73$

11.6 Amortization, page 555

New Terms Introduced in this Section

Amortization   Present value of an annuity

Level 1, page 560

1. *Amortization* is the process of paying off a debt by systematically making partial payments until the debt and accrued interest are repaid.

3. *m* is the amount of a periodic payment (usually a monthly payment);

n is the number of payments made each year;

t is the number of years;

r is the annual interest rate;

A is the future value; and

P is the present value.

For Problems 4-11 you will use the present value of an annuity formula or you will program a calculator using the computational window shown on page 556. If you are using a calculator program, input the values for the given variables and then the output should be rounded to the nearest cent.

5. $m = 50$, $r = 0.06$, $t = 5$; $P = \$2,586.28$

7. $m = 150$, $r = 0.05$, $t = 30$; $P = \$27,942.24$

9. $m = 150$, $r = 0.08$, $t = 30$; $P = \$20,442.52$

11. $m = 1,050$, $r = 0.06$, $t = 30$; $P = \$175,131.20$

For Problems 12-19 you will use the amortization formula or you will program a calculator using the computational window shown on page 558. If you are using a calculator program, input the values for the given variables and then the output should be rounded to the nearest cent.

13. $P = 14,000$, $r = 0.10$, $t = 5$; $m = \$297.46$

15. $P = 150,000$, $r = 0.08$, $t = 30$; $m = \$1,100.65$

17. $P = 150,000$, $r = 0.10$, $t = 30$; $m = \$1,316.36$

19. $P = 260,000$, $r = 0.09$, $t = 30$; $m = \$2,092.02$

For Problems 20-31 you will use the present value of an annuity formula or you will program a calculator using the computational window shown on page 556. If you are using a calculator program, input the values for the given variables and then the output should be rounded to the nearest cent.

21. $m = 500, n = 1, r = 0.06, t = 30; P = \$6,882.42$
23. $m = 600, n = 2, r = 0.02, t = 10; P = \$10,827.33$
25. $m = 100, n = 12, r = 0.04, t = 5; P = \$5,429.91$
27. $m = 400, n = 4, r = 0.11, t = 20; P = \$12,885.18$
29. $m = 75, n = 12, r = 0.04, t = 10; P = \$7,407.76$
31. $m = 100, n = 4, r = 0.03, t = 20; P = \$5,999.44$

Level 2, page 561

33. amortization; $P = 100, n = 12, t = 1.5, r = 0.18; m = \$6.38$
35. amortization; $P = 3,520, n = 12, t = \frac{30}{12}, r = 0.19; m = \$148.31$
37. amortization; $P = 12,450, n = 12, t = \frac{30}{12}, r = 0.029, m = \$430.73$
39. amortization; $P = 985, n = 12, t = \frac{15}{12}, r = 0.17; m = \$73.35$
41. amortization; $P = 108,000(0.70) = 75,600, n = 12, t = 30, r = 0.1205; m = \$780.54$
43. amortization; $P = 859,000, n = 12, t = 20, r = 0.132; m = \$10,186.47$
45. $P = \$108,000(0.70) = \$75,600$

    Total interest (monthly payment from Problem 41) is

$$I = A - P$$
$$= \$780.54(12)(30) - \$75,600$$
$$= \$205,394.40$$

    Monthly payment for $t = 15, n = 12, r = 0.1205$ is $m = \$909.76$; Total interest is

$$I = A - P$$
$$= \$909.76(12)(15) - \$75,600$$
$$= \$88,156.80$$

    Savings: $117,238 (to the nearest dollar).

47. Present value of an annuity; $m = 50,000, n = 1, t = 20, r = 0.1225; P = \$367,695.71$

Level 3, page 561

49. Present value of an annuity; $m = 250, n = 12, t = 20, r = 0.10; P = \$25,906.15$.
    Since this is greater than the lump-sum payment, the annuity is the better choice.
51. Present value of an annuity ($m = \$1,000,000, r = 0.05, n = 1$, and $t = 20$) is $12,462,210.34.
    This would be a fair price to receive for the $20,000,000 lottery prize.

53. First subtract your monthly obligations from your monthly income:
$$\$5,500 \ - \ \$625 = \$4,875$$
Next, the amount you can afford to pay is no more than 36%, so we calculate
$$m = 0.36(\$4,875) = \$1,755$$
Finally, calculate the present value of an annuity using $m = 1,755$, $n = 12$, $t = 30$, and $r = 0.0965$; $P = \$206,029.43$

Level 3 Problem Solving, page 562

55. a. If we think about the meaning of the variables, it makes sense that as the interest rate increases that the value of the money would be greater over time. Since a sinking fund calculates the monthly payment to obtain a *fixed* amount at some time, it is reasonable that the monthly payment would be less (because the return on the money is greater). Another way to reason, is to consider the sinking fund formula:

$$m = \frac{A\left(\frac{r}{n}\right)}{\left(1+\frac{r}{n}\right)^{nt} - 1}$$

We see that the numerator will increase faster than the numerator as r increases, so m decreases.

b. If we think about the meaning of the variables, it makes sense that as the interest rate increases that the future value of the money would be greater over time. This implies that the monthly value for amortization would be greater. Another way to reason is to consider the amortization formula:

$$m = \frac{P\left(\frac{r}{n}\right)}{1 - \left(1+\frac{r}{n}\right)^{-nt}}$$

We see that the denominator has a negative exponent so it becomes smaller as r increases, and so also does the numerator becomes larger, we conclude that m would also increase as r increases.

c. If we think about the meaning of the variables, it makes sense that as the interest rate increases that the future value of the money would be greater over time. This implies that the value of an annuity would also increase over time since it is the future value of periodic payments to an interest-bearing account. Finally, the present value of an annuity would need to be less since the increases (due to higher interest rates) would be greater. Another way to reason is to consider the present value of an annuity formula

Page 293

$$P = m \left[\frac{1 - \left(1 + \frac{r}{n}\right)^{-nt}}{\frac{r}{n}} \right]$$

We see that the denominator increases as r increases, so the numerator will decrease, and consequently P will decrease.

57. First, we need to calculate the amount of the loan (which we call P):

$$P = \$225,000(0.75) = \$168,750$$

Now, we use the amortization formula for $P = 168,750$, $t = 30$, $n = 12$, $r = 0.1024$ to calculate $m = \$1,510.92$. Total monthly payments are

$$\$1,510.92(12)(30) = \$543,931.20$$

For $t = 15$, the amortization gives $m = \$1,838.25$ for total payments of

$$\$1,838.25(12)(15) = \$330,885.00$$

Savings:

$$\$543,931.20 - \$330,885.00 = \$213,046.20$$

59. First we calculate the present value of an annuity for a monthly payment of $1,500: $m = 1,500$, $r = 0.11$, $t = 20$, $n = 12$; by calculator we find: $P = \$145,322.31$. This is the amount of the loan, so if we add the down payment, we obtain a price of

$$\$145,322.31 + \$30,000 = \$175,322.31$$

Next, we calculate the present value of an annuity for a monthly payment of $1,800: $m = 1,800$, $r = 0.11$, $t = 20$, $n = 12$; by calculator we find: $P = \$174,386.77$. This is the amount of the loan, so if we add the down payment, we obtain a price of

$$\$174,386.77 + \$30,000 = \$204,386.77$$

The price range is $175,322.31 to $204,386.77.

11.7 Summary of Financial Formulas, page 562

Level 1, page 565

1. Typical is 20%-30%, although many first buyer's programs offer 5% down. Cite your sources.

3. A good procedure is to ask a series of questions. **Is it a lump sum problem?** If it is, then what is the unknown? If FUTURE VALUE is the unknown, then it is a *future value* problem. If PRESENT VALUE is the unknown, then it is a *present value* problem. **Is it a periodic payment problem?** If it is, then is the periodic payment known? If the PERIODIC PAYMENT IS KNOWN and you want to find the future value then it is an *ordinary annuity* problem. If the PERIODIC PAYMENT IS KNOWN and you want to find the present value then it is a *present value of an annuity* problem. If the PERIODIC PAYMENT IS UNKNOWN and you know the future value then it is a *sinking fund* problem. If the PERIODIC PAYMENT IS UNKNOWN and you know the present value then it is an *amortization* problem.

5. $m = \dfrac{P\left(\frac{r}{n}\right)}{1 - \left(1 + \frac{r}{n}\right)^{-nt}}$; the unknown is the periodic payment

7. Since P is unknown, that is the value that is sought, so since P represents present value, this is a present value problem.

9. We are given that it is a periodic payment problem, we also know the periodic payment and seek the future value. This is the definition of an annuity problem.

11. We are given that it is a periodic payment problem, we know the present value and seek the periodic payment. This is the definition of an amortization problem.

13. The formula $P = m \left[\dfrac{1 - \left(1 + \frac{r}{n}\right)^{-nt}}{\frac{n}{r}} \right]$ is seeking the present value where the periodic payment is known, so this is a present value of an annuity problem.

15. The formula $m = \dfrac{P\left(\frac{r}{n}\right)}{1 - \left(1 + \frac{r}{n}\right)^{-nt}}$ is seeking the periodic payment where the present value is known, so this is an amortization problem.

17. This is a periodic payment problem in which we deposit a known amount to an interest-bearing account, so we recognize this an annuity problem. The given variables are: $m = \$300$, $n = 1$, $t = 10$, $r = 0.12$; use the annuity formula or use a calculator to find $A = \$5,264.62$.

19. This is a periodic payment problem in which seek to know the amount of a periodic deposit to an interest-bearing account in which the future value is known, so we recognize this as a sinking fund problem. The given variables are: $A = \$10,000$; $n = 1$, $t = 5$, $r = 0.12$; use the sinking fund formula or use a calculator to find $m = \$1,574.10$.

Level 2, page 565

21. **a**. annuity

   b. $m = \$300$, $r = 0.12$, $n = 1$, $t = 10$; $A = \$5,264.62$

Page 295

23. **a.** sinking fund
    b. $A = \$10,000, t = 5, r = 0.09, n = 1; m = \$1,670.92$
25. **a.** future value
    b. $r = 0.0625, P = \$20,000, t = 5, n = 1; A = \$27,081.62$
27. **a.** present value of an annuity
    b. $t = 33, n = 1, m = \$500, r = 0.08; P = \$5,756.94$
29. **a.** amortization
    b. $n = 12, P = \$125,000(0.80) = \$100,000, t = 30, r = 0.12; m = \$1,028.61$
31. **a.** annuity
    b. $m = \$730, n = 1, r = 0.10, t = 15; A = \$23,193.91$
33. **a.** present value
    b. $A = \$1,000,000, r = 0.09, n = 12, t = 55; P = \$7,215.46$
35. **a.** annuity
    b. $t = 5, m = \$900, n = 4, r = 0.08; A = \$21,867.63$
37. **a.** present value
    b. $t = 5, A = \$300,000, r = 0.12, n = 12; P = \$165,134.88$
39. **a.** amortization
    b. $P = \$45,000, r = 0.12, n = 12, t = \frac{7}{2}; m = \$1,317.40$
41. **a.** annuity
    b. $m = \$1,000, n = 12, r = 0.18, t = \frac{25}{3}; A = \$228,803.04$
43. **a.** future value
    b. $P = \$800, t = 1, r = 0.10, n = 360; A = \$884.12$
45. **a.** future value
    b. $P = \$5,000, n = 1, r = 0.055, t = 12; A = \$9,506.04$
47. **a.** present value
    b. $A = \$5,000, t = 3, n = 1, r = 0.11; P = \$3,655.96$
49. **a.** present value
    b. $t = 3, A = \$200,000, r = 0.10, n = 12; P = \$148,348$
51. **a.** present value
    b. $t = \frac{18}{12}, A = \$560,000, r = 0.075, n = 360; P = \$500,420$

Level 3, page 566
53. amortization; $t = 30, P = \$185,500, r = 0.0775, n = 12; m = \$1,328.94$
55. A 20-yr loan increases the monthly payment to $1,522.86.
    This loan has total payments of $A = (\$1,522.86)(240) = \$365,486.40$.

The total interest is $I = A - P$
$$= \$365{,}486.40 - \$185{,}500$$
$$= \$179{,}986.40$$

The interest savings is the difference between the total payments (see Problem 53) @ 30 years less the total payments @ 20 years:

$$\$1{,}328.94(360) - (\$1{,}522.86)(240) = \$112{,}932$$

57. The total amount paid is found by using the monthly payment (see Problem 56)

$$\$3{,}180.90(12)(30) = \$1{,}145{,}124$$

so the amount of interest is

$$I = A - P = \$1{,}145{,}124 - \$418{,}500 = \$726{,}624$$

Level 3 Problem Solving, page 566

59. We can compare present values or future values; we choose to compare present values. We find the present value given a future value, $A = \$45{,}000$. Also, $t = 1$, $r = 0.10$, and $n = 12$. Using the present value formula or a calculator, we find $P = \$40{,}734.56$. Now, add the present payment of \$10,000 to obtain \$50,734.56. This is better than the \$50,000 one-time payment. *Note*: if you compare future values the same choice is better and the future values are \$56,047.13 vs \$55,235.65.

Chapter 11 Review Questions, page 567

STOP

Studying for a chapter examination is a personal process, one which nobody else can do for you. Simply take the time to review what you have done. Here are the new terms in Chapter 11.

Add-on interest [11.2]	Average daily balance	Daily compounding [11.1]
Adjusted balance method [11.2]	method [11.2]	Dealer's cost [11.2]
Alternating series [11.4]	Closed-end loan [11.2]	e [11.1]
Amortization [11.6]	Common difference [11.3]	Euler's number [11.1]
Annual compounding [11.1]	Common ratio [11.3]	Evaluate a summation [11.4]
Annual percentage rate [11.2]	Compound interest [11.1]	Exact interest [11.1]
Annuity [11.5]	Compound interest	Expand a summation [11.4]
APR [11.2]	formula [11.1]	Fibonacci sequence [11.3]
Arithmetic sequence [11.3]	Continuous compounding [11.1]	Fibonacci-type sequence [11.3]
Arithmetic series [11.4]	Credit card [11.2]	Finance charge [11.2]

Finite series [11.4]	Lump-sum problem [11.5]	Quarterly compounding [11.1]
Five-percent offer [11.2]	Monthly compounding [11.1]	Revolving credit [11.2]
Future value [11.1]	Monthly payment [11.5]	Semiannual compounding [11.1]
Future value formula [11.1]	Natural base [11.1]	Sequence [11.3]
Geometric sequence [11.3]	Open-end loan [11.2]	Series [11.4]
Geometric series [11.4]	Ordinary interest [11.1]	Sigma notation [11.4]
Grace period [11.2]	Partial sum [11.4]	Simple interest formula [11.1]
Infinite series [11.4]	Periodic payment problem [11.5]	Sinking fund [11.5]
Inflation [11.1]	Present value [11.1]	Sticker price [11.2]
Installment loan [11.2]	Present value formula [11.1]	Summation notation [11.4]
Interest [11.1]	Present value of an annuity [11.6]	Term of a sequence [11.3]
Interest rate [11.1]	Previous balance method [11.2]	Time [11.1]
Line of credit [11.2]	Principal [11.1]	

If you can describe the term, read on to the next one; if you cannot, then look it up in the text (the section number is shown in brackets). Next, study the types of problems listed at the end of Chapter 11.

TYPES OF PROBLEMS
Be able to estimate reasonable answers to financial problems. [11.1-11.7]
Find the amount of interest if you are given the purchase price, length of loan, and the monthly
    payment. [11.1]
Find the total amount to be repaid for a simple interest loan. [11.1]
Calculate the future value. [11.1]
Calculate the present value. [11.1]
Compare the amount of interest and future value for simple and compound interest. [11.1]
Be able to calculate the future value due to inflation. [11.1]
Be able to calculate the present value due to inflation. [11.1]
Calculate the time necessary to achieve an investment goal. [11.1]
How much must be deposited into a bank account to provide a given monthly income. [11.1]

Be able to calculate the monthly payment using add-on interest. [11.2]
Make an appropriate offer on a new car. [11.2]
Calculate the APR for installment loans or for credit cards. [11.2]
Be able to calculate the amount of credit card interest by using the previous balance method,
    the adjusted balance method, and the average daily balance method; which is better from
    the consumer point of view? [11.2]

Find the amount of interest, the monthly payment, and the APR for consumer transactions. [11.2]

Determine the monthly payment for an automobile using add-on interest. [11.2]

Find the APR if you are given the price, length of time, and the add-on interest rate. [11.2]

Calculate the APR given the amount financed, the monthly payment, the length of time financed. [11.2]

Classify a given sequence as arithmetic, geometric, or Fibonacci; be able to find the next term. [11.3]

Write out the terms of a sequence when given the general term; find a specific term. [11.3]

Evaluate (expand) summation expressions. [11.4]

Find sums by classifying and using the series formulas. [11.4]

Distinguish between sequences and series in application problems. [11.4]

Calculate the amount you can save by making a periodic payment into an interest-bearing account (annuity). [11.5]

Calculate the amount you need to deposit into an interest-bearing account to have a given amount at some time in the future (sinking fund). [11.5]

Find the present value of an annuity. [11.6]

Calculate the amount of money you can borrow, given a monthly payment, interest rate, and length of time (present value of an annuity). [11.6]

Find the monthly payment for an amortized loan. [11.6]

Calculate the down payment for a home, given the price; given the amount of the down payment, calculate the price. [11.6]

Given a monthly payment, interest rate, and length of time, find the amount of loan. [11.6]

Work applied financial problems. [11.1-11.7]

Once again, see if you can verbalize (to yourself) how to do each of the listed types of problems.

   Work all of Chapter 11 Review Questions (whether they are assigned or not). Work through all of the problems before looking at the answers, and *then* correct each of the problems. The entire solution is shown in the answer section at the back of the text. If you worked the problem correctly, move on to the next problem, but if you did not work it correctly (or you did not know what to do), look back in the chapter to study the procedure, or ask your instructor.

   Finally, go back over the homework problems you have been assigned. If you worked a problem correctly, move on to the next problem, but if you missed it on your homework, then you should look back in the book or talk to your instructor about how to work the problem.

Page 299

If you follow these steps, you should be successful with your review of this chapter.

We give all of the answers to the chapter review questions (not just the odd-numbered questions).

Chapter 11 Review Questions, page 567

1. A sequence is a list of numbers having a first term, a second term, and so on; a series is the indicated sum of the terms of a sequence. An arithmetic sequence is one that has a common difference $[a_n = a_1 + (n - 1)d]$, a geometric sequence is one that has a common ratio $[r_n = g_1 r^{n-1}]$, and a Fibonacci-type sequence is one that, given the first two terms, the next is found by adding the previous two terms $[s_n = s_{n-1} + s_{n-2}]$. The sum of an arithmetic sequence is $A_n = n\left(\frac{a_1 + a_n}{2}\right)$ or $A_n = \frac{n}{2}[2a_1 + (n - 1)d]$, and the sum of a geometric sequence is $G_n = \frac{g_1(1-r^n)}{1-r}$.

2. A good procedure is to ask a series of questions.
   Is it a lump-sum problem?
   If it is, then what is the unknown? If FUTURE VALUE is the unknown, then it is a *future value* problem. If PRESENT VALUE is the unknown, then it is a *present value* problem.
   Is it a periodic payment problem?
   If it is, then is the periodic payment known?
   If the PERIODIC PAYMENT IS KNOWN and you want to find the future value, then it is an *ordinary annuity* problem.
   If the PERIODIC PAYMENT IS KNOWN and you want to find the present value, then it is a *present value of an annuity* problem.
   If the PERIODIC PAYMENT IS UNKNOWN and you know the future value, then it is a *sinking fund* problem.
   If the PERIODIC PAYMENT IS UNKNOWN and you know the present value, then it is an *amortization* problem.

3. **a.** There is a common difference:
   $10 - 5 = 5; 15 - 10 = 5; 20 - 15 = 5; d = 5$, so it is arithmetic.
   $a_n = 5n$
   b. There is a common ratio:
   $\frac{10}{5} = 2; \frac{20}{10} = 2; \frac{40}{20} = 2; r = 2$, so it is geometric.
   $g_n = 5 \cdot 2^{n-1}$
   c. There is no common difference or common ratio. Notice
   $5 + 10 = 15; 10 + 15 = 25$; it is Fibonacci-type.
   $s_1 = 5, s_2 = 10, s_n = s_{n-1} + s_{n-2}, n \geq 3$
   d. There is no common difference or common ratio. It is also not Fibonacci-type, so it is none of the above. There is a pattern, however: add $5, 10, 15, 20, \cdots$, so the next two

Page 300

terms are $35 + 20 = 55; 55 + 25 = 80$.

e. There is a common difference:

$\frac{50}{5} = 10; \frac{500}{50} = 10; \frac{5,000}{500} = 10; r = 10$, so it is geometric.

$g_n = 5 \cdot 10^{n-1}$

f. There is no common difference or common ratio. It is also not Fibonacci-type, so it is none of the above. The is a pattern, however: the terms 5 and 50 alternate. The next two terms are 5, 50.

4. a. $\sum\limits_{k=1}^{3} (k^2 - 2k + 1) = (1^2 - 2(1) + 1) + (2^2 - 2(2) + 1) + (3^2 - 2(3) + 1) = 5$

b. $\sum\limits_{k=1}^{4} \frac{k-1}{k+1} = \frac{0}{2} + \frac{1}{3} + \frac{2}{4} + \frac{3}{5} = \frac{43}{30}$ or $1.4\overline{3}$

5. parents, grandparents, great-grandparents, $\cdots$; that is, 2, 4, 8, 16, $\cdots$. In this problem, we are looking for the sum of the first 10 terms of this geometric sequence. That is, find G_{10} where $g_1 = 2$ and $r = 2$; $G_{10} = \dfrac{2(1 - 2^{10})}{1 - 2} = 2{,}046$. Thus, there are a minimum of 2,046 people. If we include the initial person, as we should, then we see there are 2,047 people on the family tree.

6. There will be 72 divisions in 24 hours. In this problem, we are looking for the 72nd term. Since g_1 is at time zero, we are looking for g_{73} where $g_1 = 1{,}024 = 2^{10}$ and $r = 2$.

$g_{73} = g_1 g^{n-1} = 2^{10} \cdot 2^{72} = 2^{82}$

7. Calculate 5% over the original price: $18{,}579(1 + 0.05) = \$19{,}507.95$.

You should offer \$19,500 for the car.

8. a. $I = Prt$ $\qquad\qquad$ Next, $A = P + I$

$ = 13{,}500(0.029)(2)$ $\qquad\quad = 13{,}500 + 783$

$ = 783.00$ $\qquad\qquad\qquad = 14{,}283$

The monthly payment is $\$14{,}283 \div 24 = \$595.13$.

The total interest is \$783, and the monthly payment is \$595.13.

b. $APR = \dfrac{2Nr}{N+1}$ $\qquad$ The APR is 5.568%.

$ = \dfrac{2(24)(0.029)}{25}$

$ \approx 0.5568$

9. $A = 48(353.04) = 16{,}945.92;$ $\quad I = A - P$

$ = 16{,}945.92 - 11{,}450.00$

$ = 5{,}495.92$

Page 301

Also, $I = Prt$ *Simple interest formula*

$5,495.92 = 11,450r(4)$ *48 months is 4 years, so t = 4.*

$1,373.98 = 11,450r$ *Divide both sides by 4.*

$0.12 \approx r$ *Divide both sides by 11,450.*

Finally, $APR = \dfrac{2Nr}{N+1}$ The APR is about 23.5%.

$= \dfrac{2(48)r}{49}$

≈ 0.235

10. The adjusted daily balance method is most advantageous to the consumer.

Previous Balance	Adjusted Balance	Average Daily Balance
$I = Prt$	$I = Prt$	$I = Prt$
$= 525(0.09)\left(\frac{1}{12}\right)$	$= (525 - 100)(0.09)\left(\frac{1}{12}\right)$	$= \frac{525\times7+425\times24}{31}(0.09)\left(\frac{31}{365}\right)$
$= 3.9375$	$= 3.1875$	$= 3.421232877$
Finance charge: $3.94.	Finance charge is $3.19.	Finance charge is $3.42.

11. $A = \$1,000,000$; $t = 50$; $r = 0.09$; $n = 12$; Thus,

$$P = A\left(1 + \frac{r}{n}\right)^{-nt} = 1,000,000\left(1 + \frac{0.09}{12}\right)^{-(50)12} \approx 11,297.10$$

Deposit $11,297.10 to have a million dollars in 50 years.

12. a. Amortization; where $P = 154,000$, $n = 12$, $r = 0.08$, and $t = 20$. Thus,

$$m = \frac{P\left(\frac{n}{r}\right)}{1 - \left(1 + \frac{r}{n}\right)^{-nt}} = \frac{154,000\left(\frac{0.08}{12}\right)}{1 - \left(1 + \frac{0.08}{12}\right)^{-(12)(20)}} \approx 1,288.117706 \quad \text{(by calculator)}$$

The monthly payments are $1,288.12.

b. 80%(PRICE OF HOME) = $154,000

PRICE OF HOME = $192,500

c. Future value (continuous compounding):

$A = Pe^{rt}$

$= \$192,500e^{0.04(30)}$

$= \$639,122.51$

13. Present value; where $A = \$100,000$; $r = 0.064$; $t = 3\frac{4}{12} = \frac{40}{12}$; $n = 12$; thus,

$$P = A\left(1 + \frac{r}{n}\right)^{-nt} = \$100,000\left(1 + \frac{0.064}{12}\right)^{-(12)(40/12)} \approx \$80,834.49$$

14. We want to find the present value where the future value, $A = \$420,000$, is given. We also know $t = 30$, $n = 1$, and $r = 0.05$. Use the present value formula or a calculator to find $P = \$97,178.53$.

15. We want to find the value of an annuity when we know the periodic payment is $m = \$20,000$, the time is $t = 29$, the period is $n = 1$, and the rate is $r = 0.05$. Use the annuity formula or a calculator to find $A = \$1,246,454.24$.

16. We want to find the present value of an annuity when we know that $m = \$20,000$, $t = 29$, $n = 1$, and $r = 0.05$. Use the present value of an annuity formula or a calculator to find $P = \$302,821.47$.

17. We need to find the present value of a future value of $A = \$420,000$ in time $t = 30$ years where $n = 1$ and $r = 0.05$. Use the present value formula or a calculator to find $P = \$97,178.53$; You need to set aside $\$97,178.53$.

18. $I = Prt$
$= \$200,000(0.05)(1)$
$= \$10,000$

19. You can compare present values of each or the future values of each. It is easier to compare present values.
Cash value: $300,000   Annuity: $302,821.47 (see Problem 16)
                            Present value of cash: $97,178.53
                            Total: $400,000
Since $400,000 is worth more than $300,000, take the installments.

20. Compare present values.
Cash value: $300,000   Present value of annuity;
                            $m = \$20,000$, $t = 29$, $n = 1$, $r = 0.10$; $P = \$187,392.12$
                            Present value of cash:
                            $A = \$420,000$, $t = 30$, $n = 1$, $r = 0.10$; $P = \$24,069.59$
                            Total: $211,461.71
Since $300,000 is worth more than $211,461.71, take the one-time payment.

CHAPTER 12 The Nature of Counting

Chapter Overview
This chapter sets the stage for probability because it develops the ideas of permutation and combination, as well as the ability to distinguish one from the other. If you are pressed for time, you could combine this chapter with the probability chapter into one unit.

12.1 Permutations, page 573

New Terms Introduced in this Section

Arrangement	Combination	Count-down property
Distinguishable permutation	Factorial	Fundamental counting principle
Multiplication property of factorials	Ordered pair	Ordered triple
Permutation	Tree diagram	

Level 1, page 580

1. A permutation of r elements selected from a set S with n elements is an ordered arrangement of those r elements selected without repetitions. The formula is $_nP_r = \frac{n!}{(n-r)!}$.

3. $_9P_1 = 9$

5. $_9P_2 = 9 \cdot 8 = 72$

7. $_9P_4 = 9 \cdot 8 \cdot 7 \cdot 6 = 3,024$

9. $_{52}P_3 = 52 \cdot 51 \cdot 50 = 132,600$

11. $_{12}P_5 = 12 \cdot 11 \cdot 10 \cdot 9 \cdot 8 = 95,040$

13. $_8P_4 = 8 \cdot 7 \cdot 6 \cdot 5 = 1,680$

15. $_{92}P_0 = 1$

17. $_7P_5 = 7 \cdot 6 \cdot 5 \cdot 4 \cdot 3 = 2,520$

19. $_7P_3 = 7 \cdot 6 \cdot 5 = 210$

21. $_{50}P_4 = 50 \cdot 49 \cdot 48 \cdot 47 = 5,527,200$

23. $_8P_3 = 8 \cdot 7 \cdot 6 = 336$

25. $_{11}P_4 = 11 \cdot 10 \cdot 9 \cdot 8 = 7,920$

27. $_gP_h = \frac{g!}{(g-h)!}$

29. $_5P_r = \frac{5!}{(5-r)!}$

31. This is a distinguishable permutation with 7 letters, each repeated once.
$$\left(\begin{matrix} 7 \\ 1, 1, 1, 1, 1, 1, 1 \end{matrix} \right) = 7! = 5,040$$

33. This is a distinguishable permutation with 6 letters, the letter "E" twice and each of the other letters repeated once.

$$\left(\begin{smallmatrix} & 6 & \\ 2,\, 1,\, 1,\, 1,\, 1 \end{smallmatrix} \right) = 360$$

35. This is a distinguishable permutation with 11 letters, the letter "M" appears once, the letters "I" and "S" each appear four times, with the letter "P" appearing twice.

$$\left(\begin{smallmatrix} & 11 & \\ 1,\, 4,\, 4,\, 2 \end{smallmatrix} \right) = 34{,}650$$

37. This is a distinguishable permutation with 11 letters, the letter "B" appears once, the letters "O," "K," and "E" each appear twice, with the other letters appearing once.

$$\left(\begin{smallmatrix} & 11 & \\ 1,\, 2,\, 2,\, 2,\, 1,\, 1,\, 1,\, 1 \end{smallmatrix} \right) = 4{,}989{,}600$$

39. This is a distinguishable permutation with 11 letters, the letter "A" appears once, the letters "P" and "O" each appear twice, the letter "S" appears three times, the letter "I" appears twice, and finally the letter "E" appears once.

$$\left(\begin{smallmatrix} & 11 & \\ 1,\, 2,\, 2,\, 3,\, 2,\, 1 \end{smallmatrix} \right) = 831{,}600$$

Level 2, page 580

41. This is an example of the election problem in which four officers are chosen from a set of five people; $_5P_4 = 120$.

43. This is an example of the election problem in which three officers are chosen from a set of ten people; $_{10}P_3 = 720$.

45. Use the fundamental counting principle: $2 \cdot 8 = 16$.

47. This is a permutation of three objects selected from a set of eight: $_8P_3 = 336$.

49. Use the fundamental counting principle: $8 \cdot 8 \cdot 10 \cdot 10 \cdot 10 \cdot 10 \cdot 10 = 6{,}400{,}000$.

51. Use the fundamental counting principle: $10 \cdot 10 \cdot 10 \cdot 10 = 10{,}000$.

53. This is a permutation of eight objects selected from a set of eight: $_8P_8 = 8! = 40{,}320$.

55. Use the fundamental counting principle: $9 \cdot 8 \cdot 10 = 720$.

Level 3, page 581

57. Use the fundamental counting principle (since repetitions are allowed): $4 \cdot 4 \cdot 4 = 64$.

59. Beverage: 7 choices; First course: 1 choice; Second course: 7; Third course: This is tricky; if the choice is steak, then one of the following is to be chosen, so we have the following choices for the third course: steak/fries, steak/mashed, steak/corn, steak/spinach, steak/broccoli, steak/onions, swordfish, rack of lamb, whitefish, grilled chicken, southern fried chicken, lobster, or vegetarian plane. There are 13 choices. Now, use the fundamental counting principle: $7 \cdot 1 \cdot 7 \cdot 13 \cdot 7 = 4{,}459$

Page 306

12.2 Combinations, page 582

New Terms Introduced in this Section

Cards, deck of   Combination

Level 1, page 588

1. A combination of r elements selected from a set S with n elements is a subset of those r elements selected without repetitions. The formula is $\binom{n}{r} = \frac{n!}{r!(n-r)!}$.

3. Using Pascal's triangle, we find the element in the 9th row, 1st column: $\binom{9}{1} = 9$

5. Using Pascal's triangle, we find the element in the 9th row, 3rd column: $\binom{9}{3} = 84$

7. The 0th entry in any row of Pascal's triangle is 1, so $\binom{9}{0} = 1$

9. Use the formula (or calculator); $\binom{52}{3} = \frac{52!}{3!(52-3)!} = \frac{52 \cdot 51 \cdot 50}{3 \cdot 2} = 22,100$

11. The last entry in any row of Pascal's triangle is 1, so $\binom{4}{4} = 1$

13. Using Pascal's triangle, we find the element in the 7th row, 3rd column: $\binom{7}{3} = 35$

15. Use the formula (or calculator); $\binom{50}{48} = \frac{50!}{48!(50-48)!} = \frac{50 \cdot 49}{2} = 1,225$

17. By formula, $\binom{g}{h} = \frac{g!}{h!(g-h)!}$.

19. By formula, $_kP_4 = \frac{k!}{(k-4)!}$.

21. By formula, $\binom{m}{n} = \frac{m!}{n!(m-n)!}$.

23. $\binom{10}{4,3,3} = \frac{10!}{4!3!3!} = 4,200$

25. $\binom{9}{2,3,4} = \frac{9!}{2!3!4!} = 1,260$

27. $\binom{7}{1}\binom{6}{4}\binom{2}{2} = \frac{7!}{1!4!2!} = 105$

Level 2, page 588

29. The order in which the pieces of candy is selected is not important, so it is a combination: $\binom{12}{5} = 792$

31. The order in which the kings are drawn is not important, so it is a combination: $\binom{4}{2} = 6$

33. The order is not important when drawing cards in poker; $\binom{13}{5} = 1,287$

35. Use the fundamental counting principle for jacks and for two, and then combinations for each of these to obtain: $\binom{4}{3}\binom{4}{2} = 24$

37. Consider the expansion of $(B + G)^5$ and look at the coefficient of B^2G^3, namely $\binom{5}{2} = 10$.

39. $2^7 = 128$

Level 3, page 589

41. The committee problem is a combination; $\binom{30}{5} = 142,506$ (by calculator).

Page 307

43. The order in which the people shake hands does not matter, and repetition is not allowed, so it is a combination; $\binom{20}{2} = 190$ (by calculator).

45. The order in which the scoops of ice cream are put into the dish does not (usually) matter, and since the scoops are different, we would call this a combination; $\binom{31}{3} = 4{,}495$ (by calculator).

47. This is neither a permutation nor a combination; it is distinguishable permutation; $\frac{7!}{2!1!2!1!1!} = 1{,}260$ (by calculator).

49. Bob (dishwasher) and Brian (bouncer) would make a big difference if Brian is a 97-lb weakling, so this is a permutation; $_{10}P_2 = 10 \cdot 9 = 90$.

51. If you are a driver, I'm sure which car you drive makes a difference, so we would classify this as a permutation; $_6P_5 = \frac{6!}{(6-5)!} = 720$.

53. The committee problem is a combination; $\binom{7}{4} = 35$.

55. Since the order in which you would select the books is important, this is a permutation; $_7P_2 = 7 \cdot 6 = 42$.

Level 3, Problem Solving, page 589

57. To form a pentagon on the circle, we need to choose five distinct (different) points, and so repetition is not allowed, so we classify this as a combination of 5 objects (points) selected from the given set of n objects (points): $\binom{n}{5}$.

59. First, use the fundamental counting principle:

| selecting the women officers | $\cdot$ | selecting the men officers | $\cdot$ | selecting the committee |

$= {_{16}P_2} \cdot {_{19}P_2} \cdot \binom{31}{6}$; approximately $6.043394448 \times 10^{10}$ (by calculator).

12.3 Counting without Counting, page 590

Level 1, page 595

1. The *fundamental counting principle* gives the number of ways of performing two or more tasks. If task A can be performed in m ways, and if, after task A is performed, a second task B, can be performed in n ways, then task A followed by task B can be performed in $m \cdot n$ ways.

3. FCP; $3 \cdot 5 = 15$.

5. FCP; $26 \cdot 10 \cdot 9 \cdot 8 \cdot 7 \cdot 6 = 786{,}240$.

7. Committee problem is a combination; $_7C_2 = 21$.

9. The order in which the flags are flown makes a difference, so it is a permutation; $_{40}P_5 = 78{,}960{,}960$.

11. Each tumbler can assume six positions, so use the FCP; $6^4 = 1{,}296$.

Page 308

13. Five flips has $2^5 = 32$ possibilities, so at least one head is $2^5 - 1 = 31$ ways.

15. Seven flips has $2^7 = 128$ possibilities, so at least one head is $2^7 - 1 = 127$ ways.

17. Election problem is a permutation; $_{15}P_3 = 2{,}730$.

19. There are 26 letters, 10 numerals, 1 space, and 4 symbols which is a total of 41 possibilities. Since there are seven spaces, the total number of plates is $41^7 \approx 1.95 \times 10^{11}$.

21. Use the FCP; $5 \cdot 3 \cdot 3 \cdot 3 = 135$.

23. Committee problem is a combination; $_{30}C_4 = 27{,}405$.

25. The order in which the ingredients is chosen is important (a sundae has ice cream first, then topping, whipped cream, and nuts on top); permutation; $_6P_3 = 120$.

27. The order in which the hearts are drawn does not matter, so it is a combination; $_{13}C_3 = 286$.

29. The order in which the TV sets are selected does not matter, so it is a combination; $_{100}C_6 = 1{,}192{,}052{,}400$

31. The order in which the liquids are mixed makes a difference, so it is a permutation; $_6P_6 = 720$

33. It is a distinguishable permutation; $\left(\begin{smallmatrix} & 4 & \\ 1, & 1, 1, & 1 \end{smallmatrix}\right) = 4! = 24$

35. First use the fundamental counting principle (with combinations for the groomsmen and bridesmaids): 1 (bride) $\times$ 1 (groom) $\times$ 2 (parents of the bride) $\times$ 2 (parents of the groom) $\times$ 1 (best man) $\times$ 1 (maid of honor) $\times$ $_5C_5$ (groomsmen) $\times$ $_5C_5$ (bridesmaids) $= 1 \times 1 \times 2 \times 2 \times 1 \times 1 \times 5! \times 5! = 57{,}600$

37. The order of digits in a number matters, so it is a permutation; $_4P_3 = 24$

39. The order in which you answer the questions is not important, so it is a combination; $_{12}C_{10} = 66$.

Level 2, page 596

41. Subsets: $\{a, b\}$; 2 arrangements: $(a, b), (b, a)$

43. Subsets: $\{a, b, c, d\}$; 24 arrangements: $(a, b, c, d), (a, b, d, c), (a, c, b, d), (a, c, d, b),$ $(a, d, b, c), (a, d, c, b), (b, a, c, d), (b, a, d, c), (b, c, a, d), (b, c, d, a), (b, d, a, c),$ $(b, d, c, a), (c, a, b, d), (c, a, d, b), (c, b, a, d), (c, b, d, a), (c, d, a, b), (c, d, b, a),$ $(d, a, b, c), (d, a, c, b), (d, c, a, b), (d, c, b, a), (d, b, a, c), (d, b, c, a)$.

45. Use the FCP; $9 \cdot 9 \cdot 10 \cdot 10 \cdot 10 \cdot 10 \cdot 10 \cdot 10 \cdot 10 = 810{,}000{,}000$.

47. There are three possibilities for each question, so the total is $3^{20} = 3{,}486{,}784{,}401$.

49. The arrangements of books on a shelf makes a difference (Vol I/Vol II is different from Vol II/Vol I), so it is a permutation; $_{26}P_{26} = 26!$

Level 3, page 597

51. 1 min = 50 jumps; 1 hr = 3,000 jumps; 1 day = 24,000 jumps; 1 year = 6×10^6 jumps;

Page 309

1 century $= 6 \times 10^8$ jumps; thus, 5,480,523,297,162 times would take about 9,134.2 centuries!

53. In Pascal's triangle, two numbers are added to obtain the number below; in this arrangement the triangle is inverted and the numbers are subtracted to find the numbers below.

Level 3 Problem Solving, page 597

55. There must be at least one bald person with 0 hairs and any number times 0 is 0; the answer is 0.

57. The number of possibilities for one pizza is:

$$\binom{11}{5} + \binom{11}{4} + \binom{11}{3} + \binom{11}{2} + \binom{11}{1} + \binom{11}{0} = 1{,}024$$

The number of possible (different) pizzas are $\binom{1{,}024}{2} + 1{,}024 = 524{,}800$.

59. *Number of 0-topping pizzas:* $\binom{11}{0} = 1$

   Number of 1-topping pizzas: $\binom{11}{1} = 11$

   Number of 2-topping pizzas: $\binom{11}{2} + \binom{11}{1} = 55 + 11 = 66$

   Note: $\binom{11}{2}$ is the number of pizzas with two different toppings; $\binom{11}{1}$ is the number of pizza with one topping taken twice.

   Number of 3-topping pizzas: $\binom{11}{3} + \binom{11}{1}\binom{10}{1} = 165 + 110 = 275$

   Note: $\binom{11}{1}\binom{10}{1}$ is the number of pizzas where 1 topping is double and the other topping is single.

   Number of 4-topping pizzas: $\binom{11}{4} + \binom{11}{2} + \binom{11}{1}\binom{10}{2} = 330 + 55 + 495 = 880$

   Note: $\binom{11}{2}$ is the number of pizzas where 2 toppings are doubled; $\binom{11}{1}\binom{10}{2}$ is the number of pizzas where 1 topping is double and the other two toppings are single.

   Number of 5-topping pizzas: $\binom{11}{5} + \binom{11}{1}\binom{10}{3} + \binom{11}{2}\binom{9}{1} = 462 + 1{,}320 + 495 = 2{,}277$

   Thus, the total is $1 + 11 + 66 + 275 + 880 + 2{,}277 = 3{,}510$. Finally, for two pizzas, there are $\binom{3{,}510}{2} + 3{,}510 = 6{,}161{,}805$ pizzas.

12.4 Rubik's Cube and Instant Insanity, page 598
New Terms Introduced in this Section

Instant Insanity   Rubik's cube

Level 1, page 601

1. 331,776 sec = 5,529.6 min = 92.16 hr, or about 92 hr, 10 min

3.   **5.**   **7.**   **9.**   **11.**

13.   **15.**   **17.**   **19.**   **21.**

23.   **25.**   **27.**   **29.**

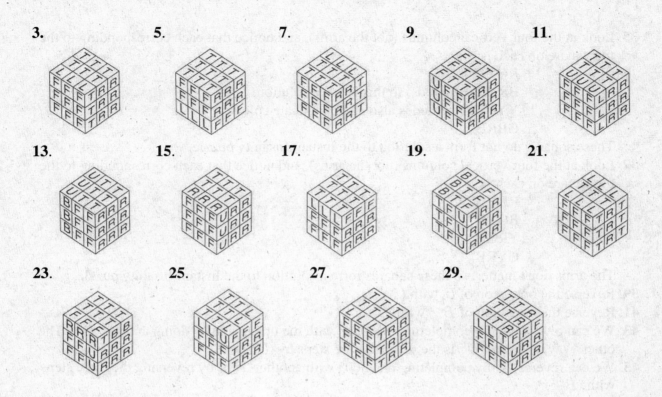

Level 2, page 601

31. Look at the four vertical columns (not the arms), and notice that each corresponding to the top square on each is:

      YRGB

      GYRB

      RBYG

      BGYR

The arms don't matter, so these squares form a solution to the Instant Insanity puzzle.

33. Look at the four vertical columns (not the arms), and notice that each corresponding to the top square on each is:

      RBGB Two blues, so this is not a solution.

      GGYY Other sides also have duplicate colors.

      RRBG

      BRGY

These squares do not form a solution to the Instant Insanity puzzle.

35. Look at the four vertical columns (not the arms), and notice that each corresponding to the top square on each is:

> RGYG
>
> BRYR Two reds, so this is not a solution.
>
> YYGB Other sides also have duplicate colors.
>
> GBRB

These squares do not form a solution to the Instant Insanity puzzle.

37. Look at the four vertical columns (not the arms), and notice that each corresponding to the top square on each is:

> RBYG
>
> BGYR
>
> YRGB
>
> GYRB

The arms don't matter, so these squares form a solution to the Instant Insanity puzzle

39. Reverse the operation of U with U^{-1}.

41. Reverse the operation of B^{-1} with B.

43. We can reverse U^2 by completing one 360° with the upper side by doing another U^2. The other way to reverse U^2 is the reverse those steps by $(U^{-1})^2$.

45. We can reverse B^3 by completing one 360° with another B or by reversing the three steps with $(B^{-1})^3$.

47. The steps BU are reversed by $U^{-1}B^{-1}$. Watch the order here.

49. The steps FT^{-1} are reversed by TF^{-1}. Watch the order here.

51. The steps $F^{-1}B^{-1}$ are reversed by BF. Watch the order here.

Level 3, page 602

53. If you take three operations such as T, R, L and carry out RL followed by T as in $T(RL)$ and then go back to the standard position and then do TR followed by L, you see they are not the same; that is $T(RL) \neq (TR)L$.

55. No; try carrying out these moves with your own cube or on the website.

Level 3 Problem Solving, page 602

57. To find the number of different possibilities we will not count the order of the cubes (certainly once we have found a solution, it doesn't matter how we switch the four cubes around — as long as we don't rotate any of them). Also, we will not count as a different solution the rotation of the final $1 \times 1 \times 4$ block around its long axis. Thus, if we do not count rotations, we see that the first cube can be placed in any one of three different ways. There is only one way of placing this correctly, but there are three ways of arranging this cube. The key here is to get the "proper faces" along the axis that is never shown. This is, if this cube looks like the one shown below, it will have an R-R that "doesn't come up."

The other two possibilities have G-B and R-Y along the axis as shown below:

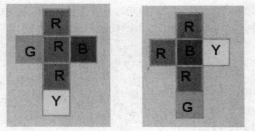

Thus, the first cube can come up three ways. Now, the other cubes do not have this much freedom of choice, since the first cube has been "fixed" in position. That is, each of the remaining cubes cannot be rotated independently, but must be rotated in conjunction with the first cube. Thus, each of the remaining cubes has 24 possibilities. If we use the fundamental counting principle, we see that there are

$$3 \cdot 24 \cdot 24 \cdot 24 = 41,472$$

possibilities, of which there is only one solution.

59.

Chapter 12 Review Questions, page 603

🛑   Studying for a chapter examination is a personal process, one which nobody else can do for you. Simply take the time to review what you have done. Here are the new terms in Chapter 12.

Arrangement [12.1]	Factorial [12.1]	Ordered triple [12.1]
Binomial theorem [12.2]	Fundamental counting	Permutation [12.1, 12.2]
Cards, deck of [12.2]	principle [12.1]	Rubik's cube [12.4]
Combination [12.1, 12.2]	Instant Insanity [12.4]	Tree diagram [12.1]
Count-down property [12.1]	Multiplication property of	
Distinguishable	factorials [12.1]	
permutation [12.1]	Ordered pair [12.1]	

If you can describe the term, read on to the next one; if you cannot, then look it up in the text (the section number is shown in brackets). Next, study the types of problems listed at the end of Chapter 12.

TYPES OF PROBLEMS

Apply the Fundamental Counting Principle. [12.1]

Simplify expressions involving factorials. [12.1]

Know the permutation formula. [12.1]

Evaluate permutations and distinguishable permutations.[12.1]

Know the combination formula. [12.2]

Evaluate combinations. [12.2]

Distinguish among a permutation, a combination, a distinguishable permutation, and the fundamental counting principle. [12.3]

Answer applied counting questions. [12.1-12.3]

Use the notation involving Rubik's cube. [12.4]

Explain the Instant Insanity problem. [12.4]

Once again, see if you can verbalize (to yourself) how to do each of the listed types of problems.

    Work all of Chapter 12 Review Questions (whether they are assigned or not). Work through all of the problems before looking at the answers, and *then* correct each of the problems. The entire solution is shown in the answer section at the back of the text. If you worked the problem correctly, move on to the next problem, but if you did not work it correctly (or you did not know what to do), look back in the chapter to study the procedure, or ask your instructor.

    Finally, go back over the homework problems you have been assigned. If you worked a problem correctly, move on to the next problem, but if you missed it on your homework, then you should look back in the book or talk to your instructor about how to work the problem.

    If you follow these steps, you should be successful with your review of this chapter.

We give all of the answers to the chapter review questions (not just the odd-numbered questions).

Chapter 12 Review Questions, page 603

1. $8! - 3! = 40{,}320 - 6$
$$= 40{,}314$$

2. $8 - 3! = 8 - 6$
$$= 2$$

3. $(8 - 3)! = 5!$
$$= 120$$

4. $\left(\dfrac{8}{2}\right)! = 4!$
$$= 24$$
Note that this is not combination notation, but rather a fraction in the parentheses.

5. $\binom{8}{2}! = 28!$ Note that this is combination notation; compare with the previous problem.

6. By Pascal's triangle, $\binom{5}{3} = 10$.

7. Write this as a combination so you can use Pascal's triangle: $\quad {}_8P_3 = 3! \cdot {}_8C_3$
$$= 6 \cdot 56$$
$$= 336$$

8. ${}_{12}P_0 = 1$

9. Use the formula (or calculator) $\quad {}_{14}C_4 = \dfrac{14!}{4!(14 - 4)!}$
$$= 1{,}001$$

10. Use the formula (or calculator) $\quad {}_{100}P_3 = \dfrac{100!}{97!}$
$$= 100 \cdot 99 \cdot 98$$
$$= 970{,}200$$

11. This is a committee problem, so it is a combination. Use the formula (or calculator): ${}_{12}C_3 = 220$; they can form 220 different committees.

12. The order in which people line up is important, so it is a permutation. Use the formula (or calculator): $_5P_5 = 5! = 120$; they can form 120 different lineups at the teller's window.

13. These are distinguishable permutations. The word "HAPPY": $\begin{pmatrix} 5 \\ 1, 1, 2, 1 \end{pmatrix} = \dfrac{5!}{1!1!2!1!}$
$$= 60$$

The word "COLLEGE": $\begin{pmatrix} 7 \\ 1, 1, 2, 2, 1 \end{pmatrix} = \dfrac{7!}{2!2!}$
$$= 1{,}260$$

14. If you draw three balls, then to have at least one red, you'd need to have:

1 red and 2 whites or 2 red and 1 white or 3 red and 0 white $= \begin{pmatrix} 4 \\ 1 \end{pmatrix}\begin{pmatrix} 6 \\ 2 \end{pmatrix} + \begin{pmatrix} 4 \\ 2 \end{pmatrix}\begin{pmatrix} 6 \\ 1 \end{pmatrix} + \begin{pmatrix} 4 \\ 3 \end{pmatrix}$
$$= 4 \cdot 15 + 6 \cdot 6 + 4$$
$$= 100$$

15. This is a committee problem, so it is a combination. We need to select five members from one hundred members of the senate: $_{100}C_5 = \frac{100!}{5!(100-5)!} = 75{,}287{,}520$ (by calculator).

16. Fundamental counting principle; each item can be included or not included, and there are 8 items, so the number of possibilities is $2^8 = 256$. The claim is correct.

17. a. There are two possibilities for each question, so (by the FCP) we have $2^{10} = 1{,}024$
  b. $3^{10} = 59{,}049$
  c. $5^{10} = 9{,}765{,}625$

18. a. $24^5 = 7{,}962{,}624$
  b. Divide by 60 to change to minutes; divide again by 60 to change to hours; then divide by 24 to change to days. The result is more than 92 days (nonstop).

19. If it is the result of a combination, then the number must appear in Pascal's triangle. The only place that 65 appears is as the first and last number of the 65th row. Since it is given that there is more than one item, we see that 65 is not in the triangle, so the answer is no.

20. The number of arrangements is $_{10}P_{10} = 10! = 3{,}628{,}800$. This is almost 10,000 years, so that day will never come for the members of this club.

Page 316

CHAPTER 13 The Nature of Probability

Chapter Overview
This chapter introduces the mathematics of uncertainty. The notion of probability is not only a major branch of mathematics, but permeates a great deal of science and social science.

13.1 Introduction to Probability, page 607

New Terms Introduced in this Section

Dice	Die	Empirical probability	Equally likely outcomes
Event	Experiment	Impossible event	Law of large numbers
Mutually exclusive	Probabilistic model	Probability	Relative frequency
Sample space	Simple event	Subjective probability	Theoretical probability

Level 1, page 616

1. Empirical probability is done by experimentation and gives the percent of occurrence, whereas a theoretical probability is defined as a ratio found without performing an experiment.

3. Since E is a subset of S the number s must be between 0 and n

$$\frac{0}{n} \leq \frac{s}{n} \leq \frac{n}{n}$$

$$0 \leq \frac{s}{n} \leq 1$$

$$0 \leq P(E) \leq 1 \qquad \textit{Since } P(E) = \frac{s}{n}$$

5. Look at the middle spinner; it looks like it is divided into three parts, labeled D, E, and F. It also looks like F is half the circle and the other half is equally divided.
   a. $P(D) = \frac{1}{4}$ **b.** $P(E) = \frac{1}{4}$ **c.** $P(F) = \frac{1}{2}$

7. Look at the right-hand spinner; it looks like it is divided into four parts, labeled G, H, and I (equally divided a half-circle), and J representing the other half-circle.
   a. $P(J) = \frac{1}{2}$ **b.** $P(G \text{ or } H) = \frac{2}{3} \cdot \frac{1}{2} = \frac{1}{3}$ **c.** $P(I \text{ or } J) = 1 - \frac{1}{3} = \frac{2}{3}$

9. This is an empirical probability; $P(A) = \frac{13}{285} \approx 0.0456$; this is about 0.05.

11. This is an empirical probability; $P(\text{winning ticket}) = \frac{285}{1,500} = 0.19$

13. $P(\text{royal flush}) = \frac{4}{2,598,960} \approx 0.000001539077169$

15. $P(\text{four of a kind}) = \frac{624}{2,598,960} \approx 0.0002400960384$

17. $P(\text{flush}) = \frac{5,108}{2,598,960} \approx 0.00196540155$

19. $P(\text{three of a kind}) = \frac{54,912}{2,598,960} \approx 0.02112845138$

21. $P(\text{one pair}) = \frac{1,098,240}{2,598,960} \approx 0.42256902761$

Level 2, page 617

23. a. There is one five of clubs in a deck of cards, so $P(\text{five of clubs}) = \frac{1}{52}$.

    b. There are four fives in a deck of cards, so $P(\text{five}) = \frac{4}{52} = \frac{1}{13}$.

    c. There are thirteen clubs in a deck of cards, so $P(\text{club}) = \frac{13}{52} = \frac{1}{4}$.

25. a. There is no card in a deck of cards that is a five and a jack, so $P(\text{five and a jack}) = 0$.

    b. $P(\text{five or a jack}) = \frac{8}{52} = \frac{2}{13}$

27. a. Since the only way to obtain a tail *and* a five is $\{T5\}$, we see $P(T5) = \frac{1}{12}$.

    b. Can obtain a tail or a five by obtaining $\{T1, T2, T3, T4, T5, T6, H5\}$, so
$P(\text{tail or a five}) = \frac{7}{12}$.

    c. Success is $\{H2\}$, so $P(H2) = \frac{1}{12}$.

29. a. Success is $\{H1, H3, H5\}$, so $P(\text{heads and an odd number}) = \frac{3}{12} = \frac{1}{4}$.

    b. Success is $\{H1, H2, H3, H4, H5, H6, T1, T3, T5\}$, so
$P(\text{heads or an odd number}) = \frac{9}{12} = \frac{3}{4}$.

31. $P(\text{five}) = \frac{4}{36} = \frac{1}{9}$

33. $P(\text{seven}) = \frac{6}{36} = \frac{1}{6}$

35. $P(\text{nine}) = \frac{4}{36} = \frac{1}{9}$

37. Count the number of possibilities from Figure 13.5 so $P(\text{four or five}) = \frac{7}{36}$.

	1	2	3	4	5	6
1	(1,1)	(1,2)	(1,3)	(1,4)	(1,5)	(1,6)
2	(2,1)	(2,2)	(2,3)	(2,4)	(2,5)	(2,6)
3	(3,1)	(3,2)	(3,3)	(3,4)	(3,5)	(3,6)
4	(4,1)	(4,2)	(4,3)	(4,4)	(4,5)	(4,6)
5	(5,1)	(5,2)	(5,3)	(5,4)	(5,5)	(5,6)
6	(6,1)	(6,2)	(6,3)	(6,4)	(6,5)	(6,6)

39. Count the number of possibilities from Figure 13.5

	1	2	3	4	5	6
1	(1,1)	(1,2)	(1,3)	(1,4)	(1,5)	(1,6)
2	(2,1)	(2,2)	(2,3)	(2,4)	(2,5)	(2,6)
3	(3,1)	(3,2)	(3,3)	(3,4)	(3,5)	(3,6)
4	(4,1)	(4,2)	(4,3)	(4,4)	(4,5)	(4,6)
5	(5,1)	(5,2)	(5,3)	(5,4)	(5,5)	(5,6)
6	(6,1)	(6,2)	(6,3)	(6,4)	(6,5)	(6,6)

$P(\text{eight } or \text{ ten}) = \frac{8}{36} = \frac{2}{9}$

Page 318

41. Player A can spin a 5 or 3 and C can spin 2 or 4; look at the sample space.

$A\backslash C$	4	2
5	A wins	A wins
3	C wins	A wins

$P(A) = \frac{3}{4}$; $P(C) = \frac{1}{4}$; so pick A; $P(A$ winning$) = \frac{3}{4}$.

43. Player D can spin a 1, 7, or 8 and E can spin a 4, 5, or 6; look at the sample space.

$D\backslash E$	4	5	6
1	E wins	E wins	E wins
7	D wins	D wins	D wins
8	D wins	D wins	D wins

$P(D) = \frac{6}{9}$; $P(E) = \frac{3}{9}$; so pick D; $P(D$ winning$) = \frac{2}{3}$.

45. Player D can spin a 1, 7, or 8 and F can spin a 2, 3, or 9; look at the sample space.

$D\backslash F$	2	3	9
1	F wins	F wins	F wins
7	D wins	D wins	F wins
8	D wins	D wins	F wins

$P(D) = \frac{4}{9}$; $P(F) = \frac{5}{9}$; so pick F; $P(F$ winning$) = \frac{5}{9}$.

47. Player C can spin a 2, or 4 and F can spin a 2, 3, or 9; look at the sample space.

$C\backslash F$	2	3	9
2	Tie	F wins	F wins
4	C wins	C wins	F wins

$P(C) = \frac{2}{6}$; $P(F) = \frac{3}{6}$; so pick F; $P(F$ winning$) = \frac{1}{2}$.

Level 3, page 617

49. Answers vary; the answers here are the theoretical probabilities, but the problem requests empirical probabilities. **a**. $P(\text{two}) = 0.0278$ **b**. $P(\text{three}) = 0.0556$ **c**. $P(\text{four}) = 0.0833$ **d**. $P(\text{five}) = 0.1111$ **e**. $P(\text{six}) = 0.1389$ **f**. $P(\text{seven}) = 0.1667$ **g**. $P(\text{eight}) = 0.1389$ **h**. $P(\text{nine}) = 0.1111$ **i**. $P(\text{ten}) = 0.0833$ **j**. $P(\text{eleven}) = 0.0556$ **k**. $P(\text{twelve}) = 0.0278$

51. You should show your table of outcomes to substantiate that they do appear to be equally likely.

53. Assume that a year is 365.25 days or 8,766 hr and it is given that you are at work 1,920 hours per year. This means that $P(\text{earthquake}) = \frac{5(174)}{8,766} \approx 10\%$.

55. a. Count the number of possibilities from Figure 13.5

	1	2	3	4	5	6
1	(1,1)	(1,2)	(1,3)	(1,4)	(1,5)	(1,6)
2	(2,1)	(2,2)	(2,3)	(2,4)	(2,5)	(2,6)
3	(3,1)	(3,2)	(3,3)	(3,4)	(3,5)	(3,6)
4	(4,1)	(4,2)	(4,3)	(4,4)	(4,5)	(4,6)
5	(5,1)	(5,2)	(5,3)	(5,4)	(5,5)	(5,6)
6	(6,1)	(6,2)	(6,3)	(6,4)	(6,5)	(6,6)

$\frac{8}{36} = \frac{2}{9}$

b. Count the number of possibilities from Figure 13.5: $\frac{4}{36} = \frac{1}{9}$

Page 319

57. a.

	1	2	3	4
1	(1,1)	(1,2)	(1,3)	(1,4)
2	(2,1)	(2,2)	(2,3)	(2,4)
3	(3,1)	(3,2)	(3,3)	(3,4)
4	(4,1)	(4,2)	(4,3)	(4,4)

b. $P(\text{two}) = \frac{1}{16}$ **c.** $P(\text{three}) = \frac{1}{8}$ **d.** $P(\text{four}) = \frac{3}{16}$ **e.** $P(\text{five}) = \frac{1}{4}$ **f.** $P(\text{six}) = \frac{3}{16}$
g. $P(\text{seven}) = \frac{1}{8}$

Level 3 Problem Solving, page 618

59. There are three possibilities. The jar has either two brains (B_1 and B_2) or a brain and a kidney (B and K).

Outside jar:	B_1	B_2	B
Inside jar:	B_2	B_1	K

In 2 out of 3 of these possibilities there is a brain in the jar, so $P(\text{brain}) = \frac{2}{3}$.

13.2 Mathematical Expectation, page 618
New Terms Introduced in this Section

Expectation   Expected value   Fair game
Mathematical expectation

Level 1, page 624

1. False; there are also green spots. $P(\text{black}) = \frac{18}{38} = \frac{9}{19}$.

3. True; play with positive expectation, do not play with negative expectation; zero expectation means that it is a fair game.

5. False; there is a cost of playing: $E = \$50,000\left(\dfrac{1}{1,000,000}\right) + (-\$1)$
$$= -\$0.95$$

7. The probability is fairly small, so answer C is way too large; if you did one flip the probability is $\frac{1}{2}$, so the expectation is close to choice A. Thus, by process of elimination, choice B is the most reasonable.

9. One hundred nickels is about $5, so choice B is the most reasonable choice.

11. You certainly know more people who have had a car accident, so that is far more likely that winning the grand prize in a lottery. There are several contestants on each week on *Jeopardy*, and that is hardly newsworthy, so that is far more likely that winning the super lottery (which usually is a newsworthy event). Thus, by process of elimination, choice B is the most reasonable.

13. $E = \dfrac{1}{1,000,000}(\$80,000)$

   $= \$0.08$

15. $E = \$1\left(\dfrac{1}{5}\right) + \$5\left(\dfrac{1}{5}\right) + \$10\left(\dfrac{1}{5}\right) + \$20\left(\dfrac{1}{5}\right) + \$100\left(\dfrac{1}{5}\right) - \$20$

   $= \$7.20$

17. $E = \dfrac{1}{4}(\$0.50)$ A fair price would be to pay \$0.25 for two plays of the game.

   $= \$0.125$

19. $E = \$100\left(\dfrac{1}{100}\right) + \$10\left(\dfrac{5}{100}\right)$

   $= \$1.50$

21. $E = 41\left(\dfrac{18}{38}\right) + (-\$1)\left(\dfrac{20}{38}\right)$

   ≈ -0.05

23. $E = 17\left(\dfrac{2}{38}\right) + (-1)\left(\dfrac{36}{38}\right)$

   $\approx -\$0.05$

25. $E = \$8\left(\dfrac{4}{38}\right) + (-\$1)\left(\dfrac{34}{38}\right)$

   $\approx -\$0.05$

27. $E = \$5\left(\dfrac{6}{38}\right) + (-\$1)\left(\dfrac{32}{38}\right)$

   $\approx -\$0.05$

29. $E = \$2\left(\dfrac{12}{38}\right) + (-\$1)\left(\dfrac{26}{38}\right)$

   $\approx -\$0.05$

Level 2, page 625

31. $E = 5\left(\frac{1}{2}\right) + 0\left(\frac{1}{4}\right) + 10\left(\frac{1}{4}\right) - 5 = 0$; fair game

33. $E = 14\left(\frac{1}{8}\right) + 8\left(\frac{1}{8}\right) + 4\left(\frac{1}{4}\right) + 2\left(\frac{1}{2}\right) - 5 = -\frac{1}{4}$; not a fair game; don't play

35. $E = \frac{1}{2}(-5) + \frac{1}{4}(16) + \frac{1}{8}(8) + \frac{1}{16}(4) + \frac{1}{32}(2) + \frac{1}{32}(1) \approx 2.84$

   You should be willing to pay \$2.84 to play the game.

37. $E = \frac{1}{20}(10,000) + \frac{2}{20}(5,000) + \frac{1}{20}(1,000) = 1,050$

   Since this is more than \$1,000, the contestant should play.

39. $E = 0.5(0.06 \cdot 185,000) + 0.3(0.03 \cdot 185,000) + 0.20(0) - 800 = 6,415$

Page 321

The expected profit from the listing is $6,415.

41. $E = 825,000(\frac{1}{40}) + 225,000(\frac{1}{20}) - 25,000 = 6,875$

They should dig the well because the expectation is positive.

43. $E = 0.50(\frac{4}{52}) + 0.25(\frac{12}{52}) + 1.00(\frac{1}{52}) - 0.10 \approx 0.02$

Yes, you should play the game.

45. $E = 25(\frac{1}{1,635}) + 3(\frac{1}{163}) + 1(\frac{1}{68}) = 0.0484013102$

The expectation is about $0.05.

47. $E = 0(0.15) + 1(0.25) + 2(0.31) + 3(0.21) + 4(0.08) = 1.82$

The expected number of tardies is 1.82.

49. $E = \$1,000(0.12) + \$800(0.38) + \$500(0.45) + \$200(0.05) = \$659$

If the cost is $500, then the expected profit is $159. This is not at least the needed $200, so the merchant should not purchase them.

Level 3, page 626

51. In this book, we are answering by maximizing the gain, whereas in real life we would often want to minimize our loses. For this problem, the contestant has $350,000, which stands to be lost half the time if the contestant continues. On the other hand, the contestant has $\frac{1}{4}$ chance to win a million, and $\frac{1}{4}$ chance to win nothing (stopper). The expectation is

$$E = \tfrac{1}{2}(\$350,000) + \tfrac{1}{4}(\$1,000,000) + \tfrac{1}{4}(\$0) = \$425,000$$

Since the expectation is greater than the $350,000, the contestant should continue.

Answers for Problems 53-56 vary, but you if your instructor assigns them, you should find the results interesting.

Level 3, Problem Solving, page 627

57 $E = \$1(\frac{1}{2}) + \$2(\frac{1}{4}) + \$4(\frac{1}{8}) = \$1.50$

59. $E = \$1(\frac{1}{2}) + \$2(\frac{1}{4}) + \$4(\frac{1}{8}) + \cdots + \$2^{999}(\frac{1}{2^{1,000}}) = \$500.00$

13.3 Probability Models, page 627

New Terms Introduced in this Section

Complementary probabilities   Conditional probability   Odds against   Odds in favor
Property of complements

Level 1, page 634

1. The *fundamental counting principle* gives the number of ways of performing two or more tasks. If task A can be performed in m ways, and if, after task A is performed, a second task B, can be performed in n ways, then task A followed by task B can be performed in $m \cdot n$ ways.

3. Since there are 9 possible squares in which to place the X (first move), the total number of possibilities is $9 \cdot 8 \cdot 7 \cdot 6 \cdot 5 = 15{,}120$.

5. $P(E) = \frac{s}{n}$ whereas the odds in favor of E are $\frac{s}{f}$ where $s + f = n$.

Problems 6-11 are to be done by estimation. These are really exercises in subjective probabilities in the sense they are asking for an opinion of which is more probable. As directed by the directions, do NOT calculate these probabilities.

7. Choice A is a normal probability; choice B is a complementary probability; choice C is the conditional probability.

9. At least 2 times means 2 times or 3 times which is obviously more likely than the probability of exactly 2 times so the correct choice is B.

11. Choice A the odds are about 1 to 10,000,000, and choice B the odds are about 1 to 2,000,000, so the correct choice is B.

13. $P(\overline{B}) = 1 - P(B)$

$$= 1 - \frac{4}{5}$$

$$= \frac{1}{5}$$

15. $P(\overline{D}) = 1 - P(D)$

$$= 1 - 0.005$$

$$= 0.995$$

17. $P(\text{not a mult. of } 5) = 1 - P(\text{mult. of } 5)$

$$= 1 - \frac{1}{5}$$

$$= \frac{4}{5}$$

19. $P(\text{at least one head}) = 1 - P(\text{no head})$

$$= 1 - \frac{1}{16}$$

$$= \frac{15}{16}$$

21. There are 16 possibilities for 4 children ($2^4 = 16$); only 1 of those possibilities is four boys. The odds against are 15 to 1.

23. odds in favor $= \dfrac{s}{f} = \dfrac{0.81}{1 - 0.81} = \dfrac{81}{18} = \dfrac{9}{2}$; odds in favor are about 9 to 2.

Page 323

25. $P(\#1) = \dfrac{1}{18 + 1} = \dfrac{1}{19}$; $P(\#2) = \dfrac{2}{3 + 2} = \dfrac{2}{5}$; $P(\#3) = \dfrac{1}{2 + 1} = \dfrac{1}{3}$; $P(\#4) = \dfrac{5}{7 + 5} = \dfrac{5}{12}$;

$P(\#5) = \dfrac{1}{1 + 1} = \dfrac{1}{2}$; *Note*: at a horse race odds are determined by track betting and the sum of the probabilities is not necessarily 1.

27. The given odds are $\frac{33}{1}$; $P(\text{lie}) = \dfrac{33}{33 + 1} = \dfrac{33}{34}$

29. Reduced sample space: HHHH; HHHT; HHTH; HHTT; HTHH; HTHT; HTTH; HTTT; THHH; THHT; THTH; THTT; TTHH; TTHT; TTTH; TTTT;

$P(\text{exactly 3H} \mid \text{at least 2H}) = \dfrac{4}{11}$

Level 2, page 635

31. It is given that the drawn card is a jack, and since this card *is* a face card, this event is certain so the probability is 1.

33. It is given that the card chosen is not a spade so this leaves 39 other card, of which 13 are a heart: $\frac{13}{39} = \frac{1}{3}$

35. It is given that the drawn card is a jack, and since we know there are two black jacks we see that the desired probability is: $\frac{2}{4} = \frac{1}{2}$

37. It is given that one of the two cards is a two, which means that there are four aces left from the 51 remaining cards: $P(\text{ace} \mid \text{two}) = \frac{4}{51}$

39. It is given that one of the two cards is a heart, which means that there are 12 hearts left from the remaining 51 cards: $P(\text{heart} \mid \text{heart}) = \frac{12}{51} = \frac{4}{17}$

41. It is given that one of the two cards is red, which means there are 26 black cards left from the remaining 51 cards: $P(\text{black} \mid \text{red}) = \frac{26}{51}$

43. $P(R) = 1 - P(\overline{R})$ About 51.5% of the plates have repetitions.

$= 1 - \dfrac{26 \times 25 \times 10 \times 9 \times 8 \times 7}{26^2 \times 10^4}$

$= 0.5153$

45. $P(R) = 1 - P(\overline{R})$

$= 1 - \dfrac{26 \times 25 \times 24 \times 10 \times 9 \times 8}{26^3 \times 10^3}$

$= 0.3609$

About 36.1% of the plates have repetitions.

47. a. $P(N) = \dfrac{25}{80}$ **b.** $P(S) = \dfrac{54}{80}$ **c.** $P(C \mid S) = \dfrac{21}{54}$ **d.** $P(S \mid C) = \dfrac{21}{30}$

         $= 0.3125$ $= 0.675$ ≈ 0.389 $= 0.70$

49. a. If the rolled number is 6, it is not odd so the probability is 0: $P(\text{odd} \mid 6) = 0$

b. If the rolled number is 5, it is odd so the probability is 1: $P(\text{odd} \mid 5) = 1$

Level 3, page 636

51. a. $P(5 \mid \text{one die is a 2}) = \frac{2}{10} = \frac{1}{5}$

b. $P(3 \mid \text{one die is a 2}) = \frac{2}{10} = \frac{1}{5}$

c. $P(2 \mid \text{one die is a 2}) = 0$

53.
$$\frac{P(\overline{E})}{P(E)} = \frac{\frac{f}{n}}{\frac{s}{n}}$$
$$= \frac{f}{n} \cdot \frac{n}{s}$$
$$= \frac{f}{s}$$
$$= \text{odds against}$$

55. a. $P(P_1) = \frac{10}{35} = \frac{2}{7}$

b. $P(\overline{P}_1) = \frac{25}{35} = \frac{5}{7}$

c. $P(P_2 \mid P_1) = \frac{9}{34}$

d. $P(\overline{P}_2 \mid P_1) = \frac{25}{34}$

e. $P(P_2 \mid \overline{P}_1) = \frac{10}{34} = \frac{5}{17}$

f. $P(\overline{P}_2 \mid \overline{P}_1) = \frac{24}{34} = \frac{12}{17}$

g. $P(P_2) = \frac{10}{35} \cdot \frac{9}{34} + \frac{25}{35} \cdot \frac{10}{34} = \frac{2}{7}$

h.

57. A typical classroom is 20 ft $\times$ 30 ft $\times$ 10 ft = 10,368,000 in.^{3} If a ping-pong ball takes up 1 in.^{3} then, the odds against winning is about the same as reaching into a classroom filled with ping-pong balls and drawing out a red ball if the room contains only one ping-pong ball painted red.

Page 325

Level 3 Problem Solving, page 636
59. Answers vary; your limited resources and the betting limit imposed on the game.

13.4 Calculated Probabilities, page 636
New Terms Introduced in this Section

Addition property of probability	Birthday problem	Dependent events
Independent events	Keno	
Multiplication property of probability	With replacement	Without replacement

Level 1, page 646
1. Events E and F are *independent* if the occurrence of one of these in no way influences the probability of the other. In terms of conditional probabilities, events E and F are independent if $P(E \mid F) = P(E)$.

3. If E and F are independent, then $P(E \cap F) = P(E \text{ and } F) = P(E) \cdot P(F)$

5. $P(\overline{A}) = 1 - P(A)$

$$= 1 - \frac{1}{2}$$

$$= \frac{1}{2}$$

7. $P(\overline{C}) = 1 - P(C)$

$$= 1 - \frac{1}{6}$$

$$= \frac{5}{6}$$

9. $P(A \cap C) = P(A) \cdot P(C)$

$$= \frac{1}{2} \cdot \frac{1}{6}$$

$$= \frac{1}{12}$$

11. $P(A \cup C) = P(A) + P(B) - P(A) \cdot P(B)$

$$= \frac{1}{2} + \frac{1}{3} - \frac{1}{2} \cdot \frac{1}{3}$$

$$= \frac{2}{3}$$

Page 326

13. $P(B \cup C) = P(B) + P(C) - P(B) \cdot P(C)$

$\qquad = \dfrac{1}{3} + \dfrac{1}{6} - \dfrac{1}{3} \cdot \dfrac{1}{6}$

$\qquad = \dfrac{4}{9}$

15. $P(\overline{A \cap C}) = 1 - P(A \cap C)$

$\qquad = 1 - P(A) \cdot P(C)$

$\qquad = 1 - \dfrac{1}{2} \cdot \dfrac{1}{6}$

$\qquad = \dfrac{11}{12}$

17. $P(\overline{A \cup B}) = 1 - P(A \cup B)$

$\qquad = 1 - [P(A) + P(B) - P(A) \cdot P(B)]$

$\qquad = 1 - \left[\dfrac{1}{2} + \dfrac{1}{3} - \dfrac{1}{2} \cdot \dfrac{1}{3}\right]$

$\qquad = 1 - \left[\dfrac{3}{6} + \dfrac{2}{6} - \dfrac{1}{6}\right]$

$\qquad = 1 - \dfrac{2}{3}$

$\qquad = \dfrac{1}{3}$

19. $P(\overline{B \cup C}) = 1 - P(B \cup C)$

$\qquad = 1 - [P(B) + P(C) - P(B) \cdot P(C)]$

$\qquad = 1 - \left[\dfrac{1}{3} + \dfrac{1}{6} - \dfrac{1}{3} \cdot \dfrac{1}{6}\right]$

$\qquad = 1 - \left[\dfrac{6}{18} + \dfrac{3}{18} - \dfrac{1}{18}\right]$

$\qquad = 1 - \dfrac{8}{18}$

$\qquad = \dfrac{5}{9}$

21. $P(\overline{A \cap B \cap C}) = 1 - P(A) \cdot P(B) \cdot P(C)$

$$= 1 - \frac{1}{2} \cdot \frac{1}{3} \cdot \frac{1}{6}$$

$$= 1 - \frac{1}{36}$$

$$= \frac{35}{36}$$

For Problems 23-34, note that $P(A) = \frac{1}{2}$, $P(B) = \frac{1}{6}$, $P(C) = \frac{1}{6}$, *and* $P(D) = \frac{1}{6}$.

23. $P(A) = \{2, 3, 5\}$; $P(B) = \{3\}$ so we see that A and B are not independent.

25. What happens on the first toss of a die has no effect on what happens on the second toss of that same die, so we see A and D are independent.

27. What happens on the first toss of a die has no effect on what happens on the second toss of that same die, so we see B and D are independent.

29. A and B are not independent, so we find $A \cap B = \{2, 3, 5\} \cap \{3\} = \{3\}$.
$P(\text{first toss is a } 3) = \frac{1}{6}$ so $P(A \cap B) = \frac{1}{6}$.

31. $P(A \cup B) = P(A) + P(B) - P(A \cap B)$

$$= \frac{1}{2} + \frac{1}{6} - \frac{1}{6}$$

$$= \frac{1}{2}$$

33. $P(B \cup D) = P(B) + P(D) - P(B \cap D)$

$$= P(B) + P(D) - P(B) \cdot P(D)$$

$$= \frac{1}{6} + \frac{1}{6} - \frac{1}{6} \cdot \frac{1}{6}$$

$$= \frac{11}{36}$$

Level 2, page 646

35. a. $P(3 \text{ bars}) = P(\text{bar}) \cdot P(\text{bar}) \cdot P(\text{bar}) = \left(\frac{1}{13}\right)^3 = \frac{1}{2,197} \approx 0.000455$

   b. $P(3 \text{ oranges}) = \frac{27}{2,197} \approx 0.01229$

   c. $P(3 \text{ plums}) = \frac{27}{2,197} \approx 0.01229$

   d. $P(1\text{st cherry}) = \frac{2}{13} \approx 0.1538$

   e. P(cherries on first two wheels) $= \left(\frac{2}{13}\right)^2 = \frac{4}{169} \approx 0.0237$

37. $E = 2\left(\dfrac{2}{20}\right) + 5\left(\dfrac{2}{20} \cdot \dfrac{5}{20}\right) + 10\left(\dfrac{2}{20} \cdot \dfrac{5}{20} \cdot \dfrac{2}{20}\right) + 10\left(\dfrac{2}{20} \cdot \dfrac{5}{20} \cdot \dfrac{8}{20}\right)$

$\qquad + 14\left(\dfrac{6}{20} \cdot \dfrac{2}{20} \cdot \dfrac{3}{20}\right) + 14\left(\dfrac{3}{20} \cdot \dfrac{2}{20} \cdot \dfrac{6}{20}\right) + 20\left(\dfrac{1}{20} \cdot \dfrac{6}{20} \cdot \dfrac{1}{20}\right)$

$\qquad + 50\left(\dfrac{2}{20} \cdot \dfrac{4}{20} \cdot \dfrac{2}{20}\right) - 1$

$\quad = -0.309$ coins

For playing \$0.25 the mathematical expectation is

$$\$0.25(-0.309) \approx -\$0.08$$

39. $E = 27{,}777 P(\text{pick } 6) - 5 = 27{,}777\left(\dfrac{{}_{20}C_6}{{}_{80}C_6}\right) - 5 \approx -\$1.42$

Here is another method:

$$E = (27{,}777 - 5)P(\text{pick } 6) + (-5)P(\text{otherwise})$$

$$= 27{,}772\left(\dfrac{{}_{20}C_6}{{}_{80}C_6}\right) + (-5)\left(1 - \dfrac{{}_{20}C_6}{{}_{80}C_6}\right)$$

$$\approx 27{,}772(1.289849391 \times 10^{-4}) - 5(0.99988710151)$$

$$\approx -\$1.42$$

41. $E = \$120 \cdot P(\text{pick } 3) + \$3 \cdot P(\text{pick } 2) - \$3$

$$= \$120\left(\dfrac{{}_{20}C_3}{{}_{80}C_3}\right) + \$3\left(\dfrac{{}_{20}C_2 \cdot {}_{60}C_1}{{}_{80}C_3}\right) - \$3$$

Here is another method:

$E = (120 - 3)P(\text{pick } 3) + (3 - 3)P(\text{pick } 2) + (0 - 3)P(\text{pick } 1) + (0 - 3)P(\text{pick } 0)$

$\approx 117(0.0138753651) + 0 - 3(0.4308666018) - 3(0.4165043817)$

$\approx -\$0.92$

43. $1 - \dfrac{{}_{365}P_5}{365^5} \approx 0.027$; the probability is about 2.7%.

45. $P(\text{five tails}) = \left(\dfrac{1}{2}\right)^5 = \dfrac{1}{32}$

47. a. With replacement: Without replacement:

$\quad P(\text{two red marbles}) = \dfrac{5}{8} \cdot \dfrac{5}{8} \qquad P(\text{two red marbles}) = \dfrac{{}_5C_2}{{}_8C_2}$

$\qquad\qquad\qquad\qquad = \dfrac{25}{64} \qquad\qquad\qquad\qquad\quad = \dfrac{10}{28}$

$\qquad\qquad\qquad\qquad\qquad\qquad\qquad\qquad\qquad\quad = \dfrac{5}{14}$

Page 329

b. With replacement:

$$P(\text{two black marbles}) = \frac{3}{8} \cdot \frac{3}{8}$$
$$= \frac{9}{64}$$

Without replacement:

$$P(\text{two black marbles}) = \frac{{}_3C_2}{{}_8C_2}$$
$$= \frac{3}{28}$$

c. With replacement:

$$P(\text{one red and one black marble}) = \frac{5}{8} \cdot \frac{3}{8} + \frac{3}{8} \cdot \frac{5}{8}$$
$$= \frac{30}{64}$$
$$= \frac{15}{32}$$

Without replacement:

$$P(\text{one red and one black marble}) = \frac{{}_5C_1 \cdot {}_3C_1}{{}_8C_2}$$
$$= \frac{15}{28}$$

d. With replacement:

$$P(\text{red then black}) = \frac{5}{8} \cdot \frac{3}{8}$$
$$= \frac{15}{64}$$

Without replacement:

$$P(\text{red then black}) = \frac{{}_5P_1 \cdot {}_3P_1}{{}_8P_2}$$
$$= \frac{5 \cdot 3}{56}$$
$$= \frac{15}{56}$$

49. a. $E = \frac{1}{4}(1) + \left(\frac{3}{4}\right)\left(\frac{1}{4}\right)(1) + \left(\frac{3}{4}\right)^2\left(\frac{1}{4}\right)(1) + \frac{27}{64}(-1)$; yes, play
$= 0.15625$

b. $E = \frac{1}{4}(1) + \left(\frac{3}{4}\right)\left(\frac{13}{51}\right)(1) + \left(\frac{3}{4}\right)\left(\frac{38}{51}\right)\left(\frac{13}{50}\right)(1) + \left(\frac{3}{4}\right)\left(\frac{38}{51}\right)\left(\frac{37}{50}\right)(-1)$
≈ 0.17294117647
yes, play

Level 3, page 647

51. First read the guest essay and then formulate an response. Use some personal experiences. You should write at least half a page.

53. $E = (20 - 1)\dfrac{_{20}C_3}{_{80}C_3} + (2 - 1)\dfrac{_{20}C_2 \cdot {_{60}C_1}}{_{80}C_3} + (-1)\dfrac{_{20}C_1 \cdot {_{60}C_2}}{_{80}C_3} + (-1)\dfrac{_{60}C_3}{_{80}C_3}$

$\approx -\$0.44$

55. $E = (250 - 1)\dfrac{_{20}C_5}{_{80}C_5} + (10 - 1)\dfrac{_{20}C_4 \cdot {_{60}C_1}}{_{80}C_5} + (2 - 1)\dfrac{_{20}C_3 \cdot {_{60}C_2}}{_{80}C_5}$

$+ (-1)\dfrac{_{60}C_4}{_{80}C_4}$

$\approx -\$0.55$

Note: $P(\text{lose } \$1) = 1 - \left[\dfrac{_{20}C_5}{_{80}C_5} + \dfrac{_{20}C_4 \cdot {_{60}C_1}}{_{80}C_5} + \dfrac{_{20}C_3 \cdot {_{60}C_2}}{_{80}C_5}\right] \approx 0.903327685$

57. You can look at the odds for each game as stated on the Keno Payout figure. Since the one spot has the best odds, we see that the one spot is the best ticket.

Level 3 Problem Solving, page 648

59. First bet: $P(\text{at least one 6 in 4 rolls}) = 1 - P(\text{no 6 in 4 rolls})$

$$= 1 - \left(\frac{5}{6}\right)^4$$

$$\approx 0.5177$$

Second bet: $P(\text{at least one 12 in 24 rolls}) = 1 - P(\text{no 12 in 24 rolls})$

$$= 1 - \left(\frac{35}{36}\right)^{24}$$

$$\approx 0.4914$$

13.5 The Binomial Distribution, page 648

New Terms Introduced in this Section

Bernoulli trial	Binomial distribution theorem	Binomial experiment
Binomial random variable	Random variable	

Level 1, page 653

1. Consider an experiment with only two outcomes, A and $\overline{A}$. Let X represent the number of times that event A has occurred, with n repetitions of the experiment. The function X is called a binomial random variable.

3. $P(X = 3) = \binom{5}{3}(0.3)^3(0.7)^2 \approx 0.132$

Page 331

5. $P(X = 6) = \binom{12}{6}(0.65)^6(0.35)^6 \approx 0.128$

7. $P(X = 6) = \binom{6}{6}(0.5)^6(0.5)^0 \approx 0.016$

9. $P(X = 5) = \binom{7}{5}(0.1)^5(0.9)^2 \approx 1.701 \times 10^{-4}$

11. Tossing coins satisfies the conditions of a binomial experiment.
$P(X = 3) = \binom{5}{3}(0.5)^3(0.5)^2 = 0.3125$

13. Boy/Girl birth patterns satisfies the conditions of a binomial experiment.
$P(X = 4) = \binom{6}{4}(0.5)^4(0.5)^2 = 0.234375$

15. Tossing a die satisfies the conditions of a binomial experiment.
$P(X = 2) = \binom{5}{2}(\frac{1}{6})^2(\frac{5}{6})^3 \approx 0.161$

17. A couple having children satisfies the conditions of a binomial experiment.
$P(X = 3) + P(X = 4) = \binom{4}{3}(0.5)^3(0.5)^1 + \binom{4}{4}(0.5)^4(0.5)^0 = 0.3125$

19. $p = 0.5, q = 0.5$; $P(X = 8) = \binom{10}{8}(0.5)^8(0.5)^2 \approx 0.044$

21. $p = 0.5, q = 0.5$; $P(X = 0) + P(X = 1) = \binom{10}{0}(0.5)^0(0.5)^{10} + \binom{10}{1}(0.5)^1(0.5)^9 \approx 0.0107$

23. $p = \frac{1}{3}, q = \frac{2}{3}$; $P(X = 0) = \binom{3}{0}(\frac{1}{3})^0(\frac{2}{3})^3 \approx 0.296$

25. $p = \frac{1}{3}, q = \frac{2}{3}$; $P(X = 2) = \binom{3}{2}(\frac{1}{3})^2(\frac{2}{3})^1 \approx 0.222$

27. $p = 0.9, q = 1 - p = 0.1$; $P(X = 4) = \binom{4}{4}(0.9)^4(0.1)^0 \approx 0.656$

29. $p = 0.9, q = 1 - p = 0.1$; $P(X = 0) = \binom{4}{0}(0.9)^0(0.1)^4 = 0.0001$

31. $p = 0.9, q = 1 - p = 0.1$;
$P(X = 3) + P(X = 4) = \binom{4}{3}(0.9)^3(0.1)^1 + \binom{4}{4}(0.9)^4(0.1)^0 = 0.9477$

33. $p = 0.2, q = 1 - p = 0.8$; $P(X = 0) = \binom{3}{0}(0.2)^0(0.8)^3 = 0.512$

35. $p = 0.2, q = 1 - p = 0.8$; $P(X = 2) = \binom{3}{2}(0.2)^2(0.8)^1 = 0.096$

37. $p = 0.2, q = 1 - p = 0.8$; $P(X \geq 1) = 1 - [P(X < 1)]$
$$= 1 - P(X = 0)$$
$$= 1 - \binom{3}{0}(0.2)^0(0.8)^3$$
$$= 1 - 0.512$$
$$= 0.488$$

Level 2, page 654

39. $p = \frac{3}{4}$; $q = 1 - p = \frac{1}{4}$; $P(X = 2) = \binom{4}{2}(\frac{3}{4})^2(\frac{1}{4})^2 \approx 0.2109$

41. $p = \frac{1}{6}$; $q = 1 - p = \frac{5}{6}$; $P(X = 1) = \binom{6}{1}(\frac{1}{6})^1(\frac{5}{6})^5 \approx 0.4019$

43. $p = 0.7, q = 1 - p = 0.3$; $P(X = 5) = \binom{6}{5}(0.7)^5(0.3)^1 \approx 0.3025$

45. $p = 0.96, q = 1 - p = 0.04$; $P(X = 4) = \binom{5}{4}(0.96)^4(0.04)^1 \approx 0.1699$

47. $p = 0.85$, $q = 1 - p = 0.15$; $P(X = 15) = \binom{20}{15}(0.85)^{15}(0.15)^5 \approx 0.1028$

49. For either team to win in 7 games, we need to calculate the probability that the series is tied in 6 games; $p = \frac{3}{5}$, $q = 1 - p = \frac{2}{5}$; $P(X = 3) = \binom{6}{3}(\frac{3}{5})^3(\frac{2}{5})^3 = 0.27648$

51. $p = 0.3$, $q = 1 - p = 0.7$; $P(X \geq 3) = P(X = 3) + P(X = 4)$

$$= \binom{4}{3}(0.3)^3(0.7)^1 + \binom{4}{4}(0.3)^4(0.7)^0$$

$$= 0.0837$$

Level 3, page 654

53. Find n such that $1 - \binom{n}{0}(0.1)^0(0.9)^n \geq 0.80$.

$$1 - \binom{n}{0}(0.1)^0(0.9)^n \geq 0.80 \qquad \text{The smallest number of missiles is 16.}$$

$$\binom{n}{0}(0.1)^0(0.9)^n < 0.20$$

$$(0.9)^n < 0.20$$

$$n \geq 15.3$$

55. a. If she is guessing then $p = 0.5$ and $q = 0.5$.

$$P(X \geq 5) = P(X = 5) + P(X = 6)$$

$$= \binom{6}{5}(0.5)^5(0.5)^1 + \binom{6}{6}(0.5)^6(0.5)^0$$

$$= 0.109375$$

b. If the claim of 90% is correct, then $p = 0.9$ and $q = 0.1$.

$$P(X \geq 5) = P(X = 5) + P(X = 6)$$

$$= \binom{6}{5}(0.9)^5(0.1)^1 + \binom{6}{6}(0.9)^6(0.1)^0$$

$$= 0.885735$$

Thus, the probability of denying the claim is $1 - P(X \geq 5) = 0.114265$.

Level 3 Problem Solving, page 654

57. $P(X = 1) + P(X = 2) = \binom{3}{1}(0.5)^1(0.5)^2 + \binom{3}{2}(0.5)^2(0.5)^1 = 0.75$

59. $P(X = 1) + P(X = 4) = \binom{5}{1}(0.5)^1(0.5)^4 + \binom{5}{4}(0.5)^4(0.5)^1 = 0.3125$

Chapter 13 Review Questions, page 655

STOP

Studying for a chapter examination is a personal process, one which nobody else can do for you. Simply take the time to review what you have done. Here are the new terms in Chapter 13.

Page 333

Addition property of
    probability [13.4]

Bernoulli trial [13.5]

Binomial distribution
    theorem [13.5]

Binomial experiment [13.5]

Binomial random variable
    [13.5]

Birthday problem [13.4]

Complementary probabilities
    [13.3]

Conditional probability [13.3]

Dependent events [13.4]

Dice [13.1]

Die [13.1]

Empirical probability [13.1]

Equally likely outcomes [13.1]

Event [13.1]

Expectation [13.2]

Expected value [13.2]

Experiment [13.1]

Fair game [13.2]

Impossible event [13.1]

Independent events [13.4]

Law of large numbers [13.1]

Keno [13.4]

Mathematical expectation
    [13.2]

Multiplication property of
    probability [13.4]

Mutually exclusive [13.1]

Odds against [13.3]

Odds in favor [13.3]

Probabilistic model [13.1]

Probability [13.1]

Property of complements
    [13.3]

Random variable [13.5]

Relative frequency [13.1]

Sample space [13.1]

Simple event [13.1]

Subjective probability [13.1]

Theoretical probability [13.1]

With replacement [13.4]

Without replacement [13.4]

If you can describe the term, read on to the next one; if you cannot, then look it up in the text (the section number is shown in brackets). Next, study the types of problems listed at the end of Chapter 13.

TYPES OF PROBLEMS

Find empirical probabilities. [13.1]

Find probabilities by looking at the sample space. [13.1]

Decide which of two events is more probable. [13.1]

Find the expectation with a cost of playing. [13.2]

Find the mathematical expectation for a simple event. [13.2]

Find the mathematical expectation for a compound event. [13.2]

Make decisions based on mathematical expectation. [13.2]

Find expectations for roulette. [13.2]

Know the procedure for using tree diagrams to find probabilities. [13.3]

Find the probability of a complement. [13.3]

Find the odds, given the probability. [13.3]

Find the probability, given the odds. [13.3]

Find conditional probabilities. [13.3]

Find the probabilities of unions, intersections, and complements. [13.4]

Find slot machine probabilities. [13.4]

Calculate Keno probabilities. [13.4]

Use tree diagrams to find probabilities [13.4]

Find binomial probabilities. [13.5]
Answer questions involving applied probabilities. [13.1-13.5]

Once again, see if you can verbalize (to yourself) how to do each of the listed types of problems.

    Work all of Chapter 13 Review Questions (whether they are assigned or not). Work through all of the problems before looking at the answers, and *then* correct each of the problems. The entire solution is shown in the answer section at the back of the text. If you worked the problem correctly, move on to the next problem, but if you did not work it correctly (or you did not know what to do), look back in the chapter to study the procedure, or ask your instructor.

    Finally, go back over the homework problems you have been assigned. If you worked a problem correctly, move on to the next problem, but if you missed it on your homework, then you should look back in the book or talk to your instructor about how to work the problem.

    If you follow these steps, you should be successful with your review of this chapter.

We give all of the answers to the chapter review questions (not just the odd-numbered questions).

Chapter 13 Review Questions, page 655

1. C; odds of winning the lottery about 1 to 5,000,000, being struck by lightning about 1 to 600,000, and of appearing on the *Tonight Show* about 1 to 490,000.

2. B; A social security number has 9 digits, so the odd of picking the right one are about 1 in 100,000,000 (too large by a factor of 100). The odds of winning a state lottery are about 1 in 5,000,000 (too large by a factor of 5). A phone number in which the first digit is known leaves six digits unknown, which gives the odds of 1 in 1,000,000.... bingo!

3. $P(\text{defective}) = \frac{4}{1,000} = 0.004$

4. The possible primes are 2, 3, and 5, so $P(\text{prime}) = \frac{3}{6} = \frac{1}{2}$.

5. Look at Figure 13.5;

	1	2	3	4	5	6
1	(1,1)	(1,2)	(1,3)	(1,4)	(1,5)	(1,6)
2	(2,1)	(2,2)	(2,3)	(2,4)	(2,5)	(2,6)
3	(3,1)	(3,2)	(3,3)	(3,4)	(3,5)	(3,6)
4	(4,1)	(4,2)	(4,3)	(4,4)	(4,5)	(4,6)
5	(5,1)	(5,2)	(5,3)	(5,4)	(5,5)	(5,6)
6	(6,1)	(6,2)	(6,3)	(6,4)	(6,5)	(6,6)

$P(\text{eight}) = \frac{5}{36}$

6. There are 4 jacks, 4 queens, 4 kings, and 4 aces, so $P(\text{jack or better}) = \frac{16}{52} = \frac{4}{13}$

7. $P(\text{ace}) = \frac{3}{51} = \frac{1}{17}$

8. Look at Figure 13.5.
    a. $P(5 \text{ on at least one of the dice}) = \frac{11}{36}$

Page 335

b. $P(\text{5 on one die or 4 on the other}) = \frac{20}{36} = \frac{5}{9}$

c. $P(\text{5 on one die and 4 on the other}) = \frac{2}{36} = \frac{1}{18}$

9. a. $P(E) = 0.01; P(\overline{E}) = 1 - P(E) = 1 - 0.01 = 0.99$

b. $P(E) = \frac{9}{10}; P(\overline{E}) = 1 - \frac{9}{10} = \frac{1}{10}$

$$\text{Odds in favor} = \frac{P(E)}{P(\overline{E})} \qquad \text{The odds are 9 to 1.}$$

$$= \frac{\frac{9}{10}}{\frac{1}{10}}$$

$$= \frac{9}{10} \cdot \frac{10}{1}$$

$$= \frac{9}{1}$$

c. Let E be the event. Given $f = 1,000$ and $s = 1$

$$P(E) = \frac{s}{s+f}$$

$$= \frac{1}{1 + 1,000}$$

$$= \frac{1}{1,001}$$

10. $E = P(\text{one}) \cdot 12 \qquad$ The expected value is \$2.

$$= \frac{1}{6}(12)$$

$$= 2$$

11. a. $P(\text{orange} \mid \text{orange}) = \frac{3}{5}$

b. $P(\text{orange} \mid \text{orange}) = \frac{2}{4} = \frac{1}{2}$

12. $P(\overline{A \cup B}) = 1 - P(A \cup B)$

$$= 1 - [P(A) + P(B) - P(A \cap B)]$$

$$= 1 - P(A) - P(B) + P(A \cap B)$$

$$= 1 - P(A) - P(B) + P(A)P(B)$$

$$= 1 - \frac{2}{3} - \frac{3}{5} + \frac{2}{3} \cdot \frac{3}{5}$$

$$= \frac{2}{15}$$

13. $P(X = x) = \binom{n}{x}(0.001)^x(0.999)^{n-x}$

14. $p = 0.85; q = 1 - p = 0.15; P(X = 3) = \binom{5}{3}(0.85)^3(0.15)^2 \approx 0.138$

15. $p = 0.85; q = 1 - p = 0.15; P(X = 4) = \binom{4}{4}(0.85)^4(0.15)^0 \approx 0.522$

Page 336

16. List the sample space:

MM ⎫
MF ⎬ Two success out of 3 possibilities; $P(\text{male}) = \frac{2}{3}$
FM ⎭

FF } Vet rules this out, so delete from the sample space.

17. $P(X \geq 1) = 1 - P(X = 0)$

$$= 1 - \binom{4}{0}(0.35)^0(0.65)^4$$

$$\approx 0.821$$

18. Write out the possible sample spaces. The best possibility is with die C:

		D							
		5	5	5	5	1	1	1	1
	2	5	5	5	5	2	2	2	2
	2	5	5	5	5	2	2	2	2
C	2	5	5	5	5	2	2	2	2
	2	5	5	5	5	2	2	2	2
	2	5	5	5	5	2	2	2	2
	2	5	5	5	5	2	2	2	2
	6	6	6	6	6	6	6	6	6
	6	6	6	6	6	6	6	6	6

We see that there are 64 possibilities and that die C wins 40 of those times; thus, $P(C \text{ wins}) = \frac{40}{64} = \frac{5}{8}$.

19. $P(\text{winning}) = \frac{1}{5}$. It is tempting to say 1 out of 25, but after you pick a key, one of five cars will fit that key.

20. $P(\text{winning a particular car}) = \frac{1}{5} \cdot \frac{1}{5} = \frac{1}{25}$

CHAPTER 14 The Nature of Statistics

> **Chapter Overview**
> This chapter gives a brief overview of statistics. You will hardly finish any college degree without some exposure to statistics. Intelligent manipulation and interpretation of large amounts of information is essential to informed knowledge.

14.1 Frequency Distributions and Graphs, page 661
New Terms Introduced in this Section

Bar graph	Circle graph	Classes	Frequency
Frequency distribution	Graph	Grouped frequency distribution	Interval
Line graph	Pictograph	Pie graph (chart)	Statistics
Stem-and-leaf plot			

Level 1, page 667

1. There are many possible types of graphs. Here are the ones mentioned in this section: A bar graph is a graph consisting of parallel bars or rectangles with lengths proportional to the frequency with which specified quantities occur in a data set. A line graph is a diagram of lines made by connected data points which represent successive changes in the value of a variable. A circle graph, or pie chart, is a circular graph having radii dividing the circle into sectors proportional in angle and area to the relative size of the quantities in a data set. A pictograph is a pictorial representation of numerical data or relationships with each value represented by a proportional number of pictures.

3. It does not mean anything at all without some further information.

5. **a.** Look at the line graph labeled Herb Smith. It is the tallest in the month labeled "O" which is October.
    b. Look at the line graph labeled Lisa Brown. It is the shortest in the second month labeled "A" which is August.

7. **a.** U.S. trade with Canada for the years 1990-2005
    b. The U.S. bars are higher, so the United States has more imports.

9. **a.** The water heater is 11% (from the graph), so $1,100(0.11) = 121$ kWh

    b. The stove is 4%; $1,100(0.04) = 44$ kWh

    c. The refrigerator is 19%; $1,100(0.19) = 209$ kWh

    d. The clothes dryer is 5%; $1,100(0.05) = 55$ kWh

11. Look for the lowest "x" on the graph which is in the C category. It is located at about 67 (grade) with a percent of 60%; the best answer is C.

13. This graph is meaningless without a scale.

15. Graphs are based on height, but the impression is that of area.

17. Three-dimensional bars are used to represent linear data.

19. Look at Figure 14.24. Find the region for 115 lbs; it is the second graph on the top row. Next, find 4 drinks and then look down that column to find 3 hours. The region is red. Now, from the written description in front of the figure, red means an accident is 25 times more likely.

21. Look at Figure 14.24. Find the region for 195 lbs; it is the second graph on the bottom row. Next, find 4 drinks and then look down that column to find 3 hours. The region is gray. Now, from the legend at the bottom, this may be illegal, and definitely is illegal if under 18 years of age.

23. Look at Figure 14.24. Find the region for 185 lbs; it is the first graph on the bottom row. Next, find 6 drinks and then look down that column to find 3 hours. The region is red. Red means that it would be illegal to drive.

Level 2, page 670

25. Data set B

Number of Cars	Tally	Frequency
0	I	1
1	IIII	4
2	HHH III	8
3	II	2
4	I	1

27. Data Set D

Temperature	Tally	Frequency
39	I	1
40	I	1
41		
42		
43	II	2
44	I	1
45	II	2
46		
47	I	1
48		
49	IIII	4
50	IIII	4
51	I	1
52	II	2
53	II	2
54	II	2
55	II	2
56		
57	II	2
58	II	2
59	I	1

Page 340

29.

```
0 | 0  1  1  1  1  2  2  2  2  2  2  2  2  2  3  3  4
```

31.

```
3 | 9
4 | 0  3  3  4  5  5  7  9  9  9
5 | 0  0  0  0  1  2  2  3  3  4  4  5  5  7  7  8  8  9
```

33.

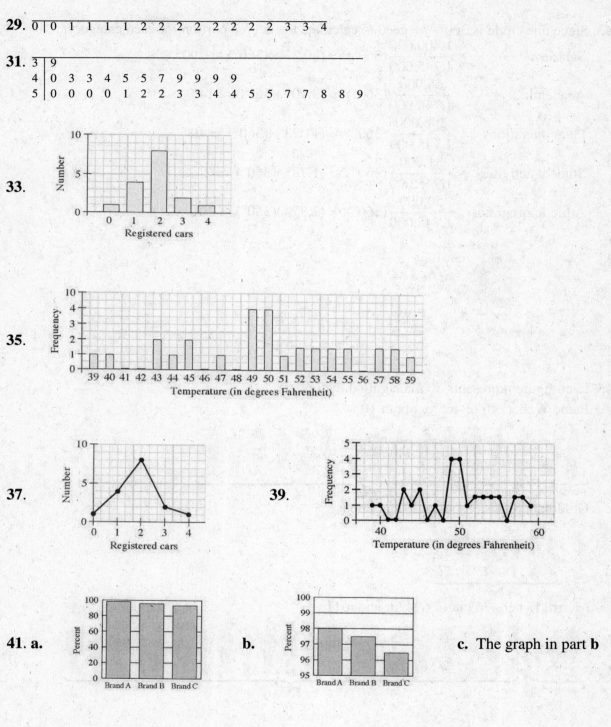

35.

37.

39.

41. a.   **b.**   **c.** The graph in part **b**

Page 341

43. Since one circle is 360°, we need to calculate the degree portion for each category;

Salaries: $\dfrac{1,400,000}{1,745,000}(360°) \approx (80.23\%)(360°) \approx 290°$

Academic $\dfrac{68,000}{1,745,000}(360°) \approx (3.90\%)(360°) \approx 14°$

Plant operations $\dfrac{196,000}{1,745,000}(360°) \approx (11.23\%)(360°) \approx 40°$

Student activities $\dfrac{31,000}{1,745,000}(360°) \approx (1.78\%)(360°) \approx 6°$

Athletic programs $\dfrac{50,000}{1,745,000}(360°) \approx (2.87\%)(360°) \approx 10°$

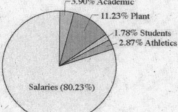

45. Each figure represents 30 managing directors.

Paine Weber: 46 of 465 or about 10%

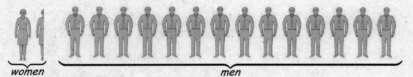

Goldman, Sachs: 9 out of 173 or about 5%

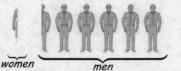

Merrill Lynch: 76 out of 694, or about 11%

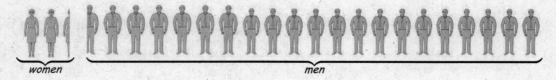

Page 342

47. Look at Figure 14.14 (Problem 8) and the set of bars labeled 2005. We estimate the heights of the bars: imports $78 billion and exports $110 billion. The difference (trade balance) is $78 − $110 = −$32. The trade balance is approximately negative $32 billion.

49. Look at Figure 14.12 (Problem 6) and notice that CBS has the maximum value; almost 15 million households in 1981.

51. You might formulate one of several possible conclusions. The most obvious is that the ratings for all the networks (except for FOX have generally declined from 1986 to the present.

Level 3, page 672

53. Look at the Social Security Reform figure. The correct part to look at is the pictograph at the bottom right of the figure. Next, find 2020 and note that this figure is labeled 2.4 persons.

55. Look at the Social Security Reform figure. The correct part to look at is the line graph at the bottom left of the figure. The graph crosses the axis at the location labeled 36. According to this graph, the social security system will go broke in 2036.

57. The line graph at the top right begins at 50 rather than 0.

59. The advertisement says that the car is 57 in. wide on the outside, but a full 5-ft across on the inside.

14.2 Descriptive Statistics, page 673

New Terms Introduced in this Section

Average	Bimodal	Box plot
Decile	Descriptive statistics	Mean
Measure of central tendency	Median	Measure of dispersion
Measure of position	Mode	Percentile
Quartile	Range	Standard deviation
Variance	Weighted mean	

Level 1, page 681

1. An *average* is a measure of central tendency and includes the mean, median, and mode.

3. A *measure of dispersion* is a single number that tells how closely the data is related to a measure of central tendency and includes the range, standard deviation, and the variance.

5. As the example shows, Andy was better than Dave in both 1989 and 1990. However, the combined batting average for Andy is 253 hits for 969 at bats for an average of 0.261; the combined batting average for Dave is 136 hits for 490 at bats for an average of 0.278; as you can see Dave beat Andy for the combined season.

7. First student's scores: 80, 80, 80, 80, 80, and 80.
   Second student's scores: 100, 100, 100, 60, 50, and 70.

9. The person making the comment scored better in English than 78% of the people taking the test and received the median score in math.

11. A variance of 0 means the data set has no dispersion. In order for this to occur, the data values must all be the same

13. By inspection, mean = 19; median = 19; no mode.

15. By inspection, mean = 767; median = 767; no mode.

17. $\overline{x} = \dfrac{\Sigma x}{n}$ The mean is 11. By inspection, the median is 9 and there is no mode.

   $= \dfrac{55}{5}$

   $= 11$

19. $\overline{x} = \dfrac{\Sigma x}{n}$ The mean is 82. By inspection, the median is 81 and there is no mode.

   $= \dfrac{410}{5}$

   $= 82$

21. $\overline{x} = \dfrac{\Sigma x}{n}$

   $= \dfrac{48}{8}$

   $= 6$

   The mean is 6. Since there is an even number of values, the median is the mean of the middle two data values: Median $= \frac{2+3}{2} = 2.5$. The mode is the most frequently occurring data value, so we see that the mode is 1.

23. First find the range, range $= 21 - 17 = 4$ and the mean, $\overline{x} = 19$ (by inspection). Find the variance and standard deviation by using a calculator or by constructing a table:

Data values	Square of the deviation from the mean
x	$(x - \overline{x})^2$
17	$(17 - 19)^2 = (-2)^2 = 4$
18	$(18 - 19)^2 = (-1)^2 = 1$
19	$(19 - 19)^2 = 0^2 = 0$
20	$(20 - 19)^2 = 1^2 = 1$
21	$(21 - 19)^2 = 2^2 = 4$
Sum:	$\Sigma(x - \overline{x})^2 = 10$

Page 344

Sample variance: $s^2 = \dfrac{\Sigma(x - \overline{x})^2}{n-1}$ Standard deviation: $s = \sqrt{\dfrac{\Sigma(x - \overline{x})^2}{n-1}}$

$\qquad\qquad\qquad = \dfrac{10}{5-1}$ $\qquad\qquad\qquad\quad = \sqrt{2.5}$

$\qquad\qquad\qquad = 2.5$ $\qquad\qquad\qquad\quad \approx 1.58$

25. First find the range, range $= 769 - 765 = 4$ and the mean, $\overline{x} = 767$ (by inspection). Find the variance and standard deviation by using a calculator or by constructing a table:

Data values	Square of the deviation from the mean
x	$(x - \overline{x})^2$
765	$(765 - 767)^2 = (-2)^2 = 4$
766	$(766 - 767)^2 = (-1)^2 = 1$
767	$(767 - 767)^2 = 0^2 = 0$
768	$(768 - 767)^2 = 1^2 = 1$
769	$(769 - 767)^2 = 2^2 = 4$
Sum:	$\Sigma(x - \overline{x})^2 = 10$

Sample variance: $s^2 = \dfrac{\Sigma(x - \overline{x})^2}{n-1}$ Standard deviation: $s = \sqrt{\dfrac{\Sigma(x - \overline{x})^2}{n-1}}$

$\qquad\qquad\qquad = \dfrac{10}{5-1}$ $\qquad\qquad\qquad\quad = \sqrt{2.5}$

$\qquad\qquad\qquad = 2.5$ $\qquad\qquad\qquad\quad \approx 1.58$

27. First find the range, range $= 25 - 1 = 24$ and the mean, $\overline{x} = \dfrac{\Sigma x}{n}$

$\qquad\qquad\qquad\qquad\qquad\qquad\qquad\qquad = \dfrac{55}{5}$

$\qquad\qquad\qquad\qquad\qquad\qquad\qquad\qquad = 11$

Find the variance and standard deviation by using a calculator or by constructing a table:

Data values	Square of the deviation from the mean
x	$(x - \overline{x})^2$
1	$(1 - 11)^2 = (-10)^2 = 100$
4	$(4 - 11)^2 = (-7)^2 = 49$
9	$(9 - 11)^2 = (-2)^2 = 4$
16	$(16 - 11)^2 = 5^2 = 25$
25	$(25 - 11)^2 = 14^2 = 196$
Sum:	$\Sigma(x - \overline{x})^2 = 374$

Page 345

Sample variance: $s^2 = \dfrac{\Sigma(x - \overline{x})^2}{n - 1}$ Standard deviation: $s = \sqrt{\dfrac{\Sigma(x - \overline{x})^2}{n - 1}}$

$\qquad\qquad\quad = \dfrac{374}{5 - 1}$ $\qquad\qquad\qquad\quad = \sqrt{93.5}$

$\qquad\qquad\quad = 93.5$ $\qquad\qquad\qquad\quad \approx 9.67$

29. First find the range, range $= 95 - 70 = 25$ and the mean, $\quad \overline{x} = \dfrac{\Sigma x}{n}$

$\qquad\qquad\qquad\qquad\qquad\qquad\qquad\qquad\qquad\qquad\qquad\qquad = \dfrac{410}{5}$

$\qquad\qquad\qquad\qquad\qquad\qquad\qquad\qquad\qquad\qquad\qquad\qquad = 82$

Find the variance and standard deviation by using a calculator or by constructing a table:

Data values	Square of the deviation from the mean
x	$(x - \overline{x})^2$
70	$(70 - 82)^2 = (-12)^2 = 144$
81	$(81 - 82)^2 = (-1)^2 = 1$
95	$(95 - 82)^2 = 13^2 = 169$
79	$(79 - 82)^2 = (-3)^2 = 9$
85	$(85 - 82)^2 = 3^2 = 9$
Sum:	$\Sigma(x - \overline{x})^2 = 332$

Sample variance: $s^2 = \dfrac{\Sigma(x - \overline{x})^2}{n - 1}$ Standard deviation: $s = \sqrt{\dfrac{\Sigma(x - \overline{x})^2}{n - 1}}$

$\qquad\qquad\quad = \dfrac{332}{5 - 1}$ $\qquad\qquad\qquad\quad = \sqrt{83}$

$\qquad\qquad\quad = 83$ $\qquad\qquad\qquad\quad \approx 9.11$

31. First find the range, range $= 21 - 0 = 21$ and the mean, $\overline{x} = \dfrac{\Sigma x}{n}$

$\qquad\qquad\qquad\qquad\qquad\qquad\qquad\qquad\qquad\qquad\qquad\qquad = \dfrac{48}{8}$

$\qquad\qquad\qquad\qquad\qquad\qquad\qquad\qquad\qquad\qquad\qquad\qquad = 6$

Find the variance and standard deviation by using a calculator or by constructing a table:

Page 346

Data values	Square of the deviation from the mean
x	$(x - \overline{x})^2$
0	$(0 - 6)^2 = (-6)^2 = 36$
1	$(1 - 6)^2 = (-5)^2 = 25$
1	$(1 - 6)^2 = (-5)^2 = 25$
2	$(2 - 6)^2 = (-4)^2 = 16$
3	$(3 - 6)^2 = (-3)^2 = 9$
4	$(4 - 6)^2 = (-2)^2 = 4$
16	$(16 - 6)^2 = 10^2 = 100$
21	$(21 - 6)^2 = 15^2 = 225$
Sum:	$\Sigma(x - \overline{x})^2 = 440$

Sample variance: $s^2 = \dfrac{\Sigma(x - \overline{x})^2}{n - 1}$ Standard deviation: $s = \sqrt{\dfrac{\Sigma(x - \overline{x})^2}{n - 1}}$

$$= \frac{440}{8 - 1}$$
$$= \frac{440}{7}$$

$$= \sqrt{\frac{440}{7}}$$
$$\approx 7.93$$

33. Note that the data are already arranged in order from least to most. First find the median, which is the middle value. since there are an even number of data values, we find the mean of the 5th and 6th values:

$$\frac{431.52 + 433.50}{2} = \$432.51$$

Thus, Q_2 = median = 432.51. Next, find the first quartile Q_1 which is the median of all items below Q_1; $Q_1 = 427.48$. Similarly, $Q_3 = 442.28$. Finally, for the box plot, we note the smallest value is \$415.24 and the largest value is \$457.94.

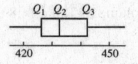

Level 2, page 682

35. mean: $\overline{x} = \dfrac{\Sigma x}{n}$

$$= \frac{12 + 9 + 10 + 16 + 10 + 21}{6}$$

$$= 13$$

For the median list the values in ascending order: $9, 10, 10, 12, 16, 21$, so the median is

Page 347

$\dfrac{10 + 12}{2} = 11$. The mode is the most frequently occurring data value; is 10.

37. We find the weighted mean: $\overline{x} = \dfrac{\Sigma(wx)}{\Sigma w}$

$$= \dfrac{1(40) + 3(50) + 4(60) + 10(70) + 6(80) + 1(90)}{25}$$

$$= 68$$

For the median list the values in ascending order and note that the median is the 13th score; we see it is 70.

The mode is the most frequently occurring data value; it is 70.

The range is $90 - 40 = 50$.

39. mean: $\overline{x} = \dfrac{\Sigma x}{n}$

$$= \dfrac{366 + 391 + 358 + 373 + 346 + 364}{6}$$

$$= \dfrac{1{,}099}{3}$$

Since the units are in thousands we note the mean (rounded to the nearest thousand) is 366,000. For the median list the values in ascending order so the median is $365,000 since $\dfrac{364 + 366}{2} = 365$.

There is no mode (since there is no most frequently occurring data value).

41. Since the mode is the largest value, namely 10, we see answer is A.

43. Since it is a symmetric distribution, we see the answer is C.

45. If you have a calculator with statistical functions, you would want to do this using a calculator. If not, then we find the results as follows.

   a. Look in the next-to-the-last column of Table 14.8 to find the weighted mean:

   $\overline{x} = \dfrac{\Sigma(wx)}{\Sigma w}$

   $$= \dfrac{35 + 3(70) + 2(90) + 100 + 6(110) + 3(120) + 2(130) + 140 + 160}{20}$$

   $$= 105.25$$

   b. Find the middle value of the last column of Table 14.8 (arranged from least to most). There are 20 values so we find the mean of the tenth and eleventh data values. Both are 12, so we see the median is 12 grams from fat.

   c. The most commonly occurring value in the first column of Table 14.8 is 42, so the mode is 42 grams.

Page 348

47. We show the detail here for those whose calculators may not have built-in statistical functions. The mean, $\bar{x} = \dfrac{\Sigma x}{n}$

$$= \frac{12 + 9 + 10 + 16 + 10 + 21}{6}$$

$$= 13$$

We organize our work in table format:

Data values	Square of the deviation from the mean
x	$(x - \bar{x})^2$
12	$(12 - 13)^2 = (-1)^2 = 1$
9	$(9 - 13)^2 = (-4)^2 = 16$
10	$(10 - 13)^2 = (-3)^2 = 9$
16	$(16 - 13)^2 = 3^2 = 9$
10	$(10 - 13)^2 = (-3)^2 = 9$
21	$(21 - 13)^2 = 8^2 = 64$
Sum:	$\Sigma(x - \bar{x})^2 = 108$

Standard deviation: $s = \sqrt{\dfrac{\Sigma(x - \bar{x})^2}{n - 1}}$

$$= \sqrt{\frac{108}{6 - 1}}$$

$$\approx 4.6476$$

Thus, the sample standard deviation (rounded to the nearest unit) is 5.

49. We show the detail here for those whose calculators may not have built-in statistical functions. The weighted mean is: $\bar{x} = \dfrac{\Sigma(wx)}{\Sigma w}$

$$= \frac{1(4) + 3(5) + 4(60) + 10(70) + 6(80) + 1(90)}{25}$$

$$= 68$$

We organize our work in table format:

Frequency	Data values	Square of the deviation from the mean
w	x	$(x - \bar{x})^2$
1	90	$(90 - 68)^2 = 22^2 = 484$
6	80	$(80 - 68)^2 = 12^2 = 144$
10	70	$(70 - 68)^2 = 2^2 = 4$
4	60	$(60 - 68)^2 = (-8)^2 = 64$
3	50	$(50 - 68)^2 = (-18)^2 = 324$
1	40	$(40 - 68)^2 = (-28)^2 = 784$

Page 349

Standard deviation: $s = \sqrt{\dfrac{\Sigma w(x - \overline{x})^2}{n - 1}}$

$= \sqrt{\dfrac{1(484) + 6(144) + 10(4) + 4(64) + 3(324) + 1(784)}{25 - 1}}$

≈ 11.9024

Thus, the sample standard deviation (rounded to the nearest unit) is 12.

51. We show the detail here for those whose calculators may not have built-in statistical functions. The mean is: $\overline{x} = \dfrac{\Sigma x}{n}$

$= \dfrac{366 + 391 + 358 + 373 + 346 + 364}{6}$

$= \dfrac{1{,}099}{3}$

We organize our work in table format:

Data values	Square of the deviation from the mean
x	$(x - \overline{x})^2$
366	$\left(366 - \dfrac{1{,}099}{3}\right)^2 = \left(-\dfrac{1}{3}\right)^2$
391	$\left(391 - \dfrac{1{,}099}{3}\right)^2 = \left(24\dfrac{2}{3}\right)^2$
358	$\left(358 - \dfrac{1{,}099}{3}\right)^2 = \left(-8\dfrac{1}{3}\right)^2$
373	$\left(373 - \dfrac{1{,}099}{3}\right)^2 = \left(6\dfrac{2}{3}\right)^2$
346	$\left(346 - \dfrac{1{,}099}{3}\right)^2 = \left(-20\dfrac{1}{3}\right)^2$
364	$\left(364 - \dfrac{1{,}099}{3}\right)^2 = \left(-20\dfrac{1}{3}\right)^2$
Sum:	$\Sigma(x - \overline{x})^2 = 1{,}141\dfrac{1}{3}$

Standard deviation: $s = \sqrt{\dfrac{\Sigma(x - \overline{x})^2}{n - 1}}$

$= \sqrt{\dfrac{1{,}141\frac{1}{3}}{6 - 1}}$

≈ 15.1085

Thus, the sample standard deviation (rounded to the nearest unit) is 15.

Page 350

53. mean: $\overline{x} = \dfrac{\Sigma x}{n}$

$$= \dfrac{17 + 2(19) + 20 + 21}{5}$$

$$= 19.2$$

Since the units are in thousands, the mean is 19,200 miles.

The range is $21,000 - 17,000 = 4,000$ miles.

The standard deviation:

$s = \sqrt{\dfrac{\Sigma(x - \overline{x})^2}{n-1}}$

$$= \sqrt{\dfrac{(17-19.2)^2 + 2(19-19.2)^2 + (20-19.2)^2 + (21-19.2)^2}{5-1}}$$

$$= \sqrt{2.2}$$

$$\approx 1.483239697$$

Since the units are in thousands, the standard deviation is 1,483 miles.

55. Go ahead and roll a pair of dice 20 times, and then answer the questions based on the results of your tally. For example, if the results are

Outcome	Tally	Frequency
2	I	1
3		0
4	II	2
5	I	1
6	HHH	5
7	IIII	4
8	III	3
9	II	2
10	I	1
11	I	1
12		0

weighted mean:

$\overline{x} = \dfrac{\Sigma wx}{w}$

$$= \dfrac{1(2) + 0(3) + 2(4) + 1(5) + 5(6) + 4(7) + 3(8) + 2(9) + 1(10) + 1(11) + 0(12)}{20}$$

$$= 6.75$$

Page 351

The median is the mean of the 10th and 11th data values; it is 7 since both the 10th and the 11th values are 7. The mode is the most frequently occurring value, which is 6. The range is $11 - 2 = 9$.

Level 3, page 684

57. a. $\overline{x} = \dfrac{\Sigma x}{n}$

$= \dfrac{2(2) + 2(5) + 7 + 2(8) + 2(9) + 10}{10}$

$= 6.5$

H.M. $= \dfrac{n}{\Sigma \frac{1}{x}}$

$= \dfrac{10}{\frac{1}{2} + \frac{1}{2} + \frac{1}{5} + \frac{1}{5} + \frac{1}{7} + \frac{1}{8} + \frac{1}{8} + \frac{1}{9} + \frac{1}{9} + \frac{1}{10}}$

$= \dfrac{10}{\frac{533}{252}}$

≈ 4.7

b. $\overline{x} = \dfrac{\Sigma x}{n}$ $\qquad$ H.M. $= \dfrac{n}{\Sigma \frac{1}{x}}$

$= \dfrac{52 + 61}{2}$ $\qquad\qquad = \dfrac{2}{\frac{1}{52} + \frac{1}{61}}$

$= 56.5$ $\qquad\qquad\qquad = \dfrac{2}{\frac{113}{3,172}}$

$\qquad\qquad\qquad\qquad\qquad \approx 56.1$

The arithmetic mean is 56.5 mph and the harmonic mean is 56.1 mph.

Level 3, Problem Solving, page 684

59. a. Population A is Player A and population B is Player B. $r_1 = 0.223$, $r_2 = 0.284$, $r = 0.257$, $R_1 = 0.232$, $R_2 = 0.296$ and $R = 0.251$.
Finally, $C_1 =$ against right-handed pitchers and $C_2 =$ against left-handed pitchers

b. Examples vary.

Page 352

14.3 The Normal Curve, page 685

New Terms Introduced in this Section

Bell-shaped curve   Continuous distribution   Cumulative frequency   Normal curve
Skewed distribution z-score

Level 1, page 692

1. **A** cumulative frequency is the sum of all preceding frequencies in which some order has been established.

3. A normal curve is shown in Figure 14.35.

5. In a normal curve that is skewed to the right, the mean is to the right of the median.

7.   Cumulative

0	5%
1	16%
2	45%
3	79%
4	94%
5	99%
6 or more	100%

$$\overline{x} = \frac{0(0.05) + 1(0.11) + 2(0.29) + 3(0.34) + 4(0.15) + 5(0.05) + 6(0.01)}{1}$$

$$\approx 2.62$$

The mean number of bedrooms per home is 2.62.

The cumulative distribution is 50% for the value 3, so the median number of bedrooms per home is 3. The most frequently occurring value is 3, so the mode is 3 bedrooms per home.

9.   Cumulative

0	1%
1	12%
2	47%
3	68%
4	88%
5	94%
6	97%
7	99%
8	100%

Page 353

$$\bar{x} = \frac{0(0.01) + 1(0.11) + 2(0.35) + 3(0.21) + 4(0.20) + 5(0.06) + 6(0.03) + 7(0.02) + 8(0.01)}{1}$$

≈ 2.94

The mean number of exemptions is 2.94.

The cumulative distribution is 50% for the value 3, so the median number of exemptions is 3. The most frequently occurring mean value is 2, so the mode is 2 exemptions.

11.

	Cumulative
2	0.28
6	0.40
8	0.75
12	0.81
16	1.00

$$\bar{x} = \frac{2(0.28) + 6(0.12) + 8(0.35) + 12(0.06) + 16(0.19)}{1}$$

≈ 7.84

The mean is 7.84 oz/tube.

The cumulative distribution is 50% for the value 8, so the median is 8 oz/tube.

The most frequently occurring mean value is 8, so the mode is 8 oz/tube.

For Problems 13-23, remember a normal curve is symmetric about the mean.

13. Use Table 14.10; to find $z = 1.4$, look at the row labeled 1.4 and the column headed 0.00 to find the entry 0.4192. The percent is 41.92%.

15. Use Table 14.10; to find $z = 2.43$, look at the row labeled 2.4 and the column headed 0.03 to find the entry 0.4925. The percent is 49.25%.

17. See the note at the bottom of Table 14.10. Since $z = 3.25$, use the value 0.4999. The percent is 49.99%.

19. Use Table 14.10; and symmetry to find $z = 2.33$. Look at the row labeled 2.3 and the column headed 0.03 to find the entry 0.4901. The percent is 49.01%.

21. Use Table 14.10; and symmetry to find $z = 0.46$. Look at the row labeled 0.4 and the column headed 0.06 to find the entry 0.1772. The percent is 17.72%.

23. Use Table 14.10; and symmetry to find $z = 2.22$. Look at the row labeled 2.2 and the column headed 0.02 to find the entry 0.4868. The percent is 48.68%.

25. The mean (μ) is 170 cm and the standard deviation (σ) is 5 cm.

a. 175 is 1 standard deviation above the mean, and 165 is 1 standard deviation below the mean, we see from Figure 14.35 that we would include 2(34.1%) = 68.2%. Out of the 50 people we would include 50(0.682) = 34.1 ≈ 34 people.

b. $x = 168 \, \text{cm}; \quad z = \dfrac{x - \mu}{\sigma}$ From Table 14.10, use $z = 0.4$ to find 15.54% is

$$= \frac{168 - 170}{5}$$

$$= -0.4$$

between $z = -0.4$ and the mean. We also know that 50% of the area under the normal curve is to the right of the mean, the total is $0.1554 + 0.500 = 0.6554$. We would expect $50(0.6554) = 32.77$, or 33 people to be taller than 168 cm.

27. $x = 163 \, \text{cm}; \quad z = \dfrac{x - \mu}{\sigma}$ From Table 14.10, use $z = 1.4$ to find 41.92% to be the

$$= \frac{163 - 170}{5}$$

$$= -1.4$$

probability that a randomly selected person's height is between 163 cm and the mean height of 170 cm. From the symmetry of the normal curve, we know 50% of the area under the curve is to the right of the mean; thus, the probability that a person is taller than 163 cm is the sum $41.92\% + 50\% = 91.92\%$.

29.

Height	Number	Cumulative
155		0.1%
160	1	2.3%
165	7	15.9%
170	17	50.0%
175	17	84.1%
180	7	97.7%
185	1	99.9%
190		100.0%

31. Grading on a curve refers to Figure 14.37, on page 687. To obtain an A, a test score that is at least two standard deviations above the mean is necessary. That is, $x \geq \mu + 2\sigma$. We know $\mu = 50$ and $\sigma = 5$, so it follows that $\mu + 2\sigma = 50 + 2(5) = 60$. A score of 60 or above is necessary.

Page 355

33. Since the standard deviation is the square root of the variance, the variance is the square of the standard deviation; that is var. $= 5^2 = 25$.

Level 2, page 693

35. A; $50\% - 6\% = 44\%$; Look at Table 14.10 to find x when $z = 1.56$, $\mu = 75$, and $\sigma = 8$.

$$z = \frac{x - \mu}{\sigma} \qquad \text{The cutoff is 87.}$$

$$1.56 = \frac{x - 75}{8}$$

$$x = 87.48$$

37. Since the bottom 6% receive Fs, and the next 16% receive Ds, it follows that 28% is

between the mean and the cutoff for a C. That is, $6\% + 16\% = 22\%$ and

$50\% - 22\% = 28\%$. Look at Table 14.10 to find the z-value which gives an area closest to

28%. It is 0.2794 which corresponds to $z = -0.77$. Since $\mu = 75$, and $\sigma = 8$, we see

$$z = \frac{x - \mu}{\sigma} \qquad \text{This means the lowest C grade is 69.}$$

$$-0.77 = \frac{x - 75}{8}$$

$$x = 68.84$$

39.

Scores	Grade	Percent	Cumulative
87 and over	A	6%	6%
81-86	B	16%	22%
69-80	C	56%	78%
63-68	D	16%	94%
62 and under	F	6%	100%

41. Answers vary. The mode is the largest; see Figure 14.41.

43. $2.55 = \dfrac{x - 85.7}{4.85}$ **45.** $-3.46 = \dfrac{x - 85.7}{4.85}$

  $x = 98.0675$ $x = 68.919$

47. One standard deviation is 0.3413, so greater than one standard deviation is
   $0.50 - 0.3413 = 0.1587$.

49. $z = \dfrac{36 - 35.5}{2.5}$ From Table 14.10, we find 0.0793, so the probability is

   $= 0.2$

Page 356

0.50 − 0.0793 = 0.4207 or 42.07%.

51. Variance of 9 is a standard deviation of $\sigma = 3$. We find the z-score:

$$z = \frac{159 - 165}{3}$$

$$\approx -2$$

From Table 14.10, we look at -2 to find the area of the left of $z = -2$ which is
$0.500 - 0.4772 = 0.0228$ so the probability is about 2.28%.

53. $z = \dfrac{0.41 - 0.4}{0.02}$ From Table 14.10, the area is 0.1915, so the probability of

$= 0.50$

exceeding this is $0.50 - 0.1915 = 0.3085$.

55. Var = 0.04; $\sigma = 0.2$

   a. 50% are less than the mean

   b. 2.3% is the percent that is greater than two standard deviations above mean:

     $12 + 2(0.2) = 12.4$ oz.

Level 3, page 694

57. The graph **a** has less variance and graph **b** has more variance.

Level 3 Problem Solving, page 694

59.

−4	0.00001
−3	0.00195
−2	0.06250
−1	0.50000
0	1.00000
1	0.50000
2	0.06250
3	0.00195
4	0.00001

14.4 Correlation and Regression, page 694

New Terms Introduced in this Section

Correlation	Least squares line	Least squares method
Linear correlation coefficient	Pearson correlation coefficient	Regression analysis
Scatter diagram	Significance level	

Level 1, page 699

1. Correlation is a number which measures the relationship between two variables.

3. The linear correlation coefficient, r, is $r = \dfrac{n\Sigma xy \ - \ (\Sigma x)(\Sigma y)}{\sqrt{n(\Sigma x^2) \ - \ (\Sigma x)^2} \ \sqrt{n(\Sigma y^2) \ - \ (\Sigma y)^2}}$

5. Compare the scatter diagram with those shown in Figure 14.44 to see that there is a strong positive correlation.

7. Look at Table 14.11 to find $n = 10$; since $|r| = 0.7,$ it is less that 0.765, so it is not significant at the 1% level.

9. Look at Table 14.11 to find $n = 30$; since $|r| = 0.4,$ it is greater than 0.361, so it is significant at the 5% level.

11. Look at Table 14.11 to find $n = 35$; since $|r| = 0.413,$ it is less that 0.430, so it is not significant at the 1% level.

13. Look at Table 14.11 to find $n = 25$; since $|r| = 0.521,$ it is greater than 0.505, so it is significant at the 1% level.

15. Look at Table 14.11 to find $n = 40$; since $|r| = 0.416,$ it is greater than 0.312, so it is significant at the 5% level. By the way, it is also significant at the 1% level.

17. Look at Table 14.11 to find $n = 10$; since $|r| = 0.56,$ it is less than 0.765, so it is not significant at the 1% level.

19. If you do not have a calculator, then $n = 4, \Sigma x = 29, \Sigma y = -40, \Sigma xy = -350,$ $\Sigma x^2 = 241,$ and $\Sigma y^2 = 600$; then

$$r = \dfrac{n\Sigma xy - (\Sigma x)(\Sigma y)}{\sqrt{n(\Sigma x^2) - (\Sigma x)^2} \ \sqrt{n(\Sigma y^2) - (\Sigma y)^2}}$$

$r = -0.765$; not significant

$$= \dfrac{4(-350) - (29)(-40)}{\sqrt{4(241) - 29^2} \sqrt{4(600) - (-40)^2}}$$

$$= -0.765$$

21. If you do not have a calculator, then $n = 5$, $\Sigma x = 20$, $\Sigma y = 96$, $\Sigma xy = 308$, $\Sigma x^2 = 108$, and $\Sigma y^2 = 2{,}070$; then

$$r = \frac{n\Sigma xy - (\Sigma x)(\Sigma y)}{\sqrt{n(\Sigma x^2) - (\Sigma x)^2}\,\sqrt{n(\Sigma y^2) - (\Sigma y)^2}}$$

$r = -0.954$, significant at 5%

$$= \frac{5(308) - (20)(96)}{\sqrt{5(108) - 20^2}\,\sqrt{5(2{,}070) - (96)^2}}$$

$$= -0.954$$

23. If you do not have a calculator, then $n = 5$, $\Sigma x = 482$, $\Sigma y = 203$, $\Sigma xy = 18{,}881$, $\Sigma x^2 = 46{,}754$, and $\Sigma y^2 = 10{,}309$; then

$$r = \frac{n\Sigma xy - (\Sigma x)(\Sigma y)}{\sqrt{n(\Sigma x^2) - (\Sigma x)^2}\,\sqrt{n(\Sigma y^2) - (\Sigma y)^2}}$$

$r = -0.890$, significant at 5%

$$= \frac{5(18{,}881) - (482)(203)}{\sqrt{5(46{,}754) - 2482^2}\,\sqrt{5(10{,}309) - (203)^2}}$$

$$= -0.890$$

25. If you do not have a calculator, then $n = 4$, $\Sigma x = 29$, $\Sigma y = -40$, $\Sigma xy = -350$, and $\Sigma x^2 = 241$; then

$$m = \frac{n(\Sigma xy) - (\Sigma x)(\Sigma y)}{n(\Sigma x^2) - (\Sigma x)^2} \qquad \text{and} \qquad b = \frac{\Sigma y - m(\Sigma x)}{n}$$

$$= \frac{4(-350) - (29)(-40)}{4(241) - 29^2} \qquad\qquad\qquad = \frac{(-40) - (-\frac{80}{41})(29)}{4}$$

$$= -\frac{80}{41} \qquad\qquad\qquad\qquad\qquad \approx 4.146$$

$$\approx -1.95$$

$$y' = -1.95x + 4.146$$

27. If you do not have a calculator, then $n = 5$, $\Sigma x = 20$, $\Sigma y = 96$, $\Sigma xy = 308$, and $\Sigma x^2 = 108$; then

$$m = \frac{n(\Sigma xy) - (\Sigma x)(\Sigma y)}{n(\Sigma x^2) - (\Sigma x)^2} \quad \text{and} \quad b = \frac{\Sigma y - m(\Sigma x)}{n}$$

$$= \frac{5(308) - (20)(96)}{5(108) - 20^2} \qquad\qquad = \frac{96 - (-\frac{19}{7})(20)}{5}$$

$$= -\frac{19}{7} \qquad\qquad\qquad\qquad \approx 30.0571$$

$$\approx -2.7143$$
$$y' = -2.7143x + 30.0571$$

29. If you do not have a calculator, then $n = 5$, $\Sigma x = 482$, $\Sigma y = 203$, $\Sigma xy = 18{,}881$, and $\Sigma x^2 = 46{,}754$; then

$$m = \frac{n(\Sigma xy) - (\Sigma x)(\Sigma y)}{n(\Sigma x^2) - (\Sigma x)^2} \quad \text{and} \quad b = \frac{\Sigma y - m(\Sigma x)}{n}$$

$$= \frac{5(18{,}881) - (482)(203)}{5(46{,}754) - 482^2} \qquad\qquad = \frac{203 - (-\frac{1{,}147}{482})(482)}{5}$$

$$= -\frac{1{,}147}{482} \qquad\qquad\qquad\qquad = 270$$

$$\approx -2.380$$
$$y' = -2.380x + 270.00$$

31. Notice that graphs A, D, and E show a positive correlation and therefore are paired with equations where $m > 0$. Considering those three possibilities, we note that all of these seem to have the same y-intercept and slopes which are not discernable. We note that this problem has the highest correlation coefficient, so we choose the one that has the least dispersion for the data points. The graph shown in D is the best choice.

33. Notice that graphs A, D, and E show a positive correlation and therefore are paired with equations where $m > 0$. Considering those three possibilities, we note that all of these seem to have the same y-intercept and slopes which are not discernable. We note that this problem has the smallest correlation coefficient, so we choose the one that has the most dispersion for the data points. The graph shown in A is the best choice.

35. Notice that graphs B, C, and F show a negative correlation and therefore are paired with equations where $m < 0$. Considering those three possibilities, we note that all of these seem to have the same y-intercept and slopes which are not discernable. We note that this problem has a correlation coefficient between the highest and the smallest, so we choose the one that is in the middle. The graph shown in C is the best choice.

Level 2, page 700

37. $m = \dfrac{8(240) - 40(48)}{8(210) - (40)^2} = 0$ $\qquad b = \dfrac{48 - m(40)}{8} = 6$; Least squares line is $y' = 6$

$$r = \frac{8(240) - 40(48)}{\sqrt{8(210) - (40)^2}\sqrt{8(288) - (48)^2}} = 0 \quad \text{Correlation coefficient is 0.}$$

39. $m = \dfrac{6(322) - 36(48)}{6(250) - (36)^2} = 1 \qquad b = \dfrac{48 - m(36)}{6} = 2\,;\ \text{Least squares line is } y' = x + 2$

$r = \dfrac{6(322) - 36(48)}{\sqrt{6(250) - (36)^2}\sqrt{6(418) - (48)^2}} = 1\,;\ \text{Correlation coefficient is 1.}$

41. Let x represent beer consumption and y represent wine consumption.

Year	1988	1989	1990	1991	1993	1995
x, beer consumption	24.7	25.4	24	23.1	22.8	22.6
y, beer consumption	2.24	2.09	2.05	1.85	1.74	1.80

If you do not have a calculator, then $n = 6, \Sigma x = 142.6, \Sigma y = 11.77, \Sigma xy = 280.701,$
$\Sigma x^2 = 3{,}395.46,$ and $\Sigma y^2 = 23.2783;$ then

$$r = \frac{n\Sigma xy - (\Sigma x)(\Sigma y)}{\sqrt{n(\Sigma x^2) - (\Sigma x)^2}\,\sqrt{n(\Sigma y^2) - (\Sigma y)^2}}$$

$$= \frac{6(280.701) - (142.6)(11.77)}{\sqrt{6(3{,}395.46) - 142.6^2}\,\sqrt{6(23.2783) - (11.77)^2}}$$

$$\approx 0.8830$$

From Table 14.11, we see that for $n = 6$ at the 5% level the critical value is 0.811 and at
1% it is 0.917. Since $|r| = 0.8830$ is greater than 0.811 but not greater than 0.917, we
conclude that there is a significant linear correlation at the 5% level between drinking beer
and drinking wine.

43. Let x represent the difficulty level and y represent the time.   If you do not have a
calculator, then $n = 8, \Sigma x = 27,\ \Sigma y = 115,\ \Sigma xy = 432,\ \Sigma x^2 = 109,$ and $\Sigma y^2 = 1{,}805;$
then

$$r = \frac{n\Sigma xy - (\Sigma x)(\Sigma y)}{\sqrt{n(\Sigma x^2) - (\Sigma x)^2}\,\sqrt{n(\Sigma y^2) - (\Sigma y)^2}}$$

$$= \frac{8(432) - (27)(115)}{\sqrt{8(109) - 27^2}\,\sqrt{8(1{,}805) - (115)^2}}$$

$$\approx 0.8421$$

From Table 14.11, we see that for $n = 8$ at the 1% level the critical value is 0.834. Since
$|r| = 0.8421$ is greater than 0.834, we conclude that there is a significant linear correlation
at the 1% level.

45. $y' = 2.4545x + 6.0909$

47. $r = 0.358$

No significant correlation

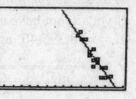

49. $r = -0.936$, significant at 1%   **51.** $y' = 0.468x + 19.235$ **53.** $y' = -6.186x + 87.179$

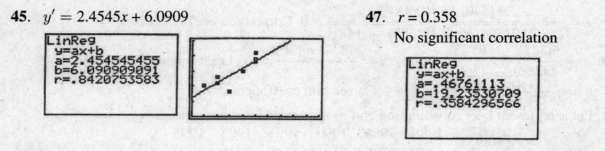

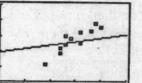

Level 3, page 701

In Problems 54-56, don't forget to change minutes to parts of a hour.

55. a. $100.1 = 3.6x + 1.04$ Time is 27:31.

       $x = 27.5167$

     The difference is 4 hr 29 minutes.

   b. The reason is that he may be getting tired.

57. a. $100.1 = 2.17x + 3.87$

       $x = 44.35$

     Change 0.35 to minutes; namely $60(0.35) \approx 21$.

   b. Time is 44:21 or 44 hr 21 min, which is 3 hours and 29 minutes less than his actual time. The reason for the slower time is that he is, no doubt, getting tired.

59. Let x represent the age (in years) and y represent the finishing time (in decimal hours, rounded to two decimal places). If you do not have a calculator, then

$n = 36, \Sigma x = 1{,}513, \Sigma y = 1{,}499.91, \Sigma xy = 63{,}579.39, \Sigma x^2 = 65{,}847$, and $\Sigma y^2 = 63{,}391.5471$; then

$$r = \frac{n\Sigma xy - (\Sigma x)(\Sigma y)}{\sqrt{n(\Sigma x^2) - (\Sigma x)^2}\,\sqrt{n(\Sigma y^2) - (\Sigma y)^2}}$$

$$= \frac{36(63{,}579.39) - (1{,}513)(1{,}499.91)}{\sqrt{36(65{,}847) - (1{,}513)^2}\sqrt{36(63{,}391.5471) - (1{,}499.91)^2}}$$

$$\approx 0.379976 \text{ or } 0.380$$

Page 362

From Table 14.11, we see that for $n = 35$ at the 5% level the critical value is 0.335 and the critical value at the 1% level is 0.430. Additionally, check $n = 40$, at the 1% level the critical value is 0.402. We know the critical values for $n = 36$ must be between those for $n = 35$ and $n = 40$. Therefore, since $|r| = 0.380$ is greater that 0.335 (for $\alpha = 0.05$ and $n = 35$) but not greater than 0.402 ($\alpha = 0.01$ and $n = 40$), we conclude there is a significant linear correlation at the 5% level (but not at the 1% level) between age and running time.

14.5 Sampling, page 702

New Terms Introduced in this Section

Fallacy of exceptions	Population	Inferential statistics	Sample
Target population	Type I error	Type II error	

Level 1, page 706

1. Dictionary: (1) facts or data of a numerical kind, assembled, classified, and tabulated so as to present significant information about a given subject (2) the science of assembling, classifying, tabulating, and analyzing such facts and data (*Webster's New World Dictionary*).

3. The fallacy of exception occurs when you form a conclusion by looking at a single case.

5. The target population is Pacific Bell's customers, so B is the correct choice.

7. The target population is the registered voters, so B is the correct choice.

9. The target population are members of the environmental group, so D is the correct choice.

11. The target population is the customers of Betty's Koi Emporium, so A is the correct choice.

13. The target population is the customers of an espresso coffee franchise, so D is the correct choice.

Level 2, page 708

Answers to Problems 14-23 may vary.

15. Survey AT&T's customers by randomly choosing from a list of AT&T's customers.

17. Survey a selection of people whose names are randomly chosen from a local list of registered voters who are party members.

19. Survey a selection of people whose names are randomly chosen from a list of all union members.

21. Survey a random selection of students chosen from a list of all students.

23. Survey a selection of people whose names are randomly chosen from a list of all customers.

Page 363

Level 3, page 708

25. The four possibilities are:
    (1) You accept that 72 is the mean, and it is the mean.
    (2) You accept that 72 is the mean, and it is not the mean.
        This is a Type II error (accept a false hypotheses).
    (3) You do not accept that 72 is the mean, and it is the mean.
        This is a Type I error (reject a true hypotheses).
    (4) You do not accept that 72 is the mean, and it is not the mean.

Level 3 Problem Solving, page 708

27. Conduct the survey by asking at least 25 people. Part of your answer needs to be how you select your respondents.

29. Conduct the survey by asking at least 25 people. Discuss how you are going to reach people in the target population. Part of your answer needs to be how you select your respondents.

Chapter 14 Review Questions, page 709

STOP

Studying for a chapter examination is a personal process, one which nobody else can do for you. Simply take the time to review what you have done. Here are the new terms in Chapter 14.

Average [14.2]	Frequency distribution [14.1]	Measures of dispersion [14.2]
Bar graph [14.1]	Graph [14.1]	Measures of position [14.2]
Bell-shaped curve [14.3]	Grouped frequency	Median [14.2]
Bimodal [14.2]	distribution [14.1]	Mode [14.2]
Box plot [14.2]	Inferential statistics [14.5]	Normal curve [14.3]
Circle graph [14.1]	Interval [14.1]	Pearson correlation
Classes [14.1]	Least squares line [14.4]	coefficient [14.4]
Continuous distribution [14.3]	Least squares method [14.4]	Percentile [14.2]
Correlation [14.4]	Line graph [14.1]	Pictograph [14.1]
Cumulative frequency [14.3]	Linear correlation	Pie chart [14.1]
Deciles [14.2]	coefficient [14.4]	Population [14.5]
Descriptive statistics [14.2]	Mean [14.2]	Quartile [14.2]
Fallacy of exceptions [14.5]	Measures of central	Range [14.2]
Frequency [14.1]	tendency [14.2]	Regression analysis [14.4]

Sample [14.5]   Statistics [overview]   Variance of a population [14.2]
Scatter diagram [14.4]   Stem-and-leaf plot [14.1]   Variance of a random
Significance level [14.4]   Target population [14.5]   sample [14.2]
Skewed distribution [14.3]   Type I error [14.5]   Weighted mean [14.2]
Standard deviation [14.2]   Type II error [14.5] z-score [14.3]

If you can describe the term, read on to the next one; if you cannot, then look it up in the text (the section number is shown in brackets). Next, study the types of problems listed at the end of Chapter 14.

TYPES OF PROBLEMS
Prepare a frequency distribution. [14.1]
Draw a bar graph. [14.1]
Draw a line graph. [14.1]
Draw a stem-and-leaf plot. [14.1]
Draw a circle graph. [14.1]
Draw a pictograph. [14.1]
Read and interpret relationships presented in graphical form. [14.1]
Recognize misuses of graphs. [14.1]
Find the mean, median, and mode for a set of data. [14.2]
Decide on an appropriate measure of central tendency. [14.2]
Find the range, standard deviation, and variance for a set of data. [14.2]
Find a cumulative distribution. [14.3]
Interpret information given in table form. [14.3]
Find the expected numbers for ranges of a normally distributed set of data. [14.3]
Determine the probability of falling within a certain range of a normally distributed set of data. [14.3]
Find and use z-scores. [14.3]
Know the meanings associated with a normal curve. [14.3]
Draw a scatter diagram for a data set. [14.4]
Decide whether there is a significant linear correlation between two given variables. [14.4]
Find a regression line for a data set.[14.4]
Discuss the type of correlation for a given data set. [14.4]
Determine whether there is a significant linear correlation, given the number of items and the
    correlation coefficient. [14.4]
Find the correlation coefficient for a given set of data. [14.4]
Choose an appropriate procedure for selecting an unbiased sample. [14.4]
Classify Type I and Type II errors. [14.5]
Make an inference about a population by taking a sample. [14.5]
Once again, see if you can verbalize (to yourself) how to do each of the listed types of problems.

Work all of Chapter 14 Review Questions (whether they are assigned or not). Work through all of the problems before looking at the answers, and *then* correct each of the problems. The entire solution is shown in the answer section at the back of the text. If you worked the problem correctly, move on to the next problem, but if you did not work it correctly (or you did not know what to do), look back in the chapter to study the procedure, or ask your instructor.

Finally, go back over the homework problems you have been assigned. If you worked a problem correctly, move on to the next problem, but if you missed it on your homework, then you should look back in the book or talk to your instructor about how to work the problem.

If you follow these steps, you should be successful with your review of this chapter.

We give all of the answers to the chapter review questions (not just the odd-numbered questions).

Chapter 14 Review Questions, page 709

1. Heads: ⦀⦀ ⦀⦀ ⦀⦀ ⦀⦀ ⦀⦀ ⦀⦀ ⦀⦀ ⦀⦀ ||| (18); Tails ⦀⦀ ⦀⦀ ⦀⦀ ⦀⦀ ⦀⦀ ⦀⦀ ⦀⦀ ⦀⦀ || (22)

2.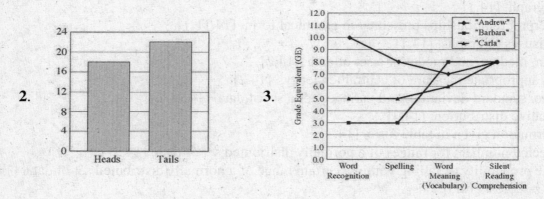

3.

4. mean = $\frac{5 + 21 + 21 + 25 + 30 + 40}{6} = 23\frac{2}{3}$

5. The median is the mid-value or the mean of the mid-values; $\frac{21 + 25}{2} = 23$.

6. The mode is the most frequently occurring value, namely 21.

7. The range is $40 - 5 = 35$.

8. The standard deviation is found:

score	(deviation from the mean)^2
5	348.44
21	7.11
21	7.11
25	1.78
30	40.11
40	266.78

Page 366

$$\frac{348.44 + 7.11 + 7.11 + 1.78 + 40.11 + 266.78}{6 - 1} = \frac{671.33}{5}$$

$$\approx 134.266$$

The standard deviation is $\sqrt{134.266} \approx 11.59$.

9. weighted mean $= \dfrac{6(5) + 10(10) + 12(35)}{5 + 10 + 35}$

$$= \frac{550}{50}$$

$$= 11$$

Since there are 50 data values, the median is the mean of the 25th and 26th values (when they are arranged in order).   median $= \dfrac{12 + 12}{2}$

$$= 12$$

The mode is the most frequently occurring data value, namely 12.

The purchase of a cola is really a popularity contest, so the mode is the most appropriate measure.

10. mean $= \dfrac{72 + 73 + 74 + 85 + 91}{5}$

$$= \frac{395}{5}$$

$$= 79$$

The mid-value, or median is 74.

There is no mode.

The most appropriate measure of a student's scores is the measure that reflects the entire distribution, so the mean is the most appropriate measure.

11. $\overline{x} = \dfrac{\Sigma x}{n}$

$$= \frac{\$2,381,876}{21}$$

$$\approx \$113,423$$

Since there are 21 data values, the median is the 11th value (counted from either the top or bottom when the data values are arranged in order); median = \$110,750.

There is no most frequently occurring value, so there is no mode.

When considering something such as salaries, there may be extremely large or small extreme values, so the median is the most appropriate measure.

Page 367

12. a. There were many more survivors among the first class, and an extraordinarily small number among the third class passengers, so the answer is yes.

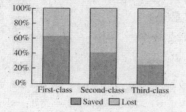

b. There were more survivors amoung the children then there were for the men, so the answer is yes.

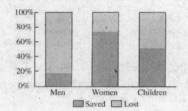

13. a. LTL cost the most in October.
   b. Courier Air generally had the lower monthly cost.
   c. Ocean and TL are about the same in October.

14.

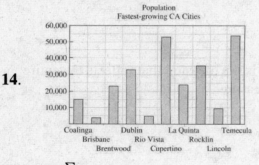

15. $\overline{x} = \dfrac{\Sigma x}{n}$

$= \dfrac{255{,}585}{10}$

$= 25{,}558.5$

There are 10 data points, so $\text{median} = \dfrac{23{,}100 + 24{,}250}{2}$

$= 23{,}675$

There is no value occurring more than once, so there is no mode.

Page 368

16. The mean, $\overline{x} = 25,558.5$.

City	(deviation from the mean)^2
Coalinga	$(15,200 - 25,558.5)^2 = (-10,358.5)^2 = 107,298,522.25$
Brisbane	$(4,060 - 25,558.5)^2 = (-21,498.5)^2 = 462,185,502.25$
Brentwood	$(23,100 - 25,558.5)^2 = (-2,458.5)^2 = 6,044,222.25$
Dublin	$(32,500 - 25,558.5)^2 = 6,941.5^2 = 48,184,422.25$
Rio Vista	$(4,850 - 25,558.5)^2 = (-20,708.5)^2 = 428,841,972.25$
Cupertino	$(52,900 - 25,558.5)^2 = 27,341.5^2 = 747,557,622.25$
La Quinta	$(24,250 - 25,558.5)^2 = (1,308.5)^2 = 1,712,172.25$
Rocklin	$(35,125 - 25,558.5)^2 = 9,566.5^2 = 91,517,922.25$
Lincoln	$(9,675 - 25,558.5)^2 = (-15,883.5)^2 = 252,285,572.25$
Temecula	$(53,800 - 25,558.5)^2 = 28,241.5^2 = 797,582,322.25$
TOTAL	$\Sigma(x - \overline{x})^2 = 2,945,617,502.5$

The standard deviation is:
$$\sigma = \sqrt{\frac{\Sigma(x - \overline{x})^2}{n - 1}}$$
$$= \sqrt{\frac{2,945,617,502.5}{9}}$$
$$\approx 18,091$$

17. a. **b.** **c.**

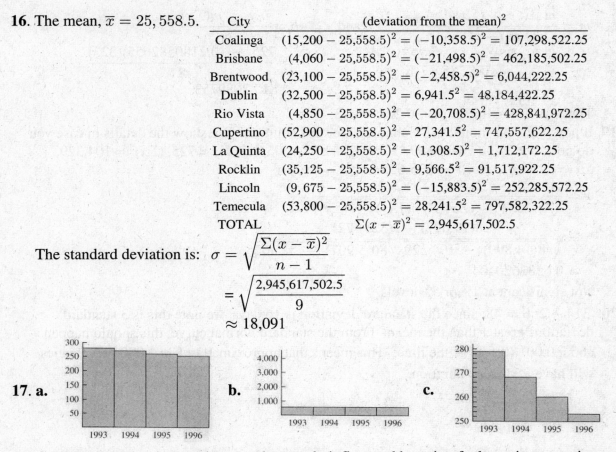

d. Answers vary; impressions can be greatly influenced by using faulty or inappropriate scale (or even worse, no scale at all).

18. You many have a calculator which will draw both the scatter diagram and the regression line.

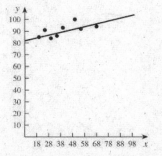

If you cannot do it using technology, we show the details:

Page 369

$$m = \frac{n(\Sigma xy) - (\Sigma x)(\Sigma y)}{n(\Sigma x^2) - (\Sigma x)^2} \qquad \text{and} \qquad b = \frac{\Sigma y - m(\Sigma x)}{n}$$

$$= \frac{8(29,677) - (323)(725)}{8(14,899) - 104,329} \qquad\qquad = \frac{725 - (0.2180582655)(323)}{8}$$

$$\approx 0.2180582655 \qquad\qquad\qquad \approx 81.82089753$$

The best-fitting line is $y' = 0.22x + 81.8$.

19. If possible, you should use technology for this solution. We show the details in case you do not. $n = 8$, $\Sigma x^2 = 14,899$, $\Sigma x = 323$, $\Sigma y^2 = 65,907$, $\Sigma y = 725$, $(\Sigma x)^2 = 104,329$, $\Sigma xy = 29,677$, $(\Sigma y)^2 = 525,625$

$$r = \frac{n\Sigma xy - (\Sigma x)(\Sigma y)}{\sqrt{n(\Sigma x^2) - (\Sigma x)^2}\ \sqrt{n(\Sigma y^2) - (\Sigma y)^2}}$$

$$= \frac{8(29,677) - (323)(725)}{\sqrt{8(14,899) - 104,329}\sqrt{8(65,907) - 525,625}}$$

$$\approx 0.6582620404$$

Not significant at 1% or 5% levels.

20. $314 - 266 = 48$; since the standard deviation is 16 days, we note this is 3 standard deviations greater than the mean. From the standard normal curve, this should happen about 0.001 (0.1 %) of the time. This means that approximately 1 in 1,000 pregnancies will have a 314-day duration.

CHAPTER 15 The Nature of Graphs and Functions

Chapter Overview
One of the most revolutionary ideas in the history of the world, not only in mathematics, is the marriage of algebra and geometry. The notion that an algebraic equation can be represented geometrically and also that a geometric graph can be represented algebraically literally changed the world. This idea is the topic of this chapter.

15.1 Cartesian Coordinates and Graphing Lines, page 715

New Terms Introduced in this Section

Abscissa	Analytic geometry	Axes	Cartesian coordinate system
Coordinates	Dependent variable	First-degree equation	Graph
Horizontal line	Independent variable	Linear equation	Ordinate
Origin	Quadrant	Satisfy	Slope
Slope-intercept form	Slope point	Solution	Standard form
Vertical line	x-intercept	y-intercept	

Level 1, page 722

1. **a**. Aphrodite **b**. Maxwell Montes **c**. Atalanta Planitia **d**. Rhea Mons **e**. Lavina Planitia

3. The y-intercept of a line is the point $(0, b)$ where the line crosses the y-axis.   The slope of a line is the steepness of a line.

5. Horizontal lines are parallel to the top of a page and have slope 0.   Vertical lines have no slope and are parallel to the edge of a page.

Ordered pairs in Problems 7-18 *may vary, but the lines will not.*

7. $(0, 5), (1, 6), (2, 7)$ plot pts.   9. $(0, 5), (1, 7), (-1, 3)$ plot pts.   11. $(0, -1), (1, 0), (2, 1)$ plot pts.

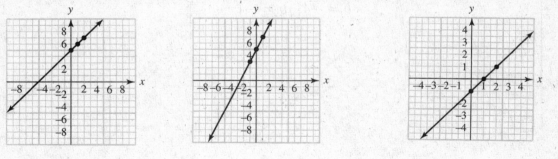

Page 371

13. $(0, 1), (1, -1), (-1, 3)$ plot pts.   **15.** $(0, 1), (1, 3), (2, 5)$ plot pts.   **17.** $(0, 2), (2, \frac{1}{2}), (4, -1)$ plot pts.

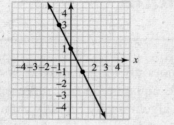
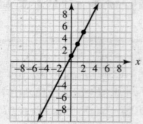
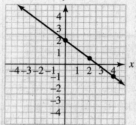

19.   **21.**   **23.**

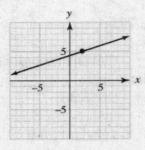
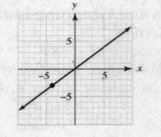
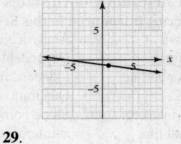

25.   **27.**   **29.**

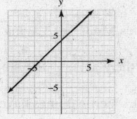
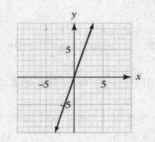
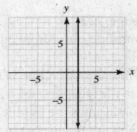

Level 2, page 722

31. $b = 3$; plot $(0, 3)$; $m = 2$ **33.** $y = -x - 4$; $b = -4, m = -1$

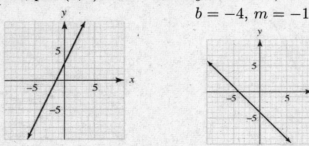

35. $2x + 3y + 6 = 0$
$$3y = -2x - 6$$
$$y = -\frac{2}{3}x - 2$$
$$b = -2; \ m = -\frac{2}{3}$$

37. $3x + 2y - 5 = 0$
$$2y = -3x + 5$$
$$y = \frac{-3}{2}x + \frac{5}{2}$$
$$b = \frac{5}{2}; \ m = \frac{-3}{2}$$

39. vertical line

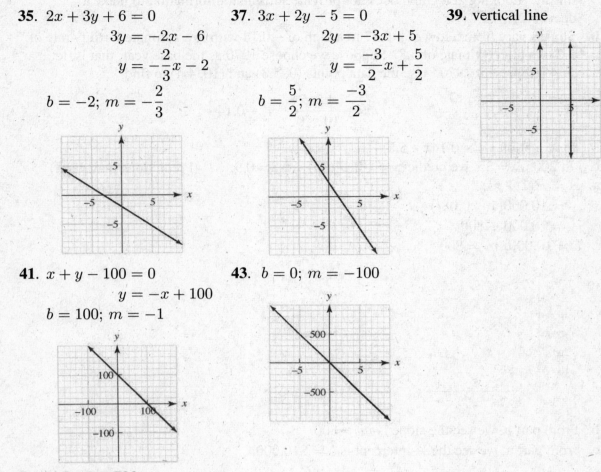

41. $x + y - 100 = 0$
$$y = -x + 100$$
$$b = 100; \ m = -1$$

43. $b = 0; \ m = -100$

Level 3, page 723

45. The y-intercept appears to be -2 and the slope $\frac{7}{10}$ (rather than $\frac{9}{10}$); C

47. The y-intercept appears to be -4 and the slope $\frac{5}{10}$ (rather than $\frac{7}{10}$); A

49. The y-intercept appears to be 2 and the slope $-\frac{9}{10}$ (rather than $-\frac{5}{10}$); E

Level 3 Problem Solving, page 723

Problems 51-52 were inspired by the Media Clips feature of the December 1996 issue of The Mathematics Teacher, *pp. 739-740. My thanks to Ron Lancaster, one of the editors of that column.*

51. a. It is declining at a rate that is nearly linear, but you can use slopes to prove that it is not linear. (Looking at a graph does not provide sufficient information to make a decision.)

b. Slope's vary from a low of -0.17 to a high of -0.14 with low estimate fertility rate of 2.2 to a high estimate of 5.8. Suppose we choose 1970 as the base year; that is, let $x = 0$ represent 1970. Use the data points $(0, 5.8)$ and $(10, 4.4)$ to find

$$m = \frac{4.4 - 5.8}{10 - 0} = -0.14$$

for equation $y = -0.14x + 5.8$.

c. In 2005, $x = 35$, we predict $y = -0.14(35) + 5.8 = 0.9$.   (Predictions may vary.)

53. a. $A = P(1 + rt)$

$= 10{,}000(1 + .0.08t)$

$= 10{,}000 + 800t$

$b = 10{,}000, \ m = 800$

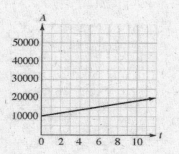

b. From part **a**, we see the slope is $m = 800$.

c. From part **a**, we see the A-intercept is $b = \$10{,}000$.

55. a. $b = -850; \ b = 1.25 = \dfrac{5}{2}$

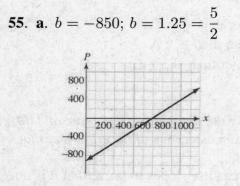

b. When $x = 0$, $P = -850$; if no items are sold there is a loss of \$850.

Page 374

c. 1.25; it is the profit increase corresponding to each unit increase in number of items sold.

57. Let $x = 0$ be the base year 1990.

$$m = \frac{19.8 - 17.0}{8 - 0} = 0.35$$

Thus, $y = 0.35x + 17.0$; for 2009, $x = 18$ so $y = 0.35(18) + 17.0 = 23.3$.
The projected population is about 23.3 million. According to Wikipedia, the 2009 population of Texas was 24.8 million. The implication is that the recent growth rate is more than the average 1990-1998 growth rate.

59. Given $y - k = m(x - h)$. Since the line passes through (x_1, y_1) and (x_2, y_2), the slope is $m = \dfrac{y_2 - y_1}{x_2 - x_1}$. Let $(h, k) = (x_1, y_1)$ be the known point. Then, by substitution,

$$y - y_1 = \left(\frac{y_2 - y_1}{x_2 - x_1} \right)(x - x_1)$$

15.2 Graphing Half-Planes, page 724

New Terms Introduced in this Section
Boundary   Closed half-plane   Half-plane   Open half-plane   Test point

Level 1, page 725
1. A *linear inequality* in two variables, say x and y, is an inequality of the form
$Ax + By + C \leq 0, Ax + By + C < 0, Ax + By + C \geq 0, Ax + By + C > 0$.

3. F; the boundary is $2x + 5y = 2$.

5. F; $(0, 0)$ does not work because it is on the boundary; choose a point not on the boundary.

7. $\quad 3x - 2y \geq -1$
$\quad 3(0) - 2(0) \geq -1$
$\qquad\qquad 0 \geq -1 \qquad$ This is true.

9. $y > 2x - 1$
$\quad 4 > 2(-2) - 1$
$\quad 4 > -5 \qquad$ This is true.

11. $4x < 3y$
$\quad 4(-2) < 3(-4)$
$\quad -8 < 12 \qquad$ This is true.

Level 2, page 725

13.   15.   17.

19.   21.   23.

25.   27.   29.

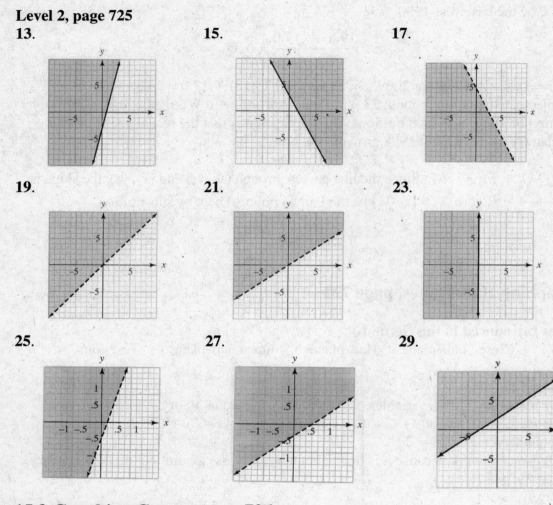

15.3 Graphing Curves, page 726
New Terms Introduced in this Section

Exponential curve   Graph   Parabola   Standard parabola   Vertex

Level 1, page 730
1. A *parabola* is a curve that is described by the path of a projectile, as shown in Figure 15.17.

3.   **5.**   **7.**

9.   **11.**   **13.**

15.   **17.**   **19.**

21.   **23.**   **25.**

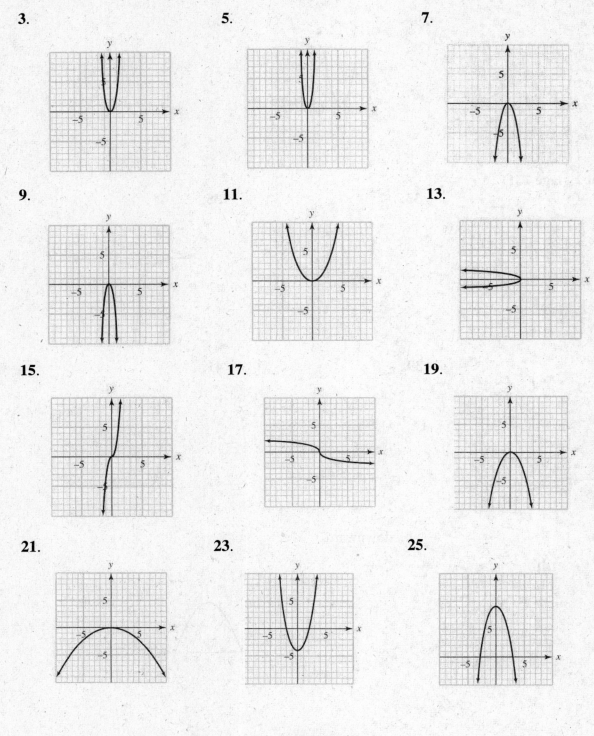

Page 377

27.   **29.**

Level 2, page 731

31.   **33.**   **35.**

37.   **39.**   **41.**

43. a.   **b.** downward   **c.**

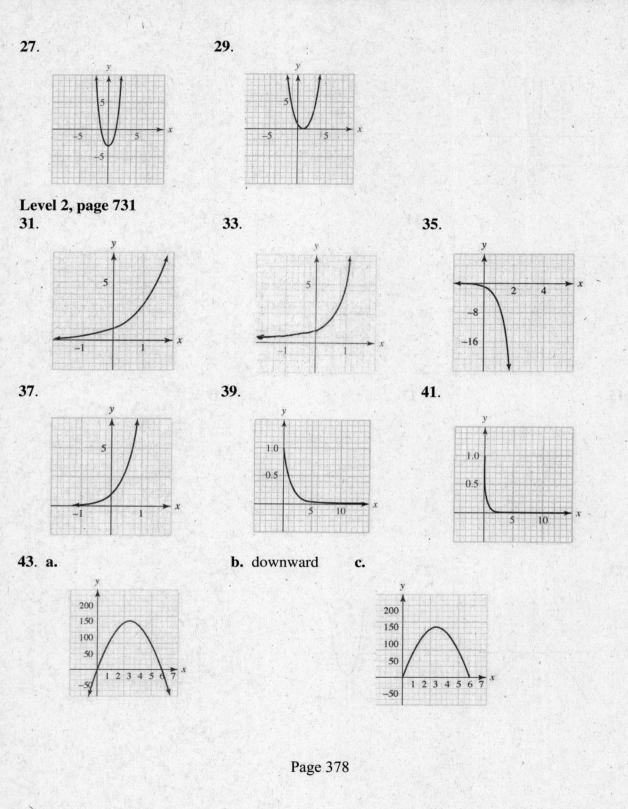

Page 378

45.

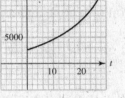

47.

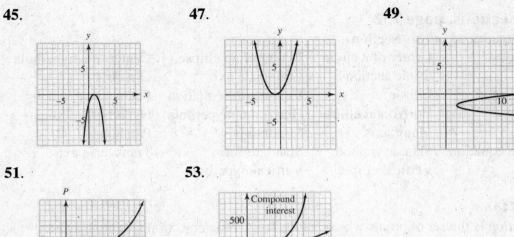

49.

51.

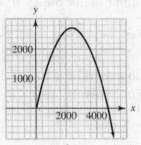

53.

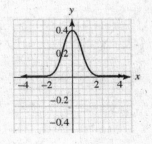

Level 3, page 731

54.

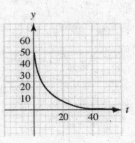

If $x = 4{,}000$, $y = 1{,}560$, positive;
$\quad x = 4{,}700$, $y = 188$, positive;
$\quad x = 5{,}000$, $y = -550$, negative.

The crossover point $(y = 0)$ is between 4,700 and 5,000, so he will make it across.

Level 3 Problem Solving, page 731

56.　　　　　　　　　　　**58.**

Page 379

15.4 Conic Sections, page 732
New Terms Introduced in this Section

Axis of parabola	Center of a circle	Center of an ellipse	Center of a hyperbola
Circle	Conic sections	Conjugate axis	Directrix
Eccentricity	Ellipse	First-degree equation	Focus
General form	Horizontal ellipse	Horizontal hyperbola	Hyperbola
Line	Major axis	Minor axis	Parabola
Second-degree equation	Slant asymptote	Standard form	Transverse axis
Vertex	Vertical ellipse	Vertical hyperbola	

Level 1, page 743

1. A *conic section* is the set of points which results from an intersection of a cone and a plane. It includes lines, parabolas, ellipses, circles, and hyperbolas.

3. An *ellipse* is the set of all points in a plane such that, for each point on the ellipse, the sum of its distances from two fixed points (called the foci) is a constant.

5. Write the equation in standard form, $\frac{x^2}{a^2} + \frac{y^2}{b^2} = 1$. The center is (0, 0); plot the intercepts on the *x*-axis and the *y*-axis. For the *x*-intercepts, plot $\pm$ the square root of the number under the x^2; for the *y*-intercepts, plot $\pm$ the square root of the number under the y^2. Finally, draw the ellipse using these intercepts.

7. **a.** Both variables are first-degree.
   b. One variable is first degree the other is second-degree.
   c. Both variables are second degree and both have the same sign (in general form).
   d. Both variables are second degree and they have opposite signs (in general form).

9. *A* and *C* have opposite signs.

11. $A = 0$ and $C \neq 0$ or $A \neq 0$ and $C = 0$

13. $A = C = 0$

15.   **17.**   **19.**

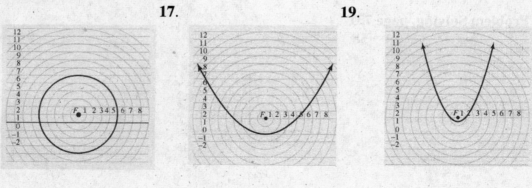

21.   **23.**

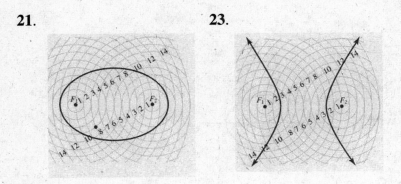

25. a. It is first degree in both x and y, so it is a line.
  b. This is second degree in both x and y and the coefficients of those square terms have the same sign, so it is an ellipse.
  c. This is first degree in x and second degree in y, so it is a parabola.
27. a. This is first degree in x and second degree in y, so it is a parabola.
  b. This is second degree in both variables and the coefficients of those square terms have different signs, so it is a hyperbola.
  c. This is second degree in both x and y and the coefficients of those square terms have the same sign, so it is an ellipse.

Level 2, page 744
29. We recognize this as a circle with center at $(0, 0)$ and radius $r = 8$.

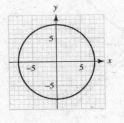

31. We recognize this as a circle with center at $(0, 0)$ and radius $r = \sqrt{250} \approx 15.8$.

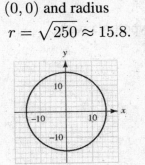

33. We recognize this an ellipse with $a = 6$ and $b = 5$.

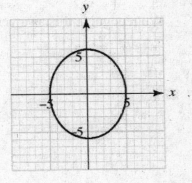

Page 381

35. $4x^2 + 9y^2 = 36$

$$\frac{4x^2}{36} + \frac{9y^2}{36} = \frac{36}{36}$$

$$\frac{x^2}{9} + \frac{y^2}{4} = 1$$

We recognize this as an ellipse with
$a = 3$ and $b = 2$.

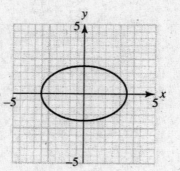

37. $16x^2 + 9y^2 = 144$

$$\frac{16x^2}{144} + \frac{9y^2}{144} = \frac{144}{144}$$

$$\frac{x^2}{9} + \frac{y^2}{16} = 1$$

We recognize this as an ellipse with
$a = 4$ and $b = 3$.

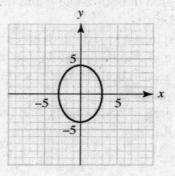

39. $y^2 - x^2 = 4$

$$\frac{y^2}{4} - \frac{x^2}{4} = 1$$

We recognize this as a vertical hyperbola with
$a = b = 2$.

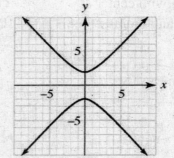

41. We recognize this as a horizontal hyperbola with $a = 2$, $b = 3$.

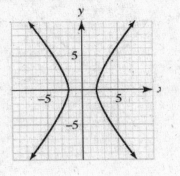

43. We recognize this as a vertical hyperbola with $a = 4$, $b = 6$.

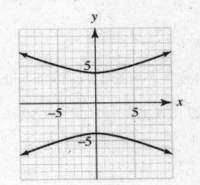

45.
$$3y^2 = 4x^2 + 12$$
$$\frac{3y^2}{12} - \frac{4x^2}{12} = \frac{12}{12}$$
$$\frac{y^2}{4} - \frac{x^2}{3} = 1$$

We recognize this as a vertical hyperbola with $a = 2$, $b = \sqrt{3}$.

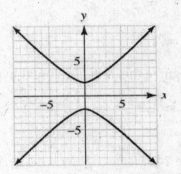

47. $x^2 - y^2 = 9$
$$\frac{x^2}{9} - \frac{y^2}{9} = 1$$

We recognize this as a horizontal hyperbola with $a = b = 3$.

Page 383

49.
$$3y^2 = 4x^2 + 5$$
$$\frac{3y^2}{5} - \frac{4x^2}{5} = \frac{5}{5}$$
$$\frac{y^2}{\frac{5}{3}} - \frac{x^2}{\frac{5}{4}} = 1$$

We recognize this as a vertical hyperbola with

$$a = \sqrt{\frac{5}{3}} \approx 1.29, \ b = \sqrt{\frac{5}{4}} = \frac{\sqrt{5}}{2} \approx 1.12$$

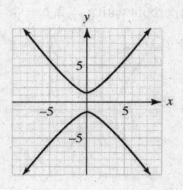

51. $3x^2 - 4y^2 = 5$
$$\frac{3x^2}{5} - \frac{4y^2}{5} = \frac{5}{5}$$
$$\frac{x^2}{\frac{5}{3}} - \frac{y^2}{\frac{5}{4}} = 1$$

We recognize this as a horizontal hyperbola with

$$a = \sqrt{\frac{5}{3}} \approx 1.29, \ b = \sqrt{\frac{5}{4}} = \frac{\sqrt{5}}{2} \approx 1.12$$

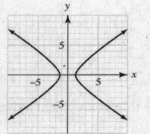

Level 3, page 744

53. $a = 1.4 \times 10^8; \ \epsilon = 0.093$

Now,

$$\epsilon = \frac{c}{a}$$
$$0.093 = \frac{c}{1.4 \times 10^8}$$
$$c = 1.302 \times 10^7$$

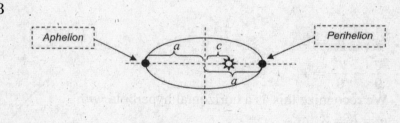

aphelion: $a + c = 1.4 \times 10^8 + 1.302 \times 10^7$ perihelion: $a - c = 1.4 \times 10^8 - 1.302 \times 10^7$
$$= 1.5 \times 10^8 \text{ mi} \qquad\qquad\qquad\qquad\qquad = 1.3 \times 10^8 \text{ mi}$$

55. apogee $= a + c = 199{,}000$ and $2a = 378{,}000$
$$189{,}000 + c = 199{,}000 \qquad \text{since } a = 189{,}000$$
$$c = 10{,}000$$

Page 384

We can now find the eccentricity since we know both c and a: $\epsilon = \dfrac{c}{a}$

$$= \frac{10,000}{189,000}$$

$$\approx 0.053$$

Level 3 Problem Solving, page 744

57. Let the vertex be $(0, 0)$. Since the parabola opens right, $y^2 = 4cx$. The parabola passes through $(4, 6)$.

$$\begin{aligned} y^2 &= 4cx \\ 6^2 &= 4c(4) \qquad \text{Substitute } (4, 6) \text{ for } (x, y). \\ 36 &= 16c \\ c &= \frac{9}{4} \text{ or } 2.25 \end{aligned}$$

The focus is 2.25 m from the vertex on the axis of the parabola.

59. The vertex is at $V(0, 0)$. The parabola passes through $(4, 6)$.

$$\begin{aligned} x^2 &= 4cy \\ 4^2 &= 4c(6) \\ c &= \frac{2}{3} \end{aligned}$$

The focus is 8 in. from the vertex on the axis of the parabola.

15.5 Functions, page 745

New Terms Introduced in this Section

Difference quotient	Domain	Exponential function	Function
function machine	Functional notation	Linear function	Logarithmic function
Probability function	Quadratic function	Range	Vertical-line test

Level 1, page 750

1. A *function* is a set of ordered pairs in which the first component is associated with exactly one second component.

3. It is a function because each first component is associated with exactly one second component.

5. It is a function because each first component is associated with exactly one second component.

7. It is not a function because the first component 4 is associated with more than one second component, namely it is associated with both 3 and 19.

Page 385

9. It is a function because each first component is associated with exactly one second component.

11. A function can consist of a single element.

13. This is not a function because a function must be a set of ordered pairs.

15. a. $f(4) = (4+5)2 = 18$
  b. $f(6) = (6+5)2 = 22$
  c. $f(-8) = (-8+5)2 = -6$
  d. $f(\frac{1}{2}) = (\frac{1}{2}+5)2 = 11$
  e. $f(t) = 2(t+5)$

17. We are given $M(x) = x^2 + 1$
  a. $M(4) = 4^2 + 1$ **b.** $M(6) = 6^2 + 1$ **c.** $M(-8) = (-8)^2 + 1$
    $= 17$ $= 37$ $= 65$
  d. $M\left(\frac{1}{2}\right) = \left(\frac{1}{2}\right)^2 + 1$ **e.** $M(t) = t^2 + 1$
    $= 1\frac{1}{4}$

19. We are given $g(x) = 2x - 1$
  a. $g(4) = 2(4) - 1$ **b.** $g(6) = 2(6) - 1$ **c.** $g(-8) = 2(-8) - 1$
    $= 7$ $= 11$ $= -17$
  d. $g\left(\frac{1}{2}\right) = 2\left(\frac{1}{2}\right) - 1$ **e.** $g(t) = 2t - 1$
    $= 0$

21. a. $f(15) = 15 - 7$ **b.** $f(-9) = -9 - 7$ **c.** $f(p) = p - 7$
    $= 8$ $= -16$

23. a. $h(0) = 3(0) - 1$ **b.** $h(-10) = 3(-10) - 1$ **c.** $h(a) = 3a - 1$
    $= -1$ $= -31$

25. a. $g(10) = \dfrac{10}{2}$ **b.** $g(-4) = \dfrac{-4}{2}$ **c.** $g(3) = \dfrac{3}{2}$
    $= 5$ $= -2$

27. Use the vertical line test to see that it is not a function because a vertical line passes through more than one point on the graph. For the domain, look at the x-scale to see that each tick mark is 1 and the graph window shows from -2 to $+2$; we find the domain: $-1 \le x \le 1$. For the range, note that the graph passes on the third tick marks, so we find the range: $-3 \le y \le 3$.

29. Use the vertical line test to see that it is not a function because a vertical line passes through more than one point on the graph. For the domain, look at the x-scale to see that each tick mark is 1; we find the domain: $x \geq -3$. For the range, note that the graph has no vertical limitations so we find the range: $\mathbb{R}$.

31. Use the vertical line test to see that it is not a function because a vertical line passes through more than one point on the graph. For the domain, look at the x-scale to see that each tick mark is 1; we find the domain $-2 \leq x \leq 3$. For the range, note that each tick on the y-axis is 1 unit, so the range is $-8 \leq y \leq 4$.

33. quadratic   **35.** logarithmic   **37.** probability

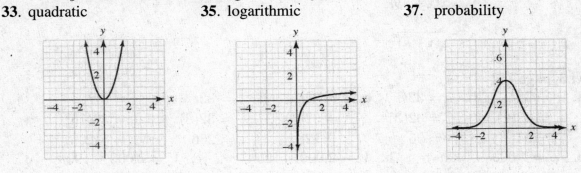

Level 2, page 752

39. $\dfrac{f(x+h) - f(x)}{h} = \dfrac{3(x+h) - 5 - (3x - 5)}{h}$

$\qquad\qquad\qquad\quad = \dfrac{3x + 3h - 5 - 3x + 5}{h}$

$\qquad\qquad\qquad\quad = 3$

41. $\dfrac{f(x+h) - f(x)}{h} = \dfrac{(x+h)^3 - x^3}{h}$

$\qquad\qquad\qquad\quad = \dfrac{x^3 + 3x^2h + 3xh^2 + h^3 - x^3}{h}$

$\qquad\qquad\qquad\quad = \dfrac{h(3x^2 + 3xh + h^2)}{h}$

$\qquad\qquad\qquad\quad = 3x^2 + 3xh + h^2$

43. $\dfrac{f(x+h) - f(x)}{h} = \dfrac{\frac{1}{x+h} - \frac{1}{x}}{h}$

$\phantom{\dfrac{f(x+h) - f(x)}{h}} = \dfrac{x - (x+h)}{x(x+h)h}$

$\phantom{\dfrac{f(x+h) - f(x)}{h}} = \dfrac{-h}{x(x+h)h}$

$\phantom{\dfrac{f(x+h) - f(x)}{h}} = \dfrac{-1}{x(x+h)}$

Level 3, page 752

45. a. $v = 32t$ **b.** $v = 32t$ **c.** $v = 32t$ **d.** $v = 32t$ **e.** $v = 32t$

$\ = 32(2)$ $\ = 32(3)$ $\ = 32(4)$ $\ = 32(8)$ $\ = 32(16)$

$\ = 64$ $\ = 96$ $\ = 128$ $\ = 256$ $\ = 512$

47. a. $V = 2{,}000 - 190n$ **b.** $V = 2{,}000 - 190n$ **c.** $V = 2{,}000 - 190n$

$\ = 2{,}000 - 190(3)$ $\ = 2{,}000 - 190(5)$ $\ = 2{,}000 - 190(7)$

$\ = 1{,}430$ $\ = 1{,}050$ $\ = 670$

$\quad$ **d.** $V = 2{,}000 - 190n$ **e.** $V = 2{,}000 - 190(10)$

$\ = 2{,}000 - 190(9)$ $\ = 2{,}000 - 190(10)$

$\ = 290$ $\ = 100$

49. A prime number is a counting number with exactly two factors. The primes less than 100 are: 2, 3, 5, 7, 11, 13, 17, 19, 23, 29, 31, 37, 41, 43, 47, 53, 59, 61, 67, 71, 73, 79, 83, 89, and 97. **a.** $P(10) = 4$ **b.** $P(-10) = 0$ **c.** $P(100) = 25$

51. $y = 3x - 5$; since any real number can be used for x, we find the domain: $\mathbb{R}$.

53. $y = \sqrt{5 - x}$; since the radicand must be nonnegative we find the domain:

$5 - x \geq 0$

$\quad 5 \geq x$

$\quad x \leq 5$

Level 3 Problem Solving, page 752

55. Toss a rock into the well and measure the time it takes to hit the bottom. We know
1 mi = 5,280 ft, so $5{,}280 = 16t^2$

$\qquad\qquad 330 = t^2$

$\qquad\qquad t \approx \pm 18.1$

Reject the negative solution, and conclude that it would be about 18 seconds to the bottom.

57. $f(x) = (x + 5)^2 - x$
$$= x^2 + 10x + 25 - x$$
$$= 10x + 25$$

59. $A = s^2$ and $P = 4s$, so $A = \left(\dfrac{P}{4}\right)^2$.

Chapter 15 Review Questions, page 753

Studying for a chapter examination is a personal process, one which nobody else can do for you. Simply take the time to review what you have done. Here are the new terms in Chapter 15.

Abscissa [15.1]
Analytic geometry [15.3]
Axes [15.1]
Axis of a parabola [15.4]
Boundary [15.2]
Cartesian coordinate system [15.1]
Center of a circle [15.4]
Center of an ellipse [15.4]
Center of a hyperbola [15.4]
Circle [15.4]
Closed half-plane [15.2]
Conic sections [15.4]
Conjugate axis [15.4]
Coordinates [15.1]
Dependent variable [15.1]
Difference quotient [15.5]
Directrix [15.4]
Domain [15.5]
Eccentricity [15.4]
Ellipse [15.4]
Exponential curve [15.3]

Exponential function [15.5]
First-degree equation [15.1, 15.4]
Focus (pl. foci) [15.4]
Function [15.5]
Function machine [15.5]
Functional notation [15.5]
General form [15.4]
Graph [15.1, 15.3]
Half-plane [15.2]
Horizontal ellipse [15.4]
Horizontal hyperbola [15.4]
Horizontal line [15.1]
Hyperbola [15.4]
Independent variable [15.1]
Line [15.4]
Linear equation [15.1]
Linear function [15.5]
Logarithmic function [15.5]
Major axis [15.4]
Minor axis [15.4]
Open half-plane [15.2]
Ordinate [15.1]

Origin [15.1]
Parabola [15.3; 15.4]
Probability function [15.5]
Quadrant [15.1]
Quadratic function [15.5]
Range [15.5]
Satisfy [15.1]
Second-degree equation [15.4]
Slant asymptotes [15.4]
Slope [15.1]
Slope-intercept form [15.1]
Slope point [15.1]
Solution [15.1]
Standard form [15.3; 15.4]
Standard parabola [15.3]
Test point [15.2]
Transverse axis [15.4]
Vertex [15.3; 15.4]
Vertical ellipse [15.4]
Vertical hyperbola [15.4]
Vertical line [15.1]
Vertical line test [15.5]

Page 389

x-intercept [15.1]   *y*-intercept [15.1]

If you can describe the term, read on to the next one; if you cannot, then look it up in the text (the section number is shown in brackets). Next, study the types of problems listed at the end of Chapter 15.

TYPES OF PROBLEMS

Graph lines by plotting points and by using the slope-intercept. [15.1]

Draw a line when given a point and the slope. [15.1]

Match a line with an equation. [15.1]

Solve applied problems involving lines, including future value, depreciation, profit, marginal
    profit, cost, and marginal cost. [15.1]

Graph first-degree inequalities with two unknowns. [15.2]

Graph parabolas and exponential curves by plotting points. [15.3]

Solve applied problems involving exponential and parabolic models. [15.3]

Identify a conic section by looking at its equation. [15.4]

Sketch a conic section using its geometric definition. [15.4]

Sketch a conic section using its standard-form equation [15.4]

Solve applied problems involving the conic sections. [15.4]

Decide whether a given set is a function. [15.5]

Use the vertical line test to decide whether a given graph represents a functions. [15.5]

Determine the output value for a function. [15.5]

Evaluate functions. [15.5]

Graph a function and then classify it as a linear, quadratic, exponential, logarithmic, or
    probability function. [15.5]

Find $\dfrac{f(x+h) - f(x)}{h}$ for a given function f. [15.5]

Once again, see if you can verbalize (to yourself) how to do each of the listed types of problems.

   Work all of Chapter 15 Review Questions (whether they are assigned or not). Work through all of the problems before looking at the answers, and *then* correct each of the problems. The entire solution is shown in the answer section at the back of the text. If you worked the problem correctly, move on to the next problem, but if you did not work it correctly (or you did not know what to do), look back in the chapter to study the procedure, or ask your instructor.

   Finally, go back over the homework problems you have been assigned. If you worked a problem correctly, move on to the next problem, but if you missed it on your homework, then you should look back in the book or talk to your instructor about how to work the problem.

   If you follow these steps, you should be successful with your review of this chapter.

Page 390

We give all of the answers to the chapter review questions (not just the odd-numbered questions).

Chapter 15 Review Questions, page 753

1. Solve for y: $5x - y = 15$
$$5x - 15 = y$$
The y-intercept is -15; the slope is 5. Plot intercept and then find slope point with rise of 5 and a run of 1.

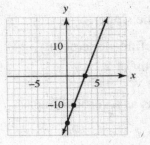

2. $y = -\dfrac{4}{5}x - 3$

The y-intercept is -3; the slope is $-\dfrac{4}{5}$. Plot intercept and then find slope point with rise of -4 and run of 5.

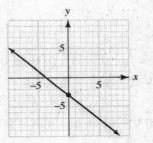

3. Solve for y: $2x + 3y = 15$
$$3y = -2x + 15$$
$$y = -\frac{2}{3}x + 5$$

The y-intercept is 5; the slope is $-\dfrac{2}{3}$. Plot intercept and then find slope point with rise of -2 and a run of 3.

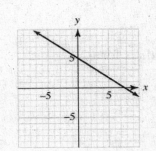

Page 391

4. Solve for y: $x = -\dfrac{2}{3}y + 1$

$$3x = -2y + 3$$
$$2y = -3x + 3$$
$$y = -\dfrac{3}{2}x + \dfrac{3}{2}$$

The y-intercept is $\dfrac{3}{2}$; the slope is $-\dfrac{3}{2}$.
Plot intercept and then find slope point
with rise of -3 and a run of 2.

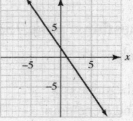

5. $x = 150$; recognize this as a vertical line.
Watch the scale.

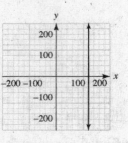

6. Graph the boundary $x = 3y$; this is a dashed line.
Choose a test point, say $(0, 3)$. $\qquad x < 3y$
$\qquad\qquad\qquad\qquad\qquad\qquad\quad 0 < 3(3)$ is true.
Shade the half-plane on the same as the test point.

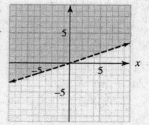

7. Formulate a table of values for $y = 1 - x^2$

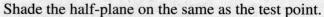

x	y
1	0
-1	0
2	-3
3	-8

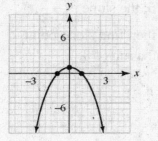

Page 392

8. Formulate a table of values for $y = -2^x$.

x	y
0	-1
1	-2
2	-4
3	-8
-1	$-1/2$
-2	$-1/4$
-3	$-1/8$

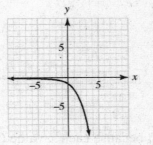

9. Recognize this as a horizontal hyperbola.

$a = \pm 4$, $b = \pm 3$

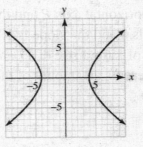

10. Recognize this as a horizontal ellipse.

$a = \pm \sqrt{20} \approx \pm 4.5$, $b = \pm \sqrt{10} \approx \pm 3.2$

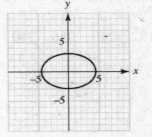

11. Recognize this as a circle centered at $(0,0)$ with $r = 1$. This is called a unit circle.

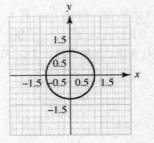

12. Recognize this as a horizontal hyperbola.

$a = \pm 1, \; b = \pm 1$

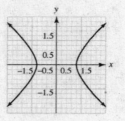

13. Yes, it is a function because each x-value (namely, 4, 5, and 6) is associated with exactly one y-value.   **14**. $f(6) = 3(6) + 2 = 20$ **15**. $g(0) = 0^2 - 3 = -3$

16. $F(10) = 5(10) + 25 = 75$ **17**. $m(10) = 5$

18. $A = 1.7P$; label the horizontal axis P and the vertical axis A. The intercept is 0 and the slope is 1.7 — that is, a rise of 1.7 and a run of 1 (or a rise of 17 and a run of 10). Note that since P represents an amount of money, it is nonnegative.

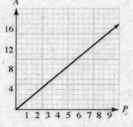

19.   Form a table of value for $y = 128t - 16t^2$.

t	0	1	2	3	4	5	6	7	8
y	0	112	192	240	256	240	192	112	0

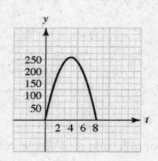

20.   Form a table of value for $P = 53{,}000e^{0.013t}$.

t	0	10	20	30	40	50
y	53,000	60,358	68,737	78,280	89,147	101,524

Page 394

CHAPTER 16 The Nature of Mathematical Systems

Chapter Overview
This chapter discusses solving simple systems of equations. The first section reviews graphing, substitution, and the addition methods of solving two linear equations with two unknowns. Most students will be familiar with these methods, but will need a brief review. The second section gives practice using these methods with particular types of problems. We then introduce matrices and Gauss-Jordan elimination as a more powerful method for solving systems of equations, which leads to a calculator method using inverse matrices. The chapter concludes with two optional sections, systems of inequalities and linear programming.

16.1 Systems of Linear Equations, page 757
New Terms Introduced in this Section

Addition method	Dependent system	Equivalent systems	Graphing method
Inconsistent system	Linear combination method	Linear system	Simultaneous solution
Substitution method	System of equations		

Level 1, page 761

1. A *system of linear equations* is two or more first degree equations considered at the same time.

3. Graph each of the equations independently and look for the points of intersection.

5. *Multiply* one or both of the equations by a constant or constants so that the coefficients of one of the variables become opposites. *Add* corresponding members of the equations to obtain a new equation in a single variable. *Solve* the derived equation for that variable. *Substitute* the value of the found variable into either of the original equations, and solve for the second variable. *State* the solution.

7.
$$3x - 4y = 16 \qquad\qquad -x + 2y = -6$$
$$3x - 16 = 4y \qquad\qquad 2y = x - 6$$
$$y = \frac{3}{4}x - 4 \qquad\qquad y = \frac{1}{2}x - 3$$

Graph this line by plotting the y-intercept $(0, -4)$ and slope $m = \frac{3}{4}$; rise of 3 and run of 4.

Graph this line by plotting the y-intercept $(0, -3)$ and slope $m = \frac{1}{2}$; rise of 1 and run of 2.

Page 395

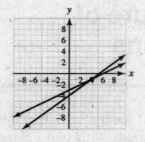

The point of intersection looks like $(4, -1)$.

9. $2x - 3y = 12$

$$y = \frac{2}{3}x - 4$$

Graph this line by plotting the
y-intercept $(0, -4)$ and slope
$m = \frac{2}{3}$; rise of 2 and run of 3.

$-4x + 6y = 18$

$6y = 18$

$$y = \frac{2}{3}x + 3$$

Graph this line by plotting the
y-intercept $(0, 3)$ and slope
$m = \frac{2}{3}$; rise of 2 and run of 3.

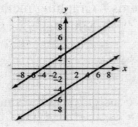

The lines are parallel, so there is no intersection point; inconsistent.

11. $6x + y = -5$

$$y = -6x - 5$$

Graph this line by plotting the
y-intercept $(0, -5)$ and slope
$m = -6$; rise of -6 and run of 1.

$x + 3y = 2$

$3y = -x + 2$

$$y = -\frac{1}{3}x + \frac{2}{3}$$

Graph this line by plotting the
y-intercept $(0, \frac{2}{3})$ and slope
$m = -\frac{1}{3}$; rise of -1 and run of 3.

Page 396

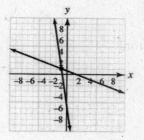

The point of intersection looks like it is $(-1, 1)$.

13. $3x + 2y = 5$

$2y = -3x + 5$

$y = -\dfrac{3}{2}x + \dfrac{5}{2}$

Graph this line by plotting the

y-intercept $(0, \dfrac{5}{2})$ and slope

$m = -\dfrac{3}{2}$; rise of -3 and run of 2.

$4x - 3y = 1$

$4x - 1 = 3y$

$y = \dfrac{4}{3}x - \dfrac{1}{3}$

Graph this line by plotting the

y-intercept $(0, -\dfrac{1}{3})$ and slope

$m = \dfrac{4}{3}$; rise of 4 and run of 3.

The point of intersection looks like it is $(1, 1)$.

15. $\begin{cases} 5x - 2y = -19 \\ x = 3y + 4 \end{cases}$ By substitution, replace x in the first equation:

$5x - 2y = -19$ *Given equation*

$5(3y + 4) - 2y = -19$ *Substitute $3y + 4$ for x.*

$15y + 20 - 2y = -19$ *Distributive property*

$13y = -39$ *Simplify and subtract 20 from both sides.*

$y = -3$ *Divide both sides by 13.*

Then $x = 3y + 4$

$= 3(-3) + 4$ *Substitute -3 for y.*

$= -5$ *Simplify.*

Page 397

The solution, by substitution, is $(-5, -3)$.

17. $\begin{cases} 3x - y = -1 \\ x = 2y + 3 \end{cases}$ By substitution, replace x in the first equation:

$3x - y = -1$ *Given equation*

$3(2y + 3) - y = -1$ *Substitute $2y + 3$ for x.*

$6y + 9 - y = -1$ *Distributive property*

$5y + 9 = -1$ *Simplify.*

$5y = -10$ *Subtract 9 from both sides.*

$y = -2$ *Divide both sides by 5.*

Then $x = 2y + 3$

$= 2(-2) + 3$ *Substitute -2 for y.*

$= -1$ *Simplify.*

The solution, by substitution, is $(-1, -2)$.

19. $\begin{cases} x + y = 12 \\ 0.6y = 0.5(12) \end{cases}$ Solve the second equation for y: $0.6y = 0.5(12)$

$$y = \frac{0.5(12)}{0.6}$$
$$= 10$$

Replace y in the first equation:

$x + y = 12$ *Given equation*

$x + 10 = 12$ *Substitute 10 for y.*

$x = 2$ *Simplify.*

The solution, by substitution, is $(2, 10)$.

21. $\begin{cases} \frac{x}{3} - y = 7 \\ x + \frac{y}{2} = 7 \end{cases}$ Solve the second equation for x: $x = -\frac{y}{2} + 7$ and multiply both sides

of the first equation by 3: $\dfrac{x}{3} - y = 7$

$x - 3y = 21$

Now replace x in this equation:

$\left(-\dfrac{y}{2} + 7\right) - 3y = 21$ *Substitute.*

$-\dfrac{7}{2}y + 7 = 21$ *Simplify.*

$-\dfrac{7}{2}y = 14$ *Subtract 7 from both sides.*

$-7y = 28$ *Multiply both sides by 2.*

$y = -4$ *Divide both sides by -7.*

Page 398

Then $\quad x = -\dfrac{y}{2} + 7$

$\qquad = -\dfrac{-4}{2} + 7$

$\qquad = 2 + 7$

$\qquad = 9$

The solution, by substitution, is $(9, -4)$.

23. $\begin{cases} x = -7y - 3 \\ 2x + 5y = 3 \end{cases}$ $\quad$ By substitution, replace x in the second equation:

$\quad 2x + 5y = 3 \qquad$ *Given equation*

$2(-7y - 3) + 5y = 3 \qquad$ *Substitute $-7y - 3$ for x.*

$-14y - 6 + 5y = 3 \qquad$ *Simplify.*

$-9y - 6 = 3$

$-9y = 9$

$y = -1$

Then $\quad x = -7y - 3$

$\qquad = -7(-1) - 3$

$\qquad = 4$

The solution, by substitution, is $(4, -1)$.

25. $\begin{cases} x + 3y = 0 \\ x = 5y + 16 \end{cases}$ $\quad$ By substitution, replace x in the first equation:

$\quad x + 3y = 0 \qquad$ *Given equation*

$(5y + 16) + 3y = 0 \qquad$ *Substitute $5y + 16$ for x.*

$8y + 16 = 0 \qquad$ *Simplify.*

$8y = -16$

$y = -2$

Then $\quad x = 5y + 16$

$\qquad = 5(-2) + 16$

$\qquad = 6$

The solution, by substitution, is $(6, -2)$.

27. $+ \begin{cases} x + y = 560 \\ x - y = 490 \end{cases}$

Add: $\quad 2x = 1{,}050$

$\qquad x = 525$

Substitute this value into either of the equations (we select the first):

Page 399

$$x + y = 560$$
$$525 + y = 560$$
$$y = 35$$

The solution, by addition, is (525, 35).

29. $2 \begin{cases} 3x + y = 13 \\ x - 2y = 9 \end{cases}$ or $\begin{cases} 6x + 2y = 26 \\ x - 2y = 9 \end{cases}$

Add: $7x = 35$
$$x = 5$$

Substitute this value into either of the equations (we select the first):
$$3x + y = 13$$
$$3(5) + y = 13$$
$$y = -2$$

The solution, by addition, is $(5, -2)$.

31. $\begin{cases} 3u + 2v = 5 \\ 4v = 10 - 6u \end{cases}$ or $\begin{cases} 3u + 2v = 5 \\ 6u + 4v = 10 \end{cases}$ so that $-2 \begin{cases} 3u + 2v = 5 \\ 6u + 4v = 10 \end{cases}$ or $\begin{cases} -6u - 4v = -10 \\ 6u + 4v = 10 \end{cases}$

Add to obtain $0 = 0$ which is true, so it is a dependent system.

33. $\begin{matrix} 2 \\ -5 \end{matrix} \begin{cases} 5s_1 + 2s_2 = 23 \\ 2s_1 + 7s_2 = 34 \end{cases}$ or $+ \begin{cases} 10s_1 + 4s_2 = 46 \\ -10s_1 - 35s_2 = -170 \end{cases}$

Add: $-31s_2 = -124$
$$s_2 = 4$$

Substitute this value into either of the two given equations (we select the first):
$$5s_1 + 2s_2 = 23$$
$$5s_1 + 2(4) = 23$$
$$5s_1 = 15$$
$$s_1 = 3$$

The solution, by addition, is $(s_1, s_2) = (3, 4)$.

35. $\begin{matrix} 2 \\ 3 \end{matrix} \begin{cases} 2u - 3v = 16 \\ 5u + 2v = 21 \end{cases}$ or $+ \begin{cases} 4u - 6v = 32 \\ 15u + 6v = 63 \end{cases}$

Add: $19u = 95$ Substitute this value into either of the two given equations
$$u = 5$$ (we select the first):
$$2u - 3v = 16$$
$$2(5) - 3v = 16$$
$$-3v = 6$$
$$v = -2$$

Page 400

The solution, by addition, is $(u, v) = (5, -2)$.

37. $2 \begin{cases} 5x + 4y = 5 \\ 15x - 2y = 8 \end{cases}$ or $+ \begin{cases} 5x + 4y = 5 \\ 30x - 4y = 16 \end{cases}$

Add: $35x = 21$

$$x = \frac{21}{35} = \frac{3}{5}$$

Substitute this value into either of the two given equations (we select the first):

$$5x + 4y = 5$$
$$5\left(\frac{3}{5}\right) + 4y = 5$$
$$3 + 4y = 5$$
$$4y = 2$$
$$y = \frac{1}{2}$$

The solution, by addition, is $\left(\frac{3}{5}, \frac{1}{2}\right)$.

Level 2, page 761

39. $+ \begin{cases} x - y = 8 \\ x + y = 2 \end{cases}$

Add: $2x = 10$ Substitute this value into either of the equations (we select the

$x = 5$

second):
$$x + y = 2$$
$$5 + y = 2$$
$$y = -3$$

The solution, by addition, is $(5, -3)$.

41. $+ 2 \begin{cases} x - 6y = -3 \\ 2x + 3y = 9 \end{cases}$ or $\begin{cases} x - 6y = -3 \\ 4x + 6y = 18 \end{cases}$

Add: $5x = 15$ Substitute this value into either of the equations (we select the

$x = 3$

second):
$$2x + 3y = 9$$
$$2(3) + 3y = 9$$
$$3y = 3$$
$$y = 1$$

The solution, by addition, is $(3, 1)$.

Page 401

43. $\begin{cases} 2x + 3y = 9 \\ x = 5y - 2 \end{cases}$ By substitution, replace x in the first equation:

$$2x + 3y = 9$$
$$2(5y - 2) + 3y = 9$$
$$10y - 4 + 3y = 9$$
$$13y = 13$$
$$y = 1$$

Then $x = 5y - 2$
$$= 5(1) - 2$$
$$= 3$$

The solution, by substitution, is (3, 1).

45. $-2 \begin{cases} 3x + 4y = 8 \\ x + 2y = 2 \end{cases}$ or $+ \begin{cases} 3x + 4y = 8 \\ -2x - 4y = -4 \end{cases}$

Add: $x = 4$; substitute this into either of the given equations (we select the second):

$$x + 2y = 2$$
$$4 + 2y = 2$$
$$2y = -2$$
$$y = -1$$

The solution, by addition, is (4, −1).

47. $-3 \begin{cases} 6x + 9y = -4 \\ 9x + 3y = 1 \end{cases}$ or $+ \begin{cases} 6x + 9y = -4 \\ -27x - 9y = -3 \end{cases}$

Add: $-21x = -7$; substitute this into either of the given equations (we select the first):

$$x = \frac{1}{3}$$

$$6x + 9y = -4$$
$$6\left(\frac{1}{3}\right) + 9y = -4$$
$$2 + 9y = -4$$
$$9y = -6$$
$$y = -\frac{2}{3}$$

The solution, by addition, is $\left(\frac{1}{3}, -\frac{2}{3}\right)$.

49. $\begin{matrix} -3 \\ 4 \end{matrix} \begin{cases} 5x + 4y = 9 \\ 9x + 3y = 12 \end{cases}$ or $+ \begin{cases} -15x - 12y = -27 \\ 36x + 12y = 48 \end{cases}$

Add: $21x = 21$; ssubstitute this into either of the given equations (we select the first):
$$x = 1$$

$$5x + 4y = 9$$
$$5(1) + 4y = 9$$
$$4y = 4$$
$$y = 1$$

The solution, by addition, is (1, 1).

51. $\begin{cases} x = \frac{3}{4}y - 2 \\ 3y - 4x = 5 \end{cases}$ Substitute the x value from the first equation into the second equation:

$$3y - 4x = 5$$
$$3y - 4\left(\frac{3}{4}y - 2\right) = 5$$
$$3y - 3y + 8 = 5$$
$$8 = 5$$

This is a false equation, so there are no values that makes this true; inconsistent.

53. $\begin{cases} x + y = 10 \\ 0.4x + 0.9y = 0.5(10) \end{cases}$ From the first equation $y = 10 - x$; substitute this into the second equation:

$$0.4x + 0.9y = 0.5(10)$$
$$0.4x + 0.9(10 - x) = 5$$
$$0.4x + 9 - 0.9x = 5$$
$$-0.5x + 9 = 5$$
$$-0.5x = -4$$
$$x = 8$$

Substitute this into the first equation to find $y = 10 - x = 10 - 8 = 2$. Thus, (8, 2).

55. $\begin{cases} y = 2x - 1 \\ y = -3x - 9 \end{cases}$

By substitution, $2x - 1 = -3x - 9$ Finally, $y = 2x - 1$
$$5x = -8$$
$$x = -\frac{8}{5}$$
$$= 2\left(-\frac{8}{5}\right) - 1$$
$$= -\frac{16}{5} - \frac{5}{5}$$
$$= -\frac{21}{5}$$

Thus, $\left(-\frac{8}{5}, -\frac{21}{5}\right)$.

Page 403

Level 3, page 761

57. Let s = Sydney's age (the cat) in days and k = Karen's age in days. We want to find the date when $4s = k$. We also are given that Karen was born on 12/7/1970 and Sydney was born on 4/18/1992. These dates are 7,803 days apart, so

$$s + 7,803 = k$$

We solve this system of equations by substituting the value for k into the second equation:

$$s + 7,803 = 4s$$
$$7,803 = 3s$$
$$s = 2,601$$

This means that the two will share a birthday 2,601 days after the cat's birth, namely on June 2, 1999.

Level 3 Problem Solving, page 762

59. $\begin{cases} xy = 1 \\ x + y = a \end{cases}$ Solve the second equation for y: $y = a - x$ and substitute this into the

first equation: $x(a - x) = 1$
$$ax - x^2 = 1$$
$$x^2 - ax + 1 = 0$$

Now, we need to use the quadratic equation: $x = \dfrac{-(-a) \pm \sqrt{(-a)^2 - 4(1)(1)}}{2(1)}$

$$= \dfrac{a \pm \sqrt{a^2 - 4}}{2}$$

Substitute the first value $x = \dfrac{a + \sqrt{a^2 - 4}}{2}$ into the equation $y = a - x$

$$= a - \dfrac{a + \sqrt{a^2 - 4}}{2}$$

$$= \dfrac{2a - a - \sqrt{a^2 - 4}}{2}$$

$$= \dfrac{a - \sqrt{a^2 - 4}}{2}$$

This is the first solution, namely $\left(\dfrac{a + \sqrt{a^2 - 4}}{2}, \dfrac{a - \sqrt{a^2 - 4}}{2} \right)$.

Page 404

Substitute the second value $x = \dfrac{a - \sqrt{a^2 - 4}}{2}$ into the equation $y = a - x$

$$= a - \frac{a - \sqrt{a^2 - 4}}{2}$$

$$= \frac{2a - a + \sqrt{a^2 + 4}}{2}$$

$$= \frac{a + \sqrt{a^2 + 4}}{2}$$

This is the second solution, namely $\left(\dfrac{a - \sqrt{a^2 - 4}}{2}, \dfrac{a + \sqrt{a^2 - 4}}{2} \right)$.

16.2 Problem Solving with Systems, page 762
New Terms Introduced in this Section

Demand   Equilibrium point   Supply

Level 1, page 770

1. $\begin{cases} \text{VALUE OF NICKELS} + \text{VALUE OF DIMES} = \text{TOTAL VALUE} \\ \text{NUMBER OF NICKELS} + 25 = \text{NUMBER OF DIMES} \end{cases}$

TOTAL VALUE = 715
↓

$\begin{cases} \text{VALUE OF NICKELS} + \text{VALUE OF DIMES} = \text{TOTAL VALUE} \\ \text{NUMBER OF NICKELS} + 25 = \text{NUMBER OF DIMES} \end{cases}$

$\quad$ *5(NUMBER OF NICKELS)* $\qquad$ *10(NUMBER OF DIMES)*
$\qquad\qquad$ ↓ $\qquad\qquad\qquad\qquad$ ↓

$\begin{cases} \text{VALUE OF NICKELS} + \text{VALUE OF DIMES} = 715 \\ \text{NUMBER OF NICKELS} + 25 = \text{NUMBER OF DIMES} \end{cases}$

$\begin{cases} 5(\text{NUMBER OF NICKELS}) + 10(\text{NUMBER OF DIMES}) = 715 \\ \text{NUMBER OF NICKELS} + 25 = \text{NUMBER OF DIMES} \end{cases}$

$n = $ NUMBER OF NICKELS and $d = $ NUMBER OF QUARTERS.

$\begin{cases} 5n + 10d = 715 \\ n + 25 = d \end{cases}$

Solve *this* system by substitution where $d = n + 25$:

$$5n + 10d = 715$$
$$5n + 10(n + 25) = 715$$
$$5n + 10n + 250 = 715$$
$$15n = 465$$
$$n = 31$$

If $n = 31$, then $d = 31 + 25 = 56$.
The box has 31 nickels and 56 dimes.

3. $\begin{cases} \text{VALUE OF NICKELS + VALUE OF QUARTERS = TOTAL VALUE} \\ \text{NUMBER OF NICKELS + NUMBER OF QUARTERS = 42} \end{cases}$

$$\text{TOTAL VALUE} = 690$$
$$\downarrow$$

$\begin{cases} \text{VALUE OF NICKELS + VALUE OF QUARTERS = TOTAL VALUE} \\ \text{NUMBER OF NICKELS + NUMBER 0F QUARTERS = 42} \end{cases}$

     5(NUMBER OF NICKELS)   *25(NUMBER OF QUARTERS)*
           $\downarrow$ $\downarrow$

$\begin{cases} \text{VALUE OF NICKELS + VALUE OF QUARTERS = 690} \\ \text{NUMBER OF NICKELS + NUMBER OF QUARTERS = 42} \end{cases}$

$\begin{cases} \text{5(NUMBER OF NICKELS) + 25(NUMBER OF QUARTERS) = 690} \\ \text{NUMBER OF NICKELS + NUMBER OF QUARTERS = 42} \end{cases}$

n = NUMBER OF NICKELS and q = NUMBER OF QUARTERS.
$\begin{cases} 5n + 25q = 690 \\ n + q = 42 \end{cases}$

Solve *this* system by substitution where $n = 42 - q$:
$$5n + 25q = 690$$
$$5(42 - q) + 25q = 690$$
$$210 - 5q + 25q = 690$$
$$20q = 480$$
$$q = 24$$

If $q = 24$, then $n = 42 - 24 = 18$.
There are 18 nickels.

5. RATE WITH THE WIND is 510 mph.

RATE AGAINST THE WIND is 270 mph.

$$\begin{cases} (\text{RATE OF PLANE}) + (\text{RATE OF WIND}) = 510 \\ (\text{RATE OF PLANE}) - (\text{RATE OF WIND}) = 270 \end{cases}$$

We assume that the rate of the plane is greater so that we obtain a positive distance. There are two unknowns; let p = RATE OF PLANE and t = RATE OF WIND.

$$+\begin{cases} p + w = 510 \\ p - w = 270 \end{cases} \quad \text{Add: } 2p = 780 \\ p = 390$$

The plane's speed is 390 mph.

7. The demand equation (which we assume is linear) passes through (110, 150) and (20, 300), where y is in thousands of units. The supply equation (which we also assume is linear) passes through (90, 300) and (10, 200). Draw the graph of these lines.

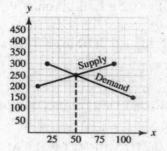

It looks like the equilibrium point for this system is $(50, 250)$. This is \$250,000 at \$50 per unit.

9. The demand equation (which we assume is linear) passes through (1, 900) and (7, 300). The supply equation (which we also assume is linear) passes through (1, 600) and (9, 1000). Draw the graph of these lines.

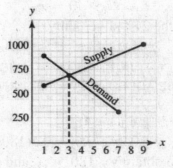

It looks like the optimum price for the items is \$3.

Page 407

11. Let x = Clint Eastwood's birth year and y = Goldie Hawn's birth year. Since Eastwood is 15 years older, you add 15 to his birth year to obtain Hawn's birth year: $x + 15 = y$. Thus, we have the following system

$$\begin{cases} x + 15 = y \\ x + y = 3{,}875 \end{cases}$$

We solve this system by substituting $x + 15$ for y in the second equation (this is easy since the first equation is already solved for y):

$$x + (x + 15) = 3{,}875$$
$$2x + 15 = 3{,}875$$
$$x = 1930$$

We are looking for Goldie Hawn's birth year, so $y = x + 15 = 1930 + 15 = 1945$.

13. Mixture **a** has $0.80(5 \text{ lb}) + 0.30(10 \text{ lb}) = 7$ lb of micoden

15. Mixture **c** has $0(8 \text{ L}) + 0.5(p\,\text{L}) = 0.5p$ L of bixon.

17. The amount of micoden is $0.8(5 \text{ lb}) + 0.3(10 \text{ lb}) = 7$ lb; as a percent $7 \text{ lb}/15 \text{ lb} = 0.46\overline{7}$ or $46\frac{2}{3}\%$.

19. AMOUNT OF BASE METAL + AMOUNT OF 21% ALLOY = AMOUNT OF 15% ALLOY

Amount of silver:

0(AMOUNT OF BASE METAL) + 0.21(AMOUNT OF 21% ALLOY) = 0.15(AMOUNT OF 15% ALLOY)

Let x = AMOUNT OF BASE METAL and we are given that AMOUNT OF 21% ALLOY = 100 oz.

$$0x + 0.21(100) = 0.15(x + 100)$$
$$21 = 0.15x + 15$$
$$6 = 0.15x$$
$$40 = x$$

It is necessary to use 40 oz of the 21% silver alloy.

21. AMOUNT OF MIXTURE = PERCENT BUTTERFAT × AMOUNT OF SOLUTION

Let x = AMOUNT OF MILK TO BE ADDED and y = AMOUNT OF CREAM TO BE ADDED. This leads to the system

$$\begin{cases} x + y = 180 \\ 0.2x + 0.6y = 0.5(180) \end{cases}$$

Solve the first equation for y to obtain $y = 180 - x$, and substitute this into the first equation to obtain

Page 408

$$0.2x + 0.6(180 - x) = 0.5(180)$$
$$0.2x + 108 - 0.6x = 90$$
$$-0.4x = -18$$
$$x = 45$$

The amount of milk to be added is 45 gallons, and the amount of cream is 135 gallons $(180 - 45 = 135)$.

Level 2, page 770

23. $\begin{cases} p = 0.005n + 12 \\ p = 150 - 0.01n \end{cases}$ Solve by substitution: $0.005n + 12 = 150 - 0.01n$
$$0.015n = 138$$
$$n = 9,200$$

Substitute this into either of the given equations (we select the second):
$$p = 150 - 0.01n$$
$$= 150 - 0.01(9,200)$$
$$= 58$$

The equilibrium point is $(58, 9\,200)$.

25. $\begin{cases} p = 0.0005n - 3 \\ p = 8 - 0.0006n \end{cases}$ Solve by substitution: $0.0005n - 3 = 8 - 0.0006n$
$$0.0011n = 11$$
$$n = 10,000$$

Substitute this into either of the given equations (we select the first):
$$p = 0.0005(10,000) - 3$$
$$= 2$$

The equilibrium point is $(2, 10,000)$.

27. $\begin{cases} q = 2d \\ 10d + 25q = 840 \end{cases}$ where d is the number of dimes and q is the number of nickels.

Substitute the first equation into the second equation:
$$10d + 25q = 840$$
$$10d + 25(2d) = 840$$
$$60d = 840$$
$$d = 14$$

Since $q = 2d = 2(14) = 28$, we see there are 14 dimes and 28 quarters.

29. $\begin{cases} x+9=y \\ x+y=3{,}851 \end{cases}$ where x = Charles Bronson's birth year and y = Clint Eastwood's birth

year. Substitute the first equation into the second equation:

$$x + y = 3{,}851$$
$$x + (x + 9) = 3{,}851$$
$$2x + 9 = 3{,}851$$
$$2x = 3{,}842$$
$$x = 1{,}921$$

Since $y = 1{,}921 + 9 = 1{,}930$, we see that Clint Eastwood was born in 1930.

31. $\begin{cases} x+6=y \\ x+y=3{,}964 \end{cases}$ where x = Sandra Bullock's birth year and y = Matt Damon's birth year.

hSubstitute the first equation into the second equation:

$$x + y = 3{,}934$$
$$x + (x + 6) = 3{,}934$$
$$2x + 6 = 3{,}934$$
$$2x = 3{,}928$$
$$x = 1{,}964$$

Matt was born in $1964 + 6 = 1970$.

33. AMOUNT OF SOLUTION: $1 + x$ where x is the amount of water added.

AMOUNT OF ANTIFREEZE: $0.80(1) + 0(x) = 0.60(1 + x)$

Solve this second equation: $\quad 0.8 = 0.60 + 0.60x$

$$0.2 = 0.60x$$
$$\frac{1}{3} = x$$

Add $\frac{1}{3}$ gal of water.

35. $\begin{cases} T=F+208{,}044 \\ T+F=316{,}224 \end{cases}$ where T is the area of Texas and F is the area of Florida.

Substitute the first equation into the second equation:

$$T + F = 316{,}224$$
$$(F + 208{,}044) + F = 316{,}224$$
$$2F = 108{,}180$$
$$F = 54{,}090$$

Then, $T = 54{,}090 + 208{,}044 = 262{,}134$, so the area of Texas is $262{,}134$ mi^2 and the area of Florida is $54{,}090$ mi^2.

37. $\begin{cases} n + q = 42 \\ 5n + 25q = 950 \end{cases}$ where n is the number of nickels and q the number of quarters. From

the first equation $n = 42 - q$; substitute this value into the second equation:

$$5n + 25q = 950$$
$$5(42 - q) + 25q = 950$$
$$210 - 5q + 25q = 950$$
$$20q = 740$$
$$q = 37$$

There are 37 quarters.

39. $\begin{cases} d = q \\ n = d + 8 \\ 5n + 10d + 25q = 1{,}240 \end{cases}$ where n is the number of nickels, d is the number of dimes,

and q the number of quarters. Substitute the first and second equations into the third equation:

$$5n + 10d + 25q = 1{,}240$$
$$5(d + 8) + 10d + 25(d) = 1{,}240$$
$$5d + 40 + 10d + 25d = 1{,}240$$
$$40d = 1{,}200$$
$$d = 30$$

There are 30 dimes.

41. $\begin{cases} 5f + 10n + 20t = 1{,}390 \\ n = 5f \\ t = 2f + 3 \end{cases}$ where n is the number of \$10 bills, f is the number of \$5

bills, and t is the number of \$20 bills. Substitute the second and third equations into the first
equation:

$$5f + 10n + 20t = 1{,}390$$
$$5f + 10(5f) + 20(2f + 3) = 1{,}390$$
$$5f + 50f + 40f + 60 = 1{,}390$$
$$95f + 60 = 1{,}390$$
$$95f = 1{,}330$$
$$f = 14$$

Use this value in the second equation: $n = 5f = 5(14) = 70$
and into the third equation: $t = 2f + 3 = 2(14) + 3 = 31$
There are fourteen \$5 bills, seventy \$10 bills, and thirty-one \$20 bills.

Page 411

43. We use $d = rt$, where $d = 870$ miles and the $t = 3\frac{1}{3}$ hr $= 200$ minutes, so

$$\begin{cases} 870 = (\text{RATE OF PLANE} + \text{RATE OF WIND})(200) \\ 870 = (\text{RATE OF PLANE} - \text{RATE OF WIND})(150) \end{cases}$$

Let $p = \text{RATE OF PLANE}$ and $w = \text{RATE OF WIND}$, we obtain

$+ \begin{cases} p + w = 4.35 & \textit{Divide both sides of the first equation by 200.} \\ p - w = 5.8 & \textit{Divide both sides of the second equation by 150} \end{cases}$

$$2p = 10.15$$
$$p = 5.075$$

This is miles per minute so the miles per hour is $60(5.075) = 304.5$.

45. $\begin{cases} x + y = 30 \\ 0.23x + 0.03y = 0.04(30) \end{cases}$ where x is the number of gallons of 23% butterfat cream and y

is the number of gallons of 3% butterfat milk. From the first equation, $y = 30 - x$, and we substitute this into the second equation:

$$0.23x + 0.03y = 1.2$$
$$0.23x + 0.03(30 - x) = 1.2$$
$$0.23x + 0.9 - 0.03x = 1.2$$
$$x = 1.5$$

$y = 30 - x = 30 - 1.5 = 28.5$, so the mixture should contain 28.5 gallons of milk with 1.5 gallons of cream.

47. $\begin{cases} t + b = 1{,}632 \\ t = b + 74 \end{cases}$ where t is the height of the Transamerica Tower and b is the height of

the Bank of America Building. Substitute the second equation into the first:

$$t + b = 1{,}632$$
$$(b + 74) + b = 1{,}632$$
$$2b + 74 = 1{,}632$$
$$2b = 1{,}558$$
$$b = 779$$

The height of the Bank of America Building is 779 ft and the height of the Transamerica Tower is 853 ft ($779 + 74 = 853$).

49. $\begin{cases} v + w = 7{,}760 \\ v = w + 760 \end{cases}$ where v is the length of the Verrazano Narrows Bridge and w is the

length of the George Washington Bridge. Substitute the second equation into the first:

Page 412

$$v + w = 7,760$$
$$(w + 760) + w = 7,760$$
$$2w = 7,000$$
$$w = 3,500$$

Then $v = 3,500 + 760 = 4,260$; the length of the Verrazano Narrows bridge is 4,260 ft.

51. The demand equation (which we assume is linear) passes through (4, 1,000) and (3, 7,000). The supply equation (which we also assume is linear) passes through (2, 1,000) and (4, 5,000). Draw the graph of these lines (number of items is in plotted in thousands)

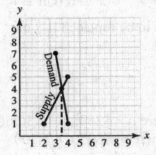

The equilibrium point is a price of $3.50 for 4,000 items.

53. $\begin{cases} p + s = 100 \\ (1)p + 0.925s = 0.94(100) \end{cases}$ where p is the number of grams of pure silver and s is the number of grams of sterling silver. From the first equation $p = 100 - s$ which we substitute into the second equation:

$$p + 0.925s = 94$$
$$(100 - s) + 0.925s = 94$$
$$-0.075s = -6$$
$$s = 80$$

Then $p = 100 - s = 100 - 80 = 20$; the mixture should contain 20 g of pure silver and 80 g of sterling silver.

55. $\begin{cases} x + 500 = y \\ 0.24x + 0.03(500) = 0.04y \end{cases}$ where x is amount of 24% butterfat solution and y is the amount of 4% butterfat mixture. Substitute the first equation into the second equation:

$$0.24x + 0.03(500) = 0.04y$$
$$0.24x + 15 = 0.04(x + 500)$$
$$0.24x + 15 = 0.04x + 20$$
$$0.20x = 5$$
$$x = 25$$

The amount of 24% butterfat that must be added is 25 gallons.

Page 413

57. AMOUNT OF PUNCH REMOVED = AMOUNT OF 7-UP ADDED

Let this amount be x. Now the amount of 7-UP in the final solution

AMOUNT OF 7-UP REMAINING AFTER x IS REMOVED + AMOUNT OF 7-UP ADDED

= AMOUNT OF 7-UP IN FINAL SOLUTION

or using variables: $0.20(5 - x) + 1.00x = 0.30(5)$

$$1 - 0.20x + x = 1.5$$
$$0.80x = 0.5$$
$$x = 0.625 = \frac{5}{8}$$

The amount of punch that needs to be replaced is $\frac{5}{8}$ qt.

Level 3, page 772

59. Supply: $n = 2.5p - 500$

Demand: $n = 200 - 0.5p$

a. Use the supply equation where $p = \$250$: $n = 2.5(250) - 500 = 125$

Use the demand equation where $p = \$250$: $n = 200 - 0.5(250) = 75$

Thus, 125 items would be supplied; 75 would be demanded.

b. Solve $n = 2.5p - 500$ for $n = 0$:

$$0 = 2.5p - 500$$
$$500 = 2.5p$$
$$200 = p$$

No items would be supplied at $200.

c. Solve $n = 200 - 0.5p$ where $n = 0$:

$$0 = 200 - 0.5p$$
$$0.5p = 200$$
$$p = 400$$

No items would be demanded at $400.

d. To solve the system, substitute the first equation into the second:

$$2.5p - 500 = 200 - 0.5p$$
$$3p = 700$$
$$p = \frac{700}{3}$$

The equilibrium price is $\frac{700}{3} \approx \$233.33$.

e. Substitute $p = \frac{700}{3}$ into either of the given equations (we select the second):

Page 414

$$n = 200 - 0.5p$$
$$= 200 - 0.5\left(\frac{700}{3}\right)$$
$$= \frac{600}{3} - \frac{350}{3}$$
$$= \frac{250}{3}$$

The number of items produced at the equilibrium price is $\frac{250}{3} \approx 83$.

16.3 Matrix Solution of a System of Equations, page 773

New Terms Introduced in this Section

Array	Augmented matrix	Column
Diagonal form	Dimension	Double subscripts
Elementary row operations	Equivalent matrices	Gauss-Jordan elimination
Matrix	Order	Pivot
Pivot row	Pivoting	Row
Row +	Row-reduced form	RowSwap
Scalar	Square matrix	Subscript
Target row	*Row	*Row +

Level 1, page 781

1. A matrix is a rectangular array of entries.
3. Select as the first pivot the element in the first row, first column, and pivot.
   The next pivot is the element in the second row, second column; pivot.
   Repeat the process until you arrive at the last row, or until the pivot element is a zero.
   If it is a zero and you can interchange that row with a row below it, so that the pivot element is no longer a zero, do so and continue. If it is zero and you cannot interchange rows so that it is not a zero, continue with the next row.
   The final matrix is called the *row-reduced form*.
5. False; the order is 4×5.
7. This statement is true.
9. This statement is true.
11. False; this notation means multiply row 4 by $\frac{1}{3}$ and add to row 2.
13. False; the first step is RowSwap([A],1,2).

Page 415

15. a. $\begin{cases} 6x + 7y + 8z = 3 \\ x + 2y + 3z = 4 \\ y + 3z = 4 \end{cases}$ **b.** $\begin{cases} x_1 = 3 \\ x_2 + 2x_3 = 4 \end{cases}$ **c.** $\begin{cases} x_1 = 32 \\ x_2 = 27 \\ x_3 = -5 \\ 0 = 3 \end{cases}$

17. RowSwap([B], 1, 2); $\begin{bmatrix} 1 & 0 & 2 & | & -8 \\ -2 & 3 & 5 & | & 9 \\ 0 & 1 & 0 & | & 5 \end{bmatrix}$

19. *Row(1/5, [D], 1); $\begin{bmatrix} 1 & 4 & 3 & | & \frac{6}{5} \\ 7 & -5 & 3 & | & 2 \\ 12 & 0 & 1 & | & 4 \end{bmatrix}$

21. *Row +(3, [B], 1, 2); $\begin{bmatrix} 1 & 3 & -5 & | & 6 \\ 0 & 13 & -14 & | & 20 \\ 0 & 5 & 1 & | & 3 \end{bmatrix}$

23. *Row +(−2, [D], 1, 2), *Row +(−3, [Ans], 1, 3); $\begin{bmatrix} 1 & 5 & 3 & | & 2 \\ 0 & -7 & -7 & | & 0 \\ 0 & -13 & -8 & | & -6 \end{bmatrix}$

25. *Row(1/3, [B], 2); $\begin{bmatrix} 1 & 5 & -3 & | & 5 \\ 0 & 1 & 3 & | & -5 \\ 0 & 2 & 1 & | & 5 \end{bmatrix}$

27. *Row+(− 1, [D], 3, 2); $\begin{bmatrix} 1 & 3 & -2 & | & 0 \\ 0 & 1 & -4 & | & 8 \\ 0 & 3 & 6 & | & 1 \end{bmatrix}$

29. *Row +(−3, [B], 2, 1), *Row +(2, [Ans], 2, 3); $\begin{bmatrix} 1 & 0 & 12 & | & 27 \\ 0 & 1 & -2 & | & -5 \\ 0 & 0 & -2 & | & -4 \end{bmatrix}$

31. *Row +(−5, [D], 2, 1), *Row +(−1, [Ans], 2, 3), *Row +(−2, [Ans], 2, 4);
$\begin{bmatrix} 1 & 0 & -26 & -8 & | & 8 \\ 0 & 1 & 5 & 2 & | & 0 \\ 0 & 0 & -1 & -2 & | & 5 \\ 0 & 0 & -13 & -3 & | & 7 \end{bmatrix}$

33. *Row(1/8, [B], 3); $\begin{bmatrix} 1 & 0 & 4 & | & -5 \\ 0 & 1 & 3 & | & 6 \\ 0 & 0 & 1 & | & 1.5 \end{bmatrix}$

Page 416

35. *Row(1/2, [D], 3); $\begin{bmatrix} 1 & 0 & -8 & 2 & | & 8 \\ 0 & 1 & -1 & 3 & | & 2 \\ 0 & 0 & 1 & 0 & | & 5 \\ 0 & 0 & -2 & 1 & | & 6 \end{bmatrix}$

37. *Row +(−4, [B], 3, 2), *Row +(3, [Ans], 3, 1); $\begin{bmatrix} 1 & 0 & 0 & | & 7 \\ 0 & 1 & 0 & | & -7 \\ 0 & 0 & 1 & | & 3 \end{bmatrix}$

39. *Row +(8, [D], 3, 1), *Row +(−4, [Ans], 3, 2), *Row +(1, [Ans], 3, 4);
$\begin{bmatrix} 1 & 0 & 0 & 2 & | & 24 \\ 0 & 1 & 0 & 2 & | & -8 \\ 0 & 0 & 1 & 0 & | & 2 \\ 0 & 0 & 0 & 1 & | & 9 \end{bmatrix}$

Level 2, page 782

41. $\begin{bmatrix} 1 & -1 & 8 \\ 1 & 1 & 2 \end{bmatrix} \rightarrow \begin{bmatrix} 1 & -1 & 8 \\ 0 & 2 & -6 \end{bmatrix} \rightarrow \begin{bmatrix} 1 & -1 & 8 \\ 0 & 1 & -3 \end{bmatrix} \rightarrow \begin{bmatrix} 1 & 0 & 5 \\ 0 & 1 & -3 \end{bmatrix}$ (5, −3)

43. $\begin{bmatrix} 1 & -6 & -3 \\ 2 & 3 & 9 \end{bmatrix} \rightarrow \begin{bmatrix} 1 & -6 & -3 \\ 0 & 15 & 15 \end{bmatrix} \rightarrow \begin{bmatrix} 1 & -6 & -3 \\ 0 & 1 & 1 \end{bmatrix} \rightarrow \begin{bmatrix} 1 & 0 & 3 \\ 0 & 1 & 1 \end{bmatrix}$ (3, 1)

45. $\begin{bmatrix} 3 & 7 & 5 \\ 4 & 9 & 7 \end{bmatrix} \rightarrow \begin{bmatrix} 3 & 7 & 5 \\ 1 & 2 & 2 \end{bmatrix} \rightarrow \begin{bmatrix} 1 & 2 & 2 \\ 3 & 7 & 5 \end{bmatrix} \rightarrow \begin{bmatrix} 1 & 2 & 2 \\ 0 & 1 & -1 \end{bmatrix} \rightarrow \begin{bmatrix} 1 & 0 & 4 \\ 0 & 1 & -1 \end{bmatrix}$ (4, −1)

47. $\begin{bmatrix} 3 & -4 & 16 \\ -1 & 2 & -6 \end{bmatrix} \rightarrow \begin{bmatrix} 1 & -2 & 6 \\ 3 & -4 & 16 \end{bmatrix} \rightarrow \begin{bmatrix} 1 & -2 & 6 \\ 0 & 2 & -2 \end{bmatrix} \rightarrow \begin{bmatrix} 1 & -2 & 6 \\ 0 & 1 & -1 \end{bmatrix}$
$\rightarrow \begin{bmatrix} 1 & 0 & 4 \\ 0 & 1 & -1 \end{bmatrix}$ (4, −1)

49. $\begin{bmatrix} 1 & 1 & 0 & 2 \\ 1 & 0 & -1 & 1 \\ 0 & -1 & 1 & 1 \end{bmatrix} \rightarrow \begin{bmatrix} 1 & 1 & 0 & 2 \\ 0 & -1 & -1 & -1 \\ 0 & -1 & 1 & 1 \end{bmatrix} \rightarrow \begin{bmatrix} 1 & 1 & 0 & 2 \\ 0 & 1 & 1 & 1 \\ 0 & -1 & 1 & 1 \end{bmatrix} \rightarrow$
$\begin{bmatrix} 1 & 1 & 0 & 2 \\ 0 & 1 & 1 & 1 \\ 0 & 0 & 2 & 2 \end{bmatrix} \rightarrow \begin{bmatrix} 1 & 0 & -1 & 1 \\ 0 & 1 & 1 & 1 \\ 0 & 0 & 1 & 1 \end{bmatrix} \rightarrow \begin{bmatrix} 1 & 0 & 0 & 2 \\ 0 & 1 & 0 & 0 \\ 0 & 0 & 1 & 1 \end{bmatrix}$ (2, 0, 1)

Page 417

51. $\begin{bmatrix} 1 & 0 & 2 & 13 \\ 2 & 1 & 0 & 8 \\ 0 & -2 & 9 & 41 \end{bmatrix} \rightarrow \begin{bmatrix} 1 & 0 & 2 & 13 \\ 0 & 1 & -4 & -18 \\ 0 & -2 & 9 & 41 \end{bmatrix} \rightarrow \begin{bmatrix} 1 & 0 & 2 & 13 \\ 0 & 1 & -4 & -18 \\ 0 & 0 & 1 & 5 \end{bmatrix} \rightarrow$

$\begin{bmatrix} 1 & 0 & 0 & 3 \\ 0 & 1 & 0 & 2 \\ 0 & 0 & 1 & 5 \end{bmatrix}$ $(3, 2, 5)$

53. $\begin{bmatrix} 5 & 0 & 1 & 9 \\ 1 & 0 & -5 & 7 \\ 1 & 1 & -1 & 0 \end{bmatrix} \rightarrow \begin{bmatrix} 1 & 1 & -1 & 0 \\ 1 & 0 & -5 & 7 \\ 5 & 0 & 1 & 9 \end{bmatrix} \rightarrow \begin{bmatrix} 1 & 1 & -1 & 0 \\ 0 & -1 & -4 & 7 \\ 0 & -5 & 6 & 9 \end{bmatrix} \rightarrow$

$\begin{bmatrix} 1 & 1 & -1 & 0 \\ 0 & 1 & 4 & -7 \\ 0 & -5 & 6 & 9 \end{bmatrix} \rightarrow \begin{bmatrix} 1 & 0 & -5 & 7 \\ 0 & 1 & 4 & -7 \\ 0 & -5 & 6 & 9 \end{bmatrix} \rightarrow \begin{bmatrix} 1 & 0 & -5 & 7 \\ 0 & 1 & 4 & -7 \\ 0 & 0 & 26 & -26 \end{bmatrix} \rightarrow$

$\begin{bmatrix} 1 & 1 & -5 & 7 \\ 0 & 1 & 4 & -7 \\ 0 & 0 & 1 & -1 \end{bmatrix} \rightarrow \begin{bmatrix} 1 & 0 & 0 & 2 \\ 0 & 1 & 0 & -3 \\ 0 & 0 & 1 & -1 \end{bmatrix}$ $(2, -3, -1)$

55. $\begin{bmatrix} 1 & 1 & 0 & -1 \\ 0 & 1 & 1 & -1 \\ 1 & 1 & 1 & 1 \end{bmatrix} \rightarrow \begin{bmatrix} 1 & 1 & 0 & -1 \\ 0 & 1 & 1 & -1 \\ 0 & 0 & 1 & 2 \end{bmatrix} \rightarrow \begin{bmatrix} 1 & 0 & -1 & 0 \\ 0 & 1 & 1 & -1 \\ 0 & 0 & 1 & 2 \end{bmatrix} \rightarrow$

$\begin{bmatrix} 1 & 0 & 0 & 2 \\ 0 & 1 & 1 & -3 \\ 0 & 0 & 1 & 2 \end{bmatrix}$ $(2, -3, 2)$

57. $\begin{bmatrix} 1 & 0 & 2 & 0 \\ 3 & -1 & 2 & 0 \\ 4 & 1 & 0 & 6 \end{bmatrix} \rightarrow \begin{bmatrix} 1 & 0 & 2 & 0 \\ 0 & -1 & -4 & 0 \\ 0 & 1 & -8 & 6 \end{bmatrix} \rightarrow \begin{bmatrix} 1 & 0 & 2 & 0 \\ 0 & 1 & 4 & 0 \\ 0 & 1 & -8 & 6 \end{bmatrix} \rightarrow$

$\begin{bmatrix} 1 & 0 & 2 & 0 \\ 0 & 1 & 4 & 0 \\ 0 & 0 & -12 & 6 \end{bmatrix} \rightarrow \begin{bmatrix} 1 & 0 & 2 & 0 \\ 0 & 1 & 4 & 0 \\ 0 & 0 & 1 & -1/2 \end{bmatrix} \rightarrow \begin{bmatrix} 1 & 0 & 0 & 1 \\ 0 & 1 & 0 & 2 \\ 0 & 0 & 1 & -1/2 \end{bmatrix}$ $\left(1, 2, -\tfrac{1}{2}\right)$

Level 3, page 782

59.

	Spray I	Spray II	Total
A	5	2	23
B	2	7	34

Let x = amount of Spray I, y = amount of Spray II

$$[M] = \begin{cases} 5x + 2y = 23 \\ 2x + 7y = 34 \end{cases}$$

$$\begin{bmatrix} 5 & 2 & 23 \\ 2 & 7 & 34 \end{bmatrix} \xrightarrow{*\text{Row} + (-2, [M], 2, 1)} \begin{bmatrix} 1 & -12 & -45 \\ 2 & 7 & 34 \end{bmatrix} \xrightarrow{*\text{Row} + (-2, [ANS], 1, 2)} \begin{bmatrix} 1 & -12 & -45 \\ 0 & 31 & 124 \end{bmatrix}$$

$$\xrightarrow{*\text{Row} (1/31, [ANS], 2)} \begin{bmatrix} 1 & -12 & -45 \\ 0 & 1 & 4 \end{bmatrix} \xrightarrow{*\text{Row} + (12, [ANS], 2, 1)} \begin{bmatrix} 1 & 0 & 3 \\ 0 & 1 & 4 \end{bmatrix} \quad \text{Solution: } (3, 4)$$

Mix 3 containers of Spray I with 4 containers of Spray II.

16.4 Inverse Matrices, page 783
New Terms Introduced in this Section

Addition of matrices	Additive inverse	Associative property
Communication matrix	Commutative property	Distributive property
Equal matrices	Identity matrix	Inverse matrix
Inverse property	Main diagonal	Matrix equation
Multiplication of matrices	Multiplicative inverse	Noncomformable matrices
Nonsingular matrix	Scalar multiplication	Singular matrix
Subtraction of matrices	Zero-one matrix	Zero matrix

Level 1, page 794

1. [M] = [N] if and only if matrices [M] and [N] are the same order and the corresponding entries are the same.

3. Let [M] be an $m \times r$ matrix and [N] an $r \times n$ matrix. The product matrix [M][N] = [P] is an $m \times n$ matrix. The entry in the ith row and jth column of [M][N] is *the sum of the products formed by multiplying each entry of the ith row of* [M] *by the corresponding element in the jth column of* [N].

5. A matrix is nonsingular if it has an inverse for multiplication.

7. If $[A][X] = [B]$, then $[X] = [A]^{-1}[B]$.

9. **a.** $[E] + [F] = \begin{bmatrix} 1 & 0 & 2 \\ 3 & -1 & 2 \\ 4 & 1 & 0 \end{bmatrix} + \begin{bmatrix} 1 & 4 & 0 \\ 3 & -1 & 2 \\ -2 & 1 & 5 \end{bmatrix}$

$$= \begin{bmatrix} 1+1 & 0+4 & 2+0 \\ 3+3 & -1-1 & 2+2 \\ 4-2 & 1+1 & 0+5 \end{bmatrix}$$

Page 419

$$= \begin{bmatrix} 2 & 4 & 2 \\ 6 & -2 & 4 \\ 2 & 2 & 5 \end{bmatrix}$$

b. $2[E] - [G] = 2\begin{bmatrix} 1 & 0 & 2 \\ 3 & -1 & 2 \\ 4 & 1 & 0 \end{bmatrix} - \begin{bmatrix} 8 & 1 & 6 \\ 3 & 5 & 7 \\ 4 & 9 & 2 \end{bmatrix}$

$$= \begin{bmatrix} 2 & 0 & 4 \\ 6 & -2 & 4 \\ 8 & 2 & 0 \end{bmatrix} - \begin{bmatrix} 8 & 1 & 6 \\ 3 & 5 & 7 \\ 4 & 9 & 2 \end{bmatrix}$$

$$= \begin{bmatrix} -6 & -1 & -2 \\ 3 & -7 & -3 \\ 4 & -7 & -2 \end{bmatrix}$$

11. a. $[F][G] = \begin{bmatrix} 1 & 4 & 0 \\ 3 & -1 & 2 \\ -2 & 1 & 5 \end{bmatrix} \begin{bmatrix} 8 & 1 & 6 \\ 3 & 5 & 7 \\ 4 & 9 & 2 \end{bmatrix}$

$$= \begin{bmatrix} 1 \cdot 8 + 4 \cdot 3 + 0 \cdot 4 & 1 \cdot 1 + 4 \cdot 5 + 0 \cdot 9 & 1 \cdot 6 + 4 \cdot 7 + 0 \cdot 2 \\ 3 \cdot 8 + (-1) \cdot 3 + 2 \cdot 4 & 3 \cdot 1 + (-1) \cdot 5 + 2 \cdot 9 & 3 \cdot 6 + (-1) \cdot 7 + 2 \cdot 2 \\ (-2) \cdot 8 + 1 \cdot 3 + 5 \cdot 4 & (-2) \cdot 1 + 1 \cdot 5 + 5 \cdot 9 & (-2) \cdot 6 + 1 \cdot 7 + 5 \cdot 2 \end{bmatrix}$$

$$= \begin{bmatrix} 20 & 21 & 34 \\ 29 & 16 & 15 \\ 7 & 48 & 5 \end{bmatrix}$$

b. $[G][F] = \begin{bmatrix} 8 & 1 & 6 \\ 3 & 5 & 7 \\ 4 & 9 & 2 \end{bmatrix} \begin{bmatrix} 1 & 4 & 0 \\ 3 & -1 & 2 \\ -2 & 1 & 5 \end{bmatrix}$

$$= \begin{bmatrix} 8 \cdot 1 + 1 \cdot 3 + 6 \cdot (-2) & 8 \cdot 4 + 1 \cdot (-1) + 6 \cdot 1 & 8 \cdot 0 + 1 \cdot 2 + 6 \cdot 5 \\ 3 \cdot 1 + 5 \cdot 3 + 7 \cdot (-2) & 3 \cdot 4 + 5 \cdot (-1) + 7 \cdot 1 & 3 \cdot 0 + 5 \cdot 2 + 7 \cdot 5 \\ 4 \cdot 1 + 9 \cdot 3 + 2 \cdot (-2) & 4 \cdot 4 + 9 \cdot (-1) + 2 \cdot 1 & 4 \cdot 0 + 9 \cdot 2 + 2 \cdot 5 \end{bmatrix}$$

$$= \begin{bmatrix} -1 & 37 & 32 \\ 4 & 14 & 45 \\ 27 & 9 & 28 \end{bmatrix}$$

13. a. $[B]^2 = \begin{bmatrix} 4 & 2 \\ -1 & 3 \end{bmatrix}^2$

$= \begin{bmatrix} 4 & 2 \\ -1 & 3 \end{bmatrix} \begin{bmatrix} 4 & 2 \\ -1 & 3 \end{bmatrix}$

$= \begin{bmatrix} 4 \cdot 4 + 2(-1) & 4 \cdot 2 + 2 \cdot 3 \\ (-1) \cdot 4 + 3(-1) & (-1) \cdot 2 + 3 \cdot 3 \end{bmatrix}$

$= \begin{bmatrix} 14 & 14 \\ -7 & 7 \end{bmatrix}$

b. $[C]^3 = \begin{bmatrix} 1 & 0 & 0 & 0 \\ 0 & 1 & 0 & 0 \\ 0 & 0 & 1 & 0 \\ 0 & 0 & 0 & 1 \end{bmatrix}^3$

$= \begin{bmatrix} 1 & 0 & 0 & 0 \\ 0 & 1 & 0 & 0 \\ 0 & 0 & 1 & 0 \\ 0 & 0 & 0 & 1 \end{bmatrix} \begin{bmatrix} 1 & 0 & 0 & 0 \\ 0 & 1 & 0 & 0 \\ 0 & 0 & 1 & 0 \\ 0 & 0 & 0 & 1 \end{bmatrix} \begin{bmatrix} 1 & 0 & 0 & 0 \\ 0 & 1 & 0 & 0 \\ 0 & 0 & 1 & 0 \\ 0 & 0 & 0 & 1 \end{bmatrix}$

$= \begin{bmatrix} 1 & 0 & 0 & 0 \\ 0 & 1 & 0 & 0 \\ 0 & 0 & 1 & 0 \\ 0 & 0 & 0 & 1 \end{bmatrix} \begin{bmatrix} 1 & 0 & 0 & 0 \\ 0 & 1 & 0 & 0 \\ 0 & 0 & 1 & 0 \\ 0 & 0 & 0 & 1 \end{bmatrix}$

$= \begin{bmatrix} 1 & 0 & 0 & 0 \\ 0 & 1 & 0 & 0 \\ 0 & 0 & 1 & 0 \\ 0 & 0 & 0 & 1 \end{bmatrix}$

15. a. not conformable   **b.** not conformable

17. $[A][X] = [B]$

$\begin{bmatrix} 1 & 2 & 4 \\ -3 & 2 & 1 \\ 2 & 0 & 1 \end{bmatrix} \begin{bmatrix} x \\ y \\ x \end{bmatrix} = \begin{bmatrix} 13 \\ 11 \\ 0 \end{bmatrix}$

$\begin{bmatrix} x + 2y + 4z \\ -3x + 2y + z \\ 2x + 0y + z \end{bmatrix} = \begin{bmatrix} 13 \\ 11 \\ 0 \end{bmatrix}$

Page 421

By the definition of equal matrices, we have $\begin{cases} x + 2y + 4z = 13 \\ -3x + 2y + z = 11 \\ 2x + z = 0 \end{cases}$

19. $[A][B] = \begin{bmatrix} 8-7 & -14+14 \\ 4-4 & -7+8 \end{bmatrix}$; $[B][A] = \begin{bmatrix} 8-7 & 28-28 \\ -2+2 & -7+8 \end{bmatrix}$

$= \begin{bmatrix} 1 & 0 \\ 0 & 1 \end{bmatrix}$ $\qquad\qquad = \begin{bmatrix} 1 & 0 \\ 0 & 1 \end{bmatrix}$

[A] and [B] are inverses since $[A][B] = [B][A] = [I]$.

Level 2, page 795

21. Use a calculator, if possible. If you do not have access to one, then:

$\begin{bmatrix} 4 & -7 & | & 1 & 0 \\ -1 & 2 & | & 0 & 1 \end{bmatrix} \rightarrow \begin{bmatrix} -1 & 2 & | & 0 & 1 \\ 4 & -7 & | & 1 & 0 \end{bmatrix}$

$\rightarrow \begin{bmatrix} 1 & -2 & | & 0 & -1 \\ 4 & -7 & | & 1 & 0 \end{bmatrix}$

$\rightarrow \begin{bmatrix} 1 & -2 & | & 0 & -1 \\ 0 & 1 & | & 1 & 4 \end{bmatrix}$

$\rightarrow \begin{bmatrix} 1 & 0 & | & 2 & 7 \\ 0 & 1 & | & 1 & 4 \end{bmatrix}$

thus, the inverse matrix is $\begin{bmatrix} 2 & 7 \\ 1 & 4 \end{bmatrix}$.

23. Use a calculator, if possible. If you do not have access to one, then:

$\begin{bmatrix} 1 & 0 & 2 & | & 1 & 0 & 0 \\ 2 & 1 & 0 & | & 0 & 1 & 0 \\ 0 & -2 & 9 & | & 0 & 0 & 1 \end{bmatrix} \rightarrow \begin{bmatrix} 1 & 0 & 2 & | & 1 & 0 & 0 \\ 0 & 1 & -4 & | & -2 & 1 & 0 \\ 0 & -2 & 9 & | & 0 & 0 & 1 \end{bmatrix}$

$\rightarrow \begin{bmatrix} 1 & 0 & 2 & | & 1 & 0 & 0 \\ 0 & 1 & -4 & | & -2 & 1 & 0 \\ 0 & 0 & 1 & | & -4 & -2 & 1 \end{bmatrix}$

$\rightarrow \begin{bmatrix} 1 & 0 & 0 & | & 9 & -4 & -2 \\ 0 & 1 & 0 & | & -18 & 9 & 4 \\ 0 & 0 & 1 & | & -4 & -2 & 1 \end{bmatrix}$

The inverse matrix is $\begin{bmatrix} 9 & -4 & -2 \\ -18 & 9 & 4 \\ -4 & 2 & 1 \end{bmatrix}$.

25.   Use a calculator, if possible. If you do not have access to one, then:

$$\left[\begin{array}{cccc|cccc} 1 & 0 & 0 & 1 & 1 & 0 & 0 & 0 \\ 0 & 2 & 0 & 0 & 0 & 1 & 0 & 0 \\ 0 & 0 & 0 & 1 & 0 & 0 & 1 & 0 \\ 2 & 0 & 1 & 0 & 0 & 0 & 0 & 1 \end{array}\right] \rightarrow \left[\begin{array}{cccc|cccc} 1 & 0 & 0 & 1 & 1 & 0 & 0 & 0 \\ 0 & 2 & 0 & 0 & 0 & 1 & 0 & 0 \\ 0 & 0 & 0 & 1 & 0 & 0 & 1 & 0 \\ 0 & 0 & 1 & -2 & -2 & 0 & 0 & 1 \end{array}\right]$$

$$\rightarrow \left[\begin{array}{cccc|cccc} 1 & 0 & 0 & 1 & 1 & 0 & 0 & 0 \\ 0 & 1 & 0 & 0 & 0 & 1/2 & 0 & 0 \\ 0 & 0 & 1 & -2 & -2 & 0 & 0 & 1 \\ 0 & 0 & 0 & 1 & 0 & 0 & 1 & 0 \end{array}\right]$$

$$\rightarrow \left[\begin{array}{cccc|cccc} 1 & 0 & 0 & 0 & 1 & 0 & -1 & 0 \\ 0 & 1 & 0 & 0 & 0 & 1/2 & 0 & 0 \\ 0 & 0 & 1 & 0 & -2 & 0 & 2 & 1 \\ 0 & 0 & 0 & 1 & 0 & 0 & 1 & 0 \end{array}\right]$$

The inverse matrix is $\begin{bmatrix} 1 & 0 & -1 & 0 \\ 0 & \frac{1}{2} & 0 & 0 \\ -2 & 0 & 2 & 1 \\ 0 & 0 & 1 & 0 \end{bmatrix}$.

27. Let $[A] = \begin{bmatrix} 4 & -7 \\ -1 & 2 \end{bmatrix}$ and $[B] = \begin{bmatrix} -2 \\ 1 \end{bmatrix}$ Use a calculator, if possible. If not,

$$[X] = [A]^{-1}[B]$$
$$= \begin{bmatrix} 2 & 7 \\ 1 & 4 \end{bmatrix}\begin{bmatrix} -2 \\ 1 \end{bmatrix}$$
$$= \begin{bmatrix} 2 \cdot (-2) + 7 \cdot 1 \\ 1(-2) + 4 \cdot 1 \end{bmatrix}$$
$$= \begin{bmatrix} 3 \\ 2 \end{bmatrix}$$

The solution is (3, 2).

29. Let $[A] = \begin{bmatrix} 4 & -7 \\ -1 & 2 \end{bmatrix}$ and $[B] = \begin{bmatrix} 48 \\ -13 \end{bmatrix}$ Use a calculator, if possible. If not,

Page 423

$$[X] = [A]^{-1}[B]$$

$$= \begin{bmatrix} 2 & 7 \\ 1 & 4 \end{bmatrix} \begin{bmatrix} 48 \\ -13 \end{bmatrix}$$

$$= \begin{bmatrix} 2 \cdot 48 + 7(-13) \\ 1 \cdot 48 + 4(-13) \end{bmatrix}$$

$$= \begin{bmatrix} 5 \\ -4 \end{bmatrix}$$

The solution is $(5, -4)$.

31. Let $[A] = \begin{bmatrix} 4 & -7 \\ -1 & 2 \end{bmatrix}$ and $[B] = \begin{bmatrix} 5 \\ 4 \end{bmatrix}$ Use a calculator, if possible. If not,

$$[X] = [A]^{-1}[B]$$

$$= \begin{bmatrix} 2 & 7 \\ 1 & 4 \end{bmatrix} \begin{bmatrix} 5 \\ 4 \end{bmatrix}$$

$$= \begin{bmatrix} 2 \cdot 5 + 7 \cdot 4 \\ 1 \cdot 5 + 4 \cdot 4 \end{bmatrix}$$

$$= \begin{bmatrix} 38 \\ 21 \end{bmatrix}$$

The solution is $(38, 21)$.

33. Let $[A] = \begin{bmatrix} 8 & 6 \\ -2 & 4 \end{bmatrix}$ and $[B] = \begin{bmatrix} 12 \\ -14 \end{bmatrix}$ Use a calculator, if possible. If not,

$$[X] = [A]^{-1}[B]$$

$$= \frac{1}{22} \begin{bmatrix} 2 & -3 \\ 1 & 4 \end{bmatrix} \begin{bmatrix} 12 \\ -14 \end{bmatrix}$$

$$= \frac{1}{22} \begin{bmatrix} 2 \cdot 12 + (-3)(-14) \\ 1 \cdot 12 + 4(-14) \end{bmatrix}$$

$$= \frac{1}{22} \begin{bmatrix} 66 \\ -44 \end{bmatrix}$$

$$= \begin{bmatrix} 3 \\ -2 \end{bmatrix}$$

The solution is $(3, -2)$.

35. Let $[A] = \begin{bmatrix} 8 & 6 \\ -2 & 4 \end{bmatrix}$ and $[B] = \begin{bmatrix} -6 \\ -26 \end{bmatrix}$ Use a calculator, if possible. If not,

$$[X] = [A]^{-1}[B]$$
$$= \frac{1}{22} \begin{bmatrix} 2 & -3 \\ 1 & 4 \end{bmatrix} \begin{bmatrix} -6 \\ -26 \end{bmatrix}$$
$$= \frac{1}{22} \begin{bmatrix} 2(-6) + (-3)(-26) \\ 1(-6) + 4(-26) \end{bmatrix}$$
$$= \frac{1}{22} \begin{bmatrix} 66 \\ -110 \end{bmatrix}$$
$$= \begin{bmatrix} 3 \\ -5 \end{bmatrix}$$

The solution is $(3, -5)$.

37. Let $[A] = \begin{bmatrix} 8 & 6 \\ -2 & 4 \end{bmatrix}$ and $[B] = \begin{bmatrix} -26 \\ 12 \end{bmatrix}$ Use a calculator, if possible. If not,

$$[X] = [A]^{-1}[B]$$
$$= \frac{1}{22} \begin{bmatrix} 2 & -3 \\ 1 & 4 \end{bmatrix} \begin{bmatrix} -26 \\ 12 \end{bmatrix}$$
$$= \frac{1}{22} \begin{bmatrix} 2(-26) + (-3)(12) \\ 1(-26) + 4 \cdot 12 \end{bmatrix}$$
$$= \frac{1}{22} \begin{bmatrix} -88 \\ 22 \end{bmatrix}$$
$$= \begin{bmatrix} -4 \\ 1 \end{bmatrix}$$

The solution is $(-4, 1)$.

Problems 39-44 all use the same inverse: $\begin{bmatrix} \frac{2}{5} & \frac{1}{5} \\ \frac{1}{15} & -\frac{2}{15} \end{bmatrix}$

39. Let $[A] = \begin{bmatrix} 2 & 3 \\ 1 & -6 \end{bmatrix}$ and $[B] = \begin{bmatrix} 9 \\ -3 \end{bmatrix}$ Use a calculator, if possible. If not,

Page 425

$$[X] = [A]^{-1}[B]$$

$$= \frac{1}{15}\begin{bmatrix} 6 & 3 \\ 1 & -2 \end{bmatrix}\begin{bmatrix} 9 \\ -3 \end{bmatrix}$$

$$= \frac{1}{15}\begin{bmatrix} 6\cdot 9 + 3(-3) \\ 1\cdot 9 + (-2)(-3) \end{bmatrix}$$

$$= \frac{1}{15}\begin{bmatrix} 45 \\ 15 \end{bmatrix}$$

$$= \begin{bmatrix} 3 \\ 1 \end{bmatrix}$$

The solution is (3, 1).

41. Let $[A] = \begin{bmatrix} 2 & 3 \\ 1 & -6 \end{bmatrix}$ and $[B] = \begin{bmatrix} 2 \\ -14 \end{bmatrix}$ Use a calculator, if possible. If not,

$$[X] = [A]^{-1}[B]$$

$$= \frac{1}{15}\begin{bmatrix} 6 & 3 \\ 1 & -2 \end{bmatrix}\begin{bmatrix} 2 \\ -14 \end{bmatrix}$$

$$= \frac{1}{15}\begin{bmatrix} 6\cdot 2 + 3(-14) \\ 1\cdot 2 + (-2)(-14) \end{bmatrix}$$

$$= \frac{1}{15}\begin{bmatrix} -30 \\ 30 \end{bmatrix}$$

$$= \begin{bmatrix} -2 \\ 2 \end{bmatrix}$$

The solution is (−2, 2).

43. Let $[A] = \begin{bmatrix} 2 & 3 \\ 1 & -6 \end{bmatrix}$ and $[B] = \begin{bmatrix} -22 \\ 49 \end{bmatrix}$ Use a calculator, if possible. If not,

$$[X] = [A]^{-1}[B]$$

$$= \frac{1}{15}\begin{bmatrix} 6 & 3 \\ 1 & -2 \end{bmatrix}\begin{bmatrix} -22 \\ 49 \end{bmatrix}$$

$$= \frac{1}{15}\begin{bmatrix} 6(-22) + 3\cdot 49 \\ 1(-22) + (-2)(49) \end{bmatrix}$$

$$= \frac{1}{15}\begin{bmatrix} 15 \\ -120 \end{bmatrix} = \begin{bmatrix} 1 \\ -8 \end{bmatrix}$$

The solution is (1, −8).

45. Let $[A] = \begin{bmatrix} 1 & 0 & 2 \\ 2 & 1 & 0 \\ 0 & -2 & 9 \end{bmatrix}$ and $[B]= \begin{bmatrix} 7 \\ 16 \\ -3 \end{bmatrix}$ Use a calculator, if possible. If not,

$[X] = [A]^{-1}[B]$

$= \begin{bmatrix} 9 & -4 & -2 \\ -18 & 9 & 4 \\ -4 & 2 & 1 \end{bmatrix} \begin{bmatrix} 7 \\ 16 \\ -3 \end{bmatrix}$

$= \begin{bmatrix} 9 \cdot 7 + (-4)16 + (-2)(-3) \\ (-18)7 + 9 \cdot 16 + 4(-3) \\ (-4)7 + 2 \cdot 16 + 1(-3) \end{bmatrix}$

$= \begin{bmatrix} 5 \\ 6 \\ 1 \end{bmatrix}$

The solution is (5, 6, 1).

47. Let $[A] = \begin{bmatrix} 1 & 0 & 2 \\ 2 & 1 & 0 \\ 0 & -2 & 9 \end{bmatrix}$ and $[B]= \begin{bmatrix} 4 \\ 0 \\ 31 \end{bmatrix}$ Use a calculator, if possible. If not,

$[X] = [A]^{-1}[B]$

$= \begin{bmatrix} 9 & -4 & -2 \\ -18 & 9 & 4 \\ -4 & 2 & 1 \end{bmatrix} \begin{bmatrix} 4 \\ 0 \\ 31 \end{bmatrix}$

$= \begin{bmatrix} 9 \cdot 4 + (-4) \cdot 0 + (-2)(31) \\ (-18)4 + 9 \cdot 0 + 4(31) \\ (-4)4 + 2 \cdot 0 + 1(31) \end{bmatrix}$

$= \begin{bmatrix} -26 \\ 52 \\ 15 \end{bmatrix}$

The solution is (−26, 52, 15).

49. Let $[A] = \begin{bmatrix} 1 & 0 & 2 \\ 2 & 1 & 0 \\ 0 & -2 & 9 \end{bmatrix}$ and $[B]= \begin{bmatrix} 12 \\ 0 \\ 10 \end{bmatrix}$ Use a calculator, if possible. If not,

$$[X] = [A]^{-1}[B]$$

$$= \begin{bmatrix} 9 & -4 & -2 \\ -18 & 9 & 4 \\ -4 & 2 & 1 \end{bmatrix} \begin{bmatrix} 12 \\ 0 \\ 10 \end{bmatrix}$$

$$= \begin{bmatrix} 9 \cdot 12 + (-4)0 + (-2)(10) \\ (-18) \cdot 12 + 9 \cdot 0 + 4(10) \\ (-4) \cdot 12 + 2 \cdot 0 + 1(10) \end{bmatrix}$$

$$= \begin{bmatrix} 88 \\ -176 \\ -38 \end{bmatrix}$$

The solution is $(88, -176, -38)$.

51. Let $[A] = \begin{bmatrix} 6 & 1 & 20 \\ 1 & -1 & 0 \\ 0 & 1 & 3 \end{bmatrix}$ and $[B] = \begin{bmatrix} 27 \\ 0 \\ 4 \end{bmatrix}$ Use a calculator, if possible. If not,

$$[X] = [A]^{-1}[B]$$

$$= \begin{bmatrix} 3 & -17 & -20 \\ 3 & -18 & -20 \\ -1 & 6 & 7 \end{bmatrix} \begin{bmatrix} 27 \\ 0 \\ 4 \end{bmatrix}$$

$$= \begin{bmatrix} 3 \cdot 27 + (-17) \cdot 0 + (-20)(4) \\ 3 \cdot 27 + (-18) \cdot 0 + (-20)(4) \\ (-1) \cdot 27 + 6 \cdot 0 + 7 \cdot 4 \end{bmatrix}$$

$$= \begin{bmatrix} 1 \\ 1 \\ 1 \end{bmatrix}$$

The solution is $(1, 1, 1)$.

53. Let $[A] = \begin{bmatrix} 6 & 1 & 20 \\ 1 & -1 & 0 \\ 0 & 1 & 3 \end{bmatrix}$ and $[B] = \begin{bmatrix} -12 \\ 6 \\ -7 \end{bmatrix}$ Use a calculator, if possible. If not,

$[X] = [A]^{-1}[B]$

$$= \begin{bmatrix} 3 & -17 & -20 \\ 3 & -18 & -20 \\ -1 & 6 & 7 \end{bmatrix} \begin{bmatrix} -12 \\ 6 \\ -7 \end{bmatrix}$$

$$= \begin{bmatrix} 3(-12) + (-17) \cdot 6 + (-20)(-7) \\ 3(-12) + (-18) \cdot 6 + (-20)(-7) \\ (-1)(-12) + 6 \cdot 6 + 7(-7) \end{bmatrix}$$

$$= \begin{bmatrix} 2 \\ -4 \\ -1 \end{bmatrix}$$

The solution is $(2, -4, -1)$.

Level 3, page 795

55. a. $P^{-1} = \begin{bmatrix} 1 & 0 & 0 & 0 & 0 \\ -1 & 1 & 0 & 0 & 0 \\ 1 & -2 & 1 & 0 & 0 \\ -1 & 3 & -3 & 1 & 0 \\ 1 & -4 & 6 & -4 & 1 \end{bmatrix}$

b. It is the same as the Pascal's matrix except that the signs of the terms alternate.

57. $[A]^3 = \begin{bmatrix} 2 & 4 & 1 & 3 \\ 4 & 2 & 3 & 4 \\ 1 & 3 & 0 & 1 \\ 3 & 4 & 1 & 2 \end{bmatrix}$ For example, the U.S. can talk to Cuba through two

intermediaries in one way, namely, U.S. to Mexico to Russia to Cuba.

Level 3 Problem Solving, page 796

59. a. $[C][W] = \begin{bmatrix} 1 & 4 & 6 \end{bmatrix} \begin{bmatrix} 2 & 1 & 3 \\ 4 & 3 & 6 \\ 1 & 2 & 4 \end{bmatrix}$

$$= [2 + 16 + 6 \quad 1 + 12 + 12 \quad 3 + 24 + 24]$$

$$= [24 \quad 25 \quad 51]$$

Riesling costs 24; Charbono, 25; and Rosé, 51

Page 429

b. $[W][D] = \begin{bmatrix} 2 & 1 & 3 \\ 4 & 3 & 6 \\ 1 & 2 & 4 \end{bmatrix} \begin{bmatrix} 40 \\ 60 \\ 30 \end{bmatrix}$

$\qquad = \begin{bmatrix} 80 + 60 + 90 \\ 160 + 180 + 180 \\ 40 + 120 + 120 \end{bmatrix}$

$\qquad = \begin{bmatrix} 230 \\ 520 \\ 280 \end{bmatrix}$

Outside bottling, 230; produced and bottled at winery, 520; estate bottles, 280

c. $([C][W])[D] = \begin{bmatrix} 24 & 25 & 51 \end{bmatrix} \begin{bmatrix} 40 \\ 60 \\ 30 \end{bmatrix}$ $\qquad [C]([W][D]) = \begin{bmatrix} 1 & 4 & 6 \end{bmatrix} \begin{bmatrix} 230 \\ 520 \\ 280 \end{bmatrix}$

$\qquad = [960 + 1,500 + 1,530]$ $\qquad\qquad = [230 + 2,080 + 1,680]$

$\qquad = [3,990]$ $\qquad\qquad\qquad\qquad = [3,990]$

It is the total cost of production of all three wines.

16.5 Modeling with Linear Programming, page 796
New Terms Introduced in this Section

Constraint	Convex set	Feasible solution	Linear programming
Objective function	Optimum solution	Superfluous constraint	System of inequalities

Level 1, page 804

1. A *system of linear inequalities* is two or more first degree inequalities considered at the same time.

3. A *linear programming problem* is a problem which seeks the maximum or minimum value of an expression, called the objective function. This solution is subject to a set of constraints, which take the form of linear inequalities.

5. a. **b.**

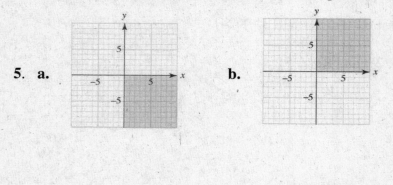

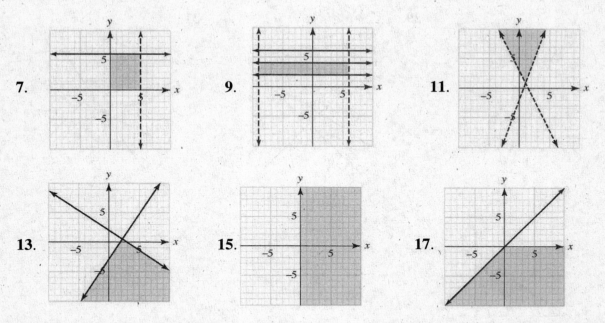

7.   9.   11.

13.   15.   17.

19. a. $(1, 3)$ does not satisfy the last inequality: $2x + 4y = 2(1) + 4(3)$ is not less than (or equal to) 8, so this is not a feasible solution.

b. $(2, 1)$ satisfies all of the inequalities so it is a feasible solution.

c. $(1, 2)$ does not satisfy the last inequality: $2x + 4y = 2(1) + 4(2)$ is not less than (or equal to) 8, so this is not a feasible solution.

d. $(-1, 4)$ does not satisfy the first inequality $-1 \geq 0$ is false, so it is not a feasible solution.

e. $(2, 2)$ does not satisfy the last inequality: $2x + 4y = 2(2) + 4(2)$ is not less than (or equal to) 8, so this is not a feasible solution.

f. $(0, 4)$ does not satisfy the last inequality: $2x + 4y = 2(0) + 4(4)$ is not less than (or equal to) 8, so this is not a feasible solution.

Level 2, page 805

21.   **23.**   **25.**

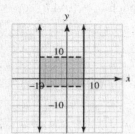   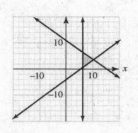   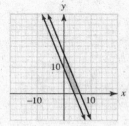

Page 431

27.

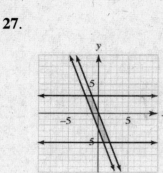

29.

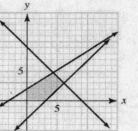

31.

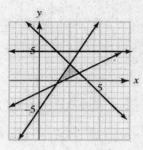

33.

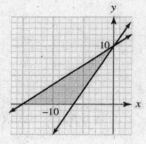

35. Graph the set of constraints. We have labeled the corner points and we find each by inspection or by solving a system of equations.

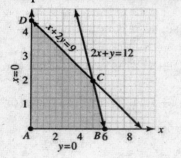

Corner point	System	Solution	Coordinates
A	$\begin{cases} x = 0 \\ y = 0 \end{cases}$	inspection	$(0, 0)$
B	$\begin{cases} 2x + y = 12 \\ y = 0 \end{cases}$	substitution	$(6, 0)$
C	$\begin{cases} 2x + y = 12 \\ x + 2y = 9 \end{cases}$	substitution	$(5, 2)$
D	$\begin{cases} x = 0 \\ x + 2y = 9 \end{cases}$	substitution	$(0, 4.5)$

Page 432

37. Graph the set of constraints.

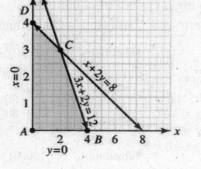

We have labeled the corner points and we find each by inspection or by solving a system of equations.

Corner point	System	Solution	Coordinates
A	$\begin{cases} x = 0 \\ y = 0 \end{cases}$	inspection	$(0, 0)$
B	$\begin{cases} 3x + 2y = 12 \\ y = 0 \end{cases}$	substitution	$(4, 0)$
C	$\begin{cases} x + 2y = 8 \\ 3x + 2y = 12 \end{cases}$	addition	$(2, 3)$
D	$\begin{cases} x = 0 \\ x + 2y = 8 \end{cases}$	substitution	$(0, 4)$

39. Graph the set of constraints.

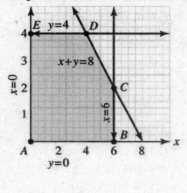

We have labeled the corner points and we find each by inspection or by solving a system of equations.

Corner point	System	Solution	Coordinate of point
A	$\begin{cases} x = 0 \\ y = 0 \end{cases}$	inspection	$(0, 0)$
B	$\begin{cases} x = 6 \\ y = 0 \end{cases}$	inspection	$(6, 0)$
C	$\begin{cases} x = 6 \\ x + y = 8 \end{cases}$	inspection	$(6, 2)$
D	$\begin{cases} x + y = 8 \\ y = 4 \end{cases}$	inspection	$(4, 4)$
E	$\begin{cases} x = 0 \\ y = 4 \end{cases}$	inspection	$(0, 4)$

41. Graph the set of constraints.

We have labeled the corner points and we find each by inspection or by solving a system of equations.

Corner point	System	Solution	Coordinates
A	$\begin{cases} x = 0 \\ y = 0 \end{cases}$	inspection	$(0, 0)$
B	$\begin{cases} 3x + 2y = 8 \\ y = 0 \end{cases}$	substitution	$\left(\dfrac{8}{3}, 0\right)$
C	$\begin{cases} 3x + 2y = 8 \\ x + 5y = 8 \end{cases}$	substitution	$\left(\dfrac{24}{13}, \dfrac{16}{13}\right)$
D	$\begin{cases} x + 5y = 8 \\ x = 4 \end{cases}$	substitution	$\left(0, \dfrac{8}{5}\right)$

Page 433

43. Graph the set of constraints.

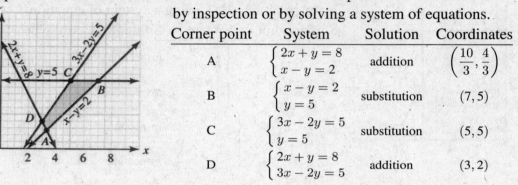

We have labeled the corner points and we find each by inspection or by solving a system of equations.

Corner point	System	Solution	Coordinates
A	$\begin{cases} x = 0 \\ 10x + 5y = 200 \end{cases}$	substitution	$(0, 40)$
B	$\begin{cases} 10x + 5y = 200 \\ 3x + 4y = 120 \end{cases}$	addition	$(8, 24)$
C	$\begin{cases} 3x + 4y = 120 \\ 2x + 5y = 100 \end{cases}$	addition	$\left(\dfrac{200}{7}, \dfrac{60}{7} \right)$
D	$\begin{cases} 2x + 5y = 100 \\ y = 0 \end{cases}$	substitution	$(50, 0)$

45. Graph the set of constraints. We have labeled the corner points and we find each by inspection or by solving a system of equations.

Corner point	System	Solution	Coordinates
A	$\begin{cases} 2x + y = 8 \\ x - y = 2 \end{cases}$	addition	$\left(\dfrac{10}{3}, \dfrac{4}{3} \right)$
B	$\begin{cases} x - y = 2 \\ y = 5 \end{cases}$	substitution	$(7, 5)$
C	$\begin{cases} 3x - 2y = 5 \\ y = 5 \end{cases}$	substitution	$(5, 5)$
D	$\begin{cases} 2x + y = 8 \\ 3x - 2y = 5 \end{cases}$	addition	$(3, 2)$

47. From your work in Problem 35, we know the corner points are $(0, 0), (6, 0), (5, 2),$ and $(0, 4.5)$. We check the value of the objective function for each of these points.

Corner point	Objective function, $W = 30x + 20y$	Conclusion
A; $(0, 0)$	$W = 30(0) + 20(0) = 0$	minimum
B; $(6, 0)$	$W = 30(6) + 20(0) = 180$	
C; $(5, 2)$	$W = 30(5) + 20(2) = 190$	maximum
D; $(0, 4.5)$	$W = 30(0) + 20(4.5) = 90$	

The maximum is $W = 190$ at $(5, 2)$.

Page 434

49. From your work in Problem 37, we know the corner points are $(0,0), (4,0), (2,3),$ and $(0,4)$. We check the value of the objective function for each of these points.

Corner point	Objective function, $P = 100x + 100y$	Conclusion
A; $(0,0)$	$P = 100(0) + 100(0) = 0$	minimum
B; $(4,0)$	$P = 100(4) + 100(0) = 400$	
C; $(2,3)$	$P = 100(2) + 100(3) = 500$	maximum
D; $(0,4)$	$P = 100(0) + 100(4) = 400$	

The maximum is $P = 500$ at $(2, 3)$.

51. From your work in Problem 39, we know the corner points are $(0,0), (6,0), (6,2), (4,4),$ and $(0,4)$. We check the value of the objective function for each of these points.

Corner point	Objective function, $A = 2x - 3y$	Conclusion
A; $(0,0)$	$A = 2(0) - 3(0) = 0$	minimum
B; $(6,0)$	$A = 2(6) - 3(0) = 12$	
C; $(6,2)$	$A = 2(6) - 3(2) = 6$	
D; $(4,4)$	$A = 2(4) - 3(4) = -4$	
E; $(0,4)$	$A = 2(0) - 3(4) = -12$	minimum

The minimum is minimum $A = -12$ at $(0, 4)$.

Level 3, page 805

53. Let $x = $ number of grams of food A

$y = $ number of grams of food B

Minimize $C = 0.29x + 0.15y$

Subject to $\begin{cases} x \geq 0, y \geq 0 \\ 10x + 5y \geq 200 \\ 2x + 5y \geq 100 \\ 3x + 4y \geq 20 \end{cases}$

55. Let $x = $ amount invested in stock (in millions of dollars)

$y = $ amount invested in bonds (in millions of dollars)

Maximize $T = 0.12x + 0.08y$

Subject to $\begin{cases} x \geq 0 \\ x \leq 8, y \geq 2 \\ x + y \leq 10 \\ x \leq 3y \end{cases}$

Level 3 Problem Solving, page 806

57. From Example 3, maximize: $P = 143x + 60y$

where x is the number of acres planted in corn and y is the number of acres planted in wheat.

Subject to:
$$\begin{cases} x \geq 0 \\ y \geq 0 \\ x + y \leq 100 \\ 120x + 210y \leq 15{,}000 \\ 110x + 30y \geq 4{,}000 \end{cases}$$

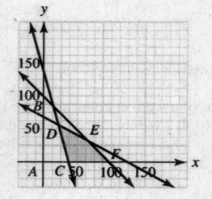

Notice that this is the same as Example 3 except the last constraint has been changed from $\leq$ to $\geq$. We draw the set of feasible solutions; note the same constraints as Example 3, except a different region is shaded.

This region is bounded by $C\left(\frac{400}{11}, 0\right)$, $D(20, 6)$, $E\left(\frac{200}{3}, \frac{100}{3}\right)$ and $F(100, 0)$. Check the corner points:

Corner point	Objective function, $P = 143x + 60y$	Conclusion
C; $\left(\dfrac{400}{11}, 0\right)$	$P = 143\left(\dfrac{400}{11}\right) + 60(0) = 5{,}200$	minimum
D; $(20, 60)$	$P = 143(20) + 60(60) = 6{,}460$	
E; $\left(\dfrac{200}{3}, \dfrac{100}{3}\right)$	$P = 143\left(\dfrac{200}{3}\right) + 60\left(\dfrac{100}{3}\right) = 11{,}533\dfrac{1}{3}$	
F; $(100, 0)$	$P = 143(100) + 60(0) = 14{,}300$	maximum

The maximum value of P is \$14,300 at $(100, 0)$. This means that, to maximize the profit subject to the constraints, the farmer should plant 100 acres in corn.

Page 436

59. Let $x =$ amount invested in Pertec stock (in thousands)

       $y =$ amount invested in Campbell Municipal Bonds (in thousands)

We wish to maximize return: $R = 0.20x + 0.10y$

Subject to: $\begin{cases} x \geq 0 \\ y \geq 0 \\ x + y \leq 100 \\ x \leq 70 \\ y \geq 20 \\ y \leq 3x \end{cases}$

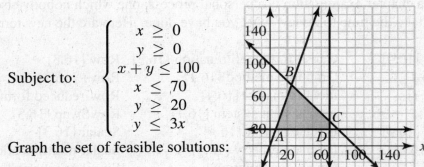

Graph the set of feasible solutions:

This region is bounded by $A\left(\frac{20}{3}, 20\right)$, $B(25, 75)$, $C(70, 30)$ and $D(70, 20)$. Check the corner points:

Corner point	Objective function, $R = 0.20x + 0.10y$	Conclusion
A; $\left(\dfrac{20}{3}, 20\right)$	$R = 0.20\left(\dfrac{20}{3}\right) + 0.10(20) = 3\dfrac{1}{3}$	minimum
B; $(25, 75)$	$R = 0.20(25) + 0.10(75) = 12.5$	
C; $(70, 30)$	$R = 0.20(70) + 0.10(30) = 17$	maximum
D; $(70, 20)$	$R = 0.20(70) + 0.10(20) = 16$	

The maximum value of R is at $(70, 30)$. This means that, to maximize the return you should invest \$70,000 in Pertec stock and \$30,000 in Campbell bonds.

Chapter 16 Review Questions, page 807

STOP

Studying for a chapter examination is a personal process, one which nobody else can do for you. Simply take the time to review what you have done. Here are the new terms in Chapter 16.

Addition of matrices [16.4]

Addition method [16.1]

Additive inverse [16.4]

Array [16.3]

Associative property [16.4]

Augmented matrix [16.3]

Column [16.3]

Communication matrix [16.4]

Commutative property [16.4]

Constraint [16.5]

Convex set [16.5]

Demand [16.2]

Dependent system [16.1]

Diagonal form [16.3]

Dimension [16.3]

Distributive property [16.4]

Double subscripts [16.3]

Elementary row
　　operations [16.3]

Equal matrices [16.4]

Equilibrium point [16.2]

Equivalent matrices [16.3]

Equivalent systems [16.1]

Feasible solution [16.5]

Gauss-Jordan elimination [16.3]

Graphing method [16.1]

Identity matrix [16.4]

Inconsistent system [16.1]

Inverse matrix [16.4]

Inverse property [16.4]

Linear combination
　　method [16.1]

Linear programming [16.5]

Linear system [16.1]

Main diagonal [16.4]

Matrix [16.3]

Matrix equation [16.4]

Multiplication of
　　matrices [16.4]

Multiplicative inverse [16.4]

Nonconformable matrices [16.4]

Nonsingular matrix [16.4]

Objective function [16.5]

Optimum solution [16.5]

Order [16.3]

Pivot [16.3]

Pivot row [16.3]

Pivoting [16.3]

Row [16.3]

Row+ [16.3]

Row-reduced form [16.3]

RowSwap [16.3]

Scalar [16.3]

Scalar multiplication [16.4]

Simultaneous solution [16.1]

Singular matrix [16.4]

Square matrix [16.3]

Subscript [16.3]

Substitution method [16.1]

Subtraction of matrices [16.4]

Superfluous constraint [16.5]

Supply [16.2]

System of equations [16.1]

System of inequalities [16.5]

Target row [16.3]

*Row [16.3]

*Row + [16.3]

Zero matrix [16.4]

Zero-one matrix [16.4]

If you can describe the term, read on to the next one; if you cannot, then look it up in the text (the section number is shown in brackets). Next, study the types of problems listed at the end of Chapter 16.

TYPES OF PROBLEMS

Solve systems of equations by graphing, substitution, or addition, as directed. [16.1]

Solve systems of equations by selecting the most appropriate method. [16.1]

Solve applied problems, including coin problems, combining rates, supply and demand, and mixture problems. [16.2]

Know the relationship between a system of equations and a corresponding matrix [16.3]

Perform elementary row operations on a given matrix. [16.3]

Solve systems of equations by the Gauss-Jordan method. [16.3]

Carry out matrix operations, including finding the inverse of a given matrix. [16.4]

Solve a system of equations using the inverse matrix method. [16.4]

Solve a system of inequalities. [16.5]

Decide whether a given point is a feasible solution for a set of constraints. [16.5]

Decide whether a given point is a corner point for a set of constraints. [16.5]

Find the corner points for a set of feasible solutions. [16.5]

Maximize or minimize an objective function subject to a set of constraints. [16.5]

Solve applied problems using a linear programming model. [16.5]

Once again, see if you can verbalize (to yourself) how to do each of the listed types of problems.

Work all of Chapter 16 Review Questions (whether they are assigned or not). Work through all of the problems before looking at the answers, and *then* correct each of the problems. The entire solution is shown in the answer section at the back of the text. If you worked the problem correctly, move on to the next problem, but if you did not work it correctly (or you did not know what to do), look back in the chapter to study the procedure, or ask your instructor.

Finally, go back over the homework problems you have been assigned. If you worked a problem correctly, move on to the next problem, but if you missed it on your homework, then you should look back in the book or talk to your instructor about how to work the problem.

If you follow these steps, you should be successful with your review of this chapter.

We give all of the answers to the chapter review questions (not just the odd-numbered questions).

Chapter 16 Review Questions, page 807

1. $[C][B] - 3[D] = \begin{bmatrix} 2 & 0 \\ 1 & 2 \\ -1 & 1 \end{bmatrix} \begin{bmatrix} 2 & -1 & 0 \\ 1 & 0 & 1 \end{bmatrix} - 3 \begin{bmatrix} 0 & 1 & 0 \\ -1 & 0 & 0 \\ 0 & 1 & -1 \end{bmatrix}$

$= \begin{bmatrix} 4 & -2 & 0 \\ 4 & -1 & 2 \\ -1 & 1 & 1 \end{bmatrix} + \begin{bmatrix} 0 & -3 & 0 \\ 3 & 0 & 0 \\ 0 & -3 & 3 \end{bmatrix}$

$= \begin{bmatrix} 4 & -5 & 0 \\ 7 & -1 & 2 \\ -1 & -2 & 4 \end{bmatrix}$

2. $[A][B][C] = \begin{bmatrix} 1 & 0 \\ 2 & -1 \end{bmatrix} \begin{bmatrix} 2 & -1 & 0 \\ 1 & 0 & 1 \end{bmatrix} \begin{bmatrix} 2 & 0 \\ 1 & 2 \\ -1 & 1 \end{bmatrix}$

$= \begin{bmatrix} 2 & -1 & 0 \\ 3 & -2 & -1 \end{bmatrix} \begin{bmatrix} 2 & 0 \\ 1 & 2 \\ -1 & 1 \end{bmatrix}$

$= \begin{bmatrix} 3 & -2 \\ 5 & -5 \end{bmatrix}$

3. [B] is a 2 × 3 matrix and [A] is 2 × 2 so these matrices are not conformable for multiplication.

4. $[C][A][B] = \begin{bmatrix} 2 & 0 \\ 1 & 2 \\ -1 & 1 \end{bmatrix} \begin{bmatrix} 1 & 0 \\ 2 & -1 \end{bmatrix} \begin{bmatrix} 2 & -1 & 0 \\ 1 & 0 & 1 \end{bmatrix}$

$= \begin{bmatrix} 2 & 0 \\ 5 & -2 \\ 1 & -1 \end{bmatrix} \begin{bmatrix} 2 & -1 & 0 \\ 1 & 0 & 1 \end{bmatrix}$

$= \begin{bmatrix} 4 & -2 & 0 \\ 8 & -5 & -2 \\ 1 & -1 & -1 \end{bmatrix}$

5. We write the augmented matrix

$\begin{bmatrix} 1 & 0 & | & 1 & 0 \\ 2 & -1 & | & 0 & 1 \end{bmatrix} \rightarrow \begin{bmatrix} 1 & 0 & | & 1 & 0 \\ 0 & -1 & | & -2 & 1 \end{bmatrix} \rightarrow \begin{bmatrix} 1 & 0 & | & 1 & 0 \\ 0 & 1 & | & 2 & -1 \end{bmatrix}$

$[A]^{-1} = \begin{bmatrix} 1 & 0 \\ 2 & -1 \end{bmatrix}^{-1}$ Note, this matrix is its own inverse.

$= \begin{bmatrix} 1 & 0 \\ 2 & -1 \end{bmatrix}$

Page 440

6. $\begin{bmatrix} 2 & 1 & | & 1 & 0 \\ -\frac{3}{2} & -\frac{1}{2} & | & 0 & 1 \end{bmatrix} \rightarrow \begin{bmatrix} 1 & \frac{1}{2} & | & \frac{1}{2} & 0 \\ -\frac{3}{2} & -\frac{1}{2} & | & 0 & 1 \end{bmatrix} \rightarrow \begin{bmatrix} 1 & \frac{1}{2} & | & \frac{1}{2} & 0 \\ 0 & \frac{1}{4} & | & \frac{3}{4} & 1 \end{bmatrix}$

$\rightarrow \begin{bmatrix} 1 & \frac{1}{2} & | & \frac{1}{2} & 0 \\ 0 & 1 & | & 3 & 4 \end{bmatrix} \rightarrow \begin{bmatrix} 1 & 0 & | & -1 & -2 \\ 0 & 1 & | & 3 & 4 \end{bmatrix}$ Inverse is $\begin{bmatrix} -1 & -2 \\ 3 & 4 \end{bmatrix}$.

7. $\begin{bmatrix} 1 & 3 & 3 & | & 1 & 0 & 0 \\ 1 & 4 & 3 & | & 0 & 1 & 0 \\ 1 & 3 & 4 & | & 0 & 0 & 1 \end{bmatrix} \rightarrow \begin{bmatrix} 1 & 3 & 3 & | & 1 & 0 & 0 \\ 0 & 1 & 0 & | & -1 & 1 & 0 \\ 0 & 0 & 1 & | & -1 & 0 & 1 \end{bmatrix} \rightarrow \begin{bmatrix} 1 & 0 & 3 & | & 4 & -3 & 0 \\ 0 & 1 & 0 & | & -1 & 1 & 0 \\ 0 & 0 & 1 & | & -1 & 0 & 1 \end{bmatrix}$

$\rightarrow \begin{bmatrix} 1 & 0 & 0 & | & 7 & -3 & -3 \\ 0 & 1 & 0 & | & -1 & 1 & 0 \\ 0 & 0 & 1 & | & -1 & 0 & 1 \end{bmatrix}$ Inverse is $\begin{bmatrix} 7 & -3 & -3 \\ -1 & 1 & 0 \\ -1 & 0 & 1 \end{bmatrix}$.

8. Graph $2x - y = 2$ by writing $y = 2x - 2$; y-intercept is -2 and slope is $2 = \frac{2}{1}$.

Graph $3x - 2y = 1$ by writing $y = \frac{3}{2}x - \frac{1}{2}$; y-intercept is $-\frac{1}{2}$ and slope is $\frac{3}{2}$.

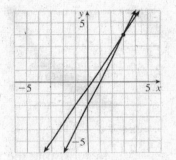

The intersection point appears to be $(3, 4)$. Check:

$2(3) - 4 = 6 - 4 = 2$

$3(3) - 2(4) = 9 - 8 = 1$

9. $2 \begin{cases} x + 3y = 3 \\ 4x - 6y = -6 \end{cases}$

$+ \begin{cases} 2x + 6y = 6 \\ 4x - 6y = -6 \end{cases}$

$6x = 0$

$x = 0$

If $x = 0$, then $0 + 3y = 3$ or $y = 1$.   The solution is $(0, 1)$.

10. Substitute the first equation into the second:

$$5x + 2(\mathbf{1 - 2x}) = 1$$
$$5x + 2 - 4x = 1$$
$$x = -1$$

If $x = -1$, then $y = 1 - 2(-1) = 1 + 2 = 3$; the solution is $(-1, 3)$.

Page 441

11. $\begin{bmatrix} 1 & 1 & 1 & | & 2 \\ 1 & 2 & -2 & | & 1 \\ 1 & 1 & 3 & | & 4 \end{bmatrix} \rightarrow \begin{bmatrix} 1 & 1 & 1 & | & 2 \\ 0 & 1 & -3 & | & -1 \\ 0 & 0 & 2 & | & 2 \end{bmatrix} \rightarrow \begin{bmatrix} 1 & 0 & 4 & | & 3 \\ 0 & 1 & -3 & | & -1 \\ 0 & 0 & 2 & | & 2 \end{bmatrix}$

$\rightarrow \begin{bmatrix} 1 & 0 & 4 & | & 3 \\ 0 & 1 & -3 & | & -1 \\ 0 & 0 & 1 & | & 1 \end{bmatrix} \rightarrow \begin{bmatrix} 1 & 0 & 0 & | & -1 \\ 0 & 1 & 0 & | & 2 \\ 0 & 0 & 1 & | & 1 \end{bmatrix}$ The solution is $(-1, 2, 1)$.

12. Let $[A] = \begin{bmatrix} 2 & 1 \\ -\frac{3}{2} & -\frac{1}{2} \end{bmatrix}$, $[X] = \begin{bmatrix} x \\ y \end{bmatrix}$, $[B] = \begin{bmatrix} 13 \\ 10 \end{bmatrix}$; The solution is $[X] = [A]^{-1}[B]$.

We found the inverse in Problem 6: $[X] = \begin{bmatrix} -1 & -2 \\ 3 & 4 \end{bmatrix}\begin{bmatrix} 13 \\ 10 \end{bmatrix} = \begin{bmatrix} -33 \\ 79 \end{bmatrix}$.

The solution is $(-33, 79)$.

13. Graphing, addition, or substitution would all be appropriate methods for this problem. By

addition $+\begin{cases} 3x - y = -10 \\ x + y = -2 \end{cases}$

$$4x = -12$$
$$x = -3$$

If $x = -3$, then $\quad x + y = -2$
$$-3 + y = -2$$
$$y = 1$$

The solution is $(-3, 1)$.

14. We use substitution, $\begin{cases} x = 5 - 3y \\ 5x + 7y = 25 \end{cases}$ to substitute the first equation into the second:

$$5x + 7y = 25$$
$$5(5 - 3y) + 7y = 25$$
$$25 - 15y + 7y = 25$$
$$25 - 8y = 25$$
$$y = 0$$

If $y = 0$, then $x = 5$ and the solution is $(5, 0)$.

15. We use Gauss-Jordan on the system $\begin{cases} x + y + z = 5 \\ x - 2y + z = -1 \\ 3x + y - 2z = 16 \end{cases}$

$\begin{bmatrix} 1 & 1 & 1 & | & 5 \\ 1 & -2 & 1 & | & -1 \\ 3 & 1 & -2 & | & 16 \end{bmatrix} \rightarrow \begin{bmatrix} 1 & 1 & 1 & | & 5 \\ 0 & -3 & 0 & | & -6 \\ 0 & -2 & -5 & | & 1 \end{bmatrix} \rightarrow \begin{bmatrix} 1 & 1 & 1 & | & 5 \\ 0 & 1 & 0 & | & 2 \\ 0 & -2 & -5 & | & 1 \end{bmatrix}$

$\rightarrow \begin{bmatrix} 1 & 0 & 1 & | & 3 \\ 0 & 1 & 0 & | & 2 \\ 0 & 0 & -5 & | & 5 \end{bmatrix} \rightarrow \begin{bmatrix} 1 & 0 & 1 & | & 3 \\ 0 & 1 & 0 & | & 2 \\ 0 & 0 & 1 & | & -1 \end{bmatrix} \rightarrow \begin{bmatrix} 1 & 0 & 0 & | & 4 \\ 0 & 1 & 0 & | & 2 \\ 0 & 0 & 1 & | & -1 \end{bmatrix}$

The solution is $(4, 2, -1)$.

16. We use the addition method: $2\begin{cases} 5x + 3y = 82 \\ 4x - 6y = -10 \end{cases} = +\begin{cases} 10x + 6y = 164 \\ 4x - 6y = -10 \end{cases}$

$$14x = 154$$
$$x = 11$$

If $x = 11$, then substitute this value into either equation (we choose the second):

$4x - 6y = -10$
$4(11) - 6y = -10$
$-6y = -54$
$y = 9$

The solution is $(11, 9)$.

17. The first boundary line: $2x - y + 2 = 0$
$$y = 2x + 2$$
Plot y-intercept $(0, 2)$, and slope of 2.

Second boundary line: $2x - y + 12 = 0$
$$y = 2x + 12$$
Plot y-intercept $(0, 12)$, and slope of 2.

Notice that the first two lines are parallel.

Third boundary line: $3x + 2y + 10 = 0$
$$2y = -3x - 10$$
$$y = -\frac{3}{2}x - 5$$
Plot y-intercept $(0, -5)$, and slope of $-\frac{3}{2}$.

Fourth boundary line: $3x + 2y - 18 = 0$
$$2y = -3x + 18$$
$$y = -\frac{3}{2}x + 9$$
Plot y-intercept $(0, 9)$, and slope of $-\frac{3}{2}$.

Notice that the last two lines are also parallel. We draw the graph.

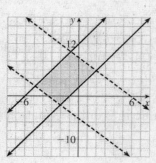

18. Let $x =$ number of bars of product I.

  $y =$ number of bars of product II.

	A	B
Product I	$3x$	$5x$
Product II	$4y$	$7y$
TOTAL	33	56

Thus, $\begin{cases} 3x + 4y = 33 \\ 5x + 7y = 56 \end{cases}$ The solution to this system is (7, 3). This means she uses 7 bars

of Product I and 3 bars of Product II.

19. Let x, y, and z be the number of products of I, II, and III that can be manufactured. This leads to the following system:

$$\begin{cases} 2x + 2y + 3z = 1{,}250 \\ x + 2y + 2z = 900 \\ x + y + 2z = 750 \end{cases}$$

The solution to this system is (100, 150, 250). This means the number of product I to be manufactured is 100; product II, 150; and product III, 250.

20. Let $x =$ number of acres of corn; $y =$ number of acres of wheat.

Maximize profit, $P = 2.10(100)x + 2.50(40)y = 210x + 100y$. The constraints are:

$$\begin{cases} x \geq 0 \\ y \geq 0 \\ x + y \leq 500 \\ 100x + 40y \leq 18{,}000 \\ 120x + 60y \leq 24{,}000 \end{cases}$$

We graph the set of feasible solutions:

Page 444

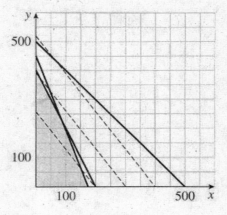

The corner points are (100, 200), (0, 400), and (180, 0).

Corner point	Objective function, $P = 210x + 100y$	Conclusion
(100, 200)	$P = 210(100) + 100(200) = 41,000$	maximum
(0, 400)	$P = 210(0) + 100(400) = 40,000$	
(180, 0)	$P = 210(180) + 100(0) = 37,800$	minimum

This means the farmer will maximize the profit if 100 acres of corn and 200 acres of wheat are planted (200 acres left unplanted).

CHAPTER 17 The Nature of Voting and Apportionment

Chapter Overview
This chapter is a fairly new branch of mathematics, one with a great deal of popularity. Different ways of making selections is much more complex than one would think, and a consideration of this topic is designed to open this topic to the layperson who may never have considered the implications of the topics of this chapter.

17.1 Voting, page 813

New Terms Introduced in this Section

Approval voting method	Binary voting	Borda count
Decisiveness (principle of)	Dictatorship	Hare method
Independence of irrelevant alternatives	Majority rule	Pairwise comparison method
Pareto principle	Plurality method	Runoff election
Sequential voting	Social choice theory	Symmetry
Tournament method	Vote	

Level 1, page 822

1. Each voter votes for one candidate. The candidate receiving the most votes is declared the winner.

3. Each voter votes for one candidate. If a candidate receives a majority of the votes, that candidate is declared to be a *first round winner*. If no candidate receives a majority of the votes, then with the **Hare method**, also known as the *plurality with an elimination runoff method*, then the candidate(s) with the fewest number of first-place votes is (are) eliminated. Each voter votes for one candidate in the second round. If a candidate receives a majority, that candidate is declared to be a *second round winner*. If no candidate receives a majority of the second round votes, then eliminate the candidate(s) with the fewest number of votes. Repeat this process until a candidate receives a majority.

5. With the **tournament method** candidates are teamed head-to-head or one-to-one, with the winner of one pairing facing a new opponent for the next election.

7. You probably used plurality voting when selecting a class president.

9. Your state convention voting for their president or leader of the body probably used the Hare voting method.

11. An example might be selecting priorities in the master plan for an organization.

13. Since the total number of votes is even, a majority is $\dfrac{22}{2} + 1 = 12$.

15. **a.** "(ACB)" means that the voter ranks three candidates in the order of A first, C next, and in last place is candidate B.
    b. In the election there were 4 voters who ranked the candidates in the order ACB.

17. **a.** "(BCA)" means that the voter preferences are for B as the first choice, C next, with the last choice being A.
    b. No people voted for this ranking.

19. Since the total number of votes is odd, a majority is $\dfrac{19 + 1}{2} = 10$.

21. **a.** "(ADBC)" means the voter picks A as the first choice, followed by D, then B, and C is in last place.
    b. It means that in the election there were 8 voters who ranked the candidates in the order ADBC.

23. **a.** $4! = 24$
    b. (ABCD), (ABDC), (ACBD), (ACDB), (ADCB), (BACD), (BCAD), (BCDA), (BDAC), (CABD), (CADB), (CBAD), (CBDA), (CDAB), (CDBA), (DABC), (DBAC), (DBCA), (DCBA)

25. **a.** $3! = 6$ **b.** $4! = 24$ **c.** $5! = 120$

Level 2, page 822

27. The number of different ways a voter can rank n candidates (with no ties allowed) is found by using the fundamental counting principle: $n(n-1)(n-2)\cdots 3 \cdot 2 \cdot 1 = n!$ ways.

29. The votes are A: $8 + 4 = 12$; B: $3 + 0 = 3$, and C: $2 + 5 = 7$. The total vote is 22, and this is even so $\frac{22}{2} + 1 = 12$, so 12 votes are necessary for a majority; thus, A wins according to the majority rule.

31.

	A	B	C		Total A	B	C
(ABC), 8:	8(3	2	1) =		24	16	8
(ACB), 4:	4(3	1	2) =		12	4	8
(BAC), 3:	3(2	3	1) =		6	9	3
(BCA), 0:	0(1	3	2) =		0	0	0
(CAB), 2:	2(2	1	3) =		4	2	6
(CBA), 5:	5(1	2	3) =		5	10	15
TOTAL:					51	41	40

A is the winner.

Page 448

33. AB → A; 1 point; AC → A; 1 point BC → tie; $\frac{1}{2}$ point for each; A wins

35. Snacks has 10 votes; drinks has 8 votes; travel slides has 3 votes; and a guest speaker has 9 votes. Snacks is the winner.

37. A majority winner is $\dfrac{293,472}{2} + 1 = 146,737$. Howard Dean is the majority winner.

39. A: 2 + 4 = 6
    B: 2 + 1 = 3
    C: 2 + 1 = 3
    There is no winner by using the majority rule.

41.

				Total		
	A	B	C	A	B	C
(ABC), 2:	2(3	2	1) =	6	4	2
(ACB), 4:	4(3	1	2) =	12	4	8
(BAC), 2:	2(2	3	1) =	4	6	2
(BCA), 1:	1(1	3	2) =	1	3	2
(CAB), 2:	2(2	1	3) =	4	2	6
(CBA), 1:	1(1	2	3) =	1	2	3
TOTAL:				28	21	23

A wins.

43. (AB): 2 + 4 + 2 = 8 and (BA): 2 + 1 + 1 = 4; A gets 1 point.
    (AC): 2 + 4 + 2 = 8 and (CA): 1 + 2 + 1 = 4; A gets 1 point.
    (BC): 2 + 2 + 1 = 5 and (CB): 4 + 2 + 1 = 7; C gets 1 point.
    This means that A has 2 points, B has no points, and C has 1 point.
    A is the winner.

45. We note that A has 1 + 3 = 4 votes; B has 4 + 3 = 7 votes; and C has 5 + 1 = 6 votes. A majority requires $\frac{17+1}{2} = 9$ votes; there is no winner by using the majority rule.

47.

				Total		
	A	B	C	A	B	C
(ABC), 1:	1(3	2	1) =	3	2	1
(ACB), 3:	3(3	1	2) =	9	3	6
(BAC), 4:	4(2	3	1) =	8	12	4
(BCA), 3:	3(1	3	2) =	3	9	6
(CAB), 5:	5(2	1	3) =	10	5	15
(CBA), 1:	1(1	2	3) =	1	2	3
TOTAL:				34	33	35

C wins.

Page 449

49. AB $\rightarrow$ A (9 to 8)

AC $\rightarrow$ C (8 to 9)

BC $\rightarrow$ C (8 to 9)

This means that A has 1 point, B has no points, and C has 2 points. C is the winner.

Level 3, page 823

51. a. The most common grade is a B.

   b. plurality vote

53. a. The most common grade is a C.

   b. plurality vote

55. There are five candidates, so the number of possibilities is 5! = 120.

57. Drop the person with the fewest first-place votes, which is E. Now the votes are:

$$A: 360 = 360$$
$$B: 240 + 80 = 320$$
$$C: 200 + 40 = 240$$
$$D: 180 = 180$$

No majority winner, so we now, eliminate D for the third ballot:

$$A: 360 + 0 = 360$$
$$B: 240 + 80 + 0 = 320$$
$$C: 200 + 180 + 40 = 420$$

For the fourth ballot, eliminate B:

$$A: 360 = 360$$
$$C: 240 + 200 + 180 + 80 + 40 = 740$$

The winner is C; this is commonly called the Hare method.

59.

					Total				
A	B	C	D	E	A	B	C	D	E
360(5	2	1	4	3) =	1,800	720	360	1,440	1,080
240(1	5	2	3	4) =	240	1,200	480	720	960
200(1	4	5	2	3) =	200	800	1,000	400	600
180(1	2	4	5	3) =	180	360	720	900	540
80(1	4	2	3	5) =	80	320	160	240	400
40(1	2	4	3	5) =	40	80	160	120	200
TOTAL:					2,540	3,480	2,880	3,820	3,780

D wins

17.2 Voting Dilemmas, page 824

New Terms Introduced in this Section

Arrow's impossibility theorem	Condorcet candidate	Condorcet criterion
Condorcet's paradox	Fairness criteria	Fair voting principles
Insincere voting	Irrelevant alternatives criterion	Majority criterion
Monotonicity criterion	Straw vote	Transitive law

Level 1, page 836

1. The *majority criterion* says that if a candidate receives a majority of the first-place votes, then that candidate should be declared the winner.

3. The *monotonicity criterion* says that a candidate who wins a first election and then gains additional support, without losing any of the original support, should also win a second election.

5. A voting plan satisfies the *fairness criteria* if it satisfies all of the following properties: majority criterion, Condorcet criterion, monotonicity criterion and *irrelevant alternatives criterion*.

7. Answers vary; be sure to support your opinion.

9. It tells us that we will not be able to make up, invent, or discover a voting plan without violating at least one of these conditions: unrestricted domain, decisiveness, symmetry and transitivity, independence of irrelevant alternatives, Pareto principle, or dictatorship.

11. **a.** The vote for winner is:
    A: $11 + 1 = 12$;
    B: $3 + 6 = 9$;
    C: $3 + 7 = 10$.
    There is no majority.
    Proceed with the Hare method of voting. The fewest first-place votes were cast for B, so eliminate B. The vote is now:
    A: $11 + 1 + 3 = 15$;
    C: $6 + 3 + 7 = 16$.
    The winner is C, so the California Teachers Association will be the collective bargaining for the faculty association.

    b. No, this does not violate the majority criterion because there is no majority winner.

13. **a.** The vote for winner is: A: 6, B: 5, C: 3. Since there are 14 voters, a majority would be 8 votes. There is no majority winner, but the plurality vote goes to A.

b.

					Total		
Outcome	A	B	C	A	B	C
(ABC), 0:	0(3	2	1) =	0	0	0
(ACB), 6:	6(3	1	2) =	18	6	12
(BAC), 5:	5(2	3	1) =	10	15	5
(BCA), 0:	0(1	3	2) =	0	0	0
(CAB), 3:	3(2	1	3) =	6	3	9
(CBA) 0:	0(1	2	3) =	0	0	0
TOTAL:				34	24	26

A is the winner.

c. No, this does not violate the majority criteria, since there is no majority winner.

15. a. The vote for winner is:

A: 2, B: 2, C: 1.

\he candidate with the fewest first-place votes is C. Delete C and have a runoff election between A and B:

A: $2 + 0 + 1 = 3$,

B: $0 + 2 + 0 = 2$,

A wins the runoff election.

b. If B withdraws before the election the votes would be:

    (AC)   (CA)

$2 + 0 + 0 = 2$ $2 + 1 + 0 = 3$

There are 5 votes, so a majority winner is C.

c. Yes, this violates the irrelevant alternative criterion. Candidate B was not the winner in part **a**, where A won the election. However, if candidate B were to withdraw from the election, that would affect the outcome and cause C to win.

17. a. A Condorcet candidate is the one that wins all the one-on-one match-ups.

A over B: $8 + 6 = 14$ and B over A: $5 + 8 = 13$; A wins

A over C: $8 + 8 = 16$ and C over A: $5 + 6 = 11$; A wins

B over C: $8 + 5 + 8 = 21$ and C over B: 6; B wins

	A	B	C
A	—	A	A
B	A	—	B
C	A	B	—

We can see from the table that A is the Condorcet candidate.

Page 452

b. Now, look at the original vote; the results are:

A: 8,

B: 5 + 8 = 13,

C: 6.

The total number of votes is 27, so there is no majority. The plurality vote goes to to B with a total of 13 first-place votes. This does violate the Condorcet criterion.

19. a. The vote is:

A: 7,

B: 3,

C: 5,

D: 2.

There is no majority winner, but A wins by the plurality method.

b. If Carol drops out the voting preferences are: (ADB)   (BAD)   (BDA)   (DBA)

A: 7   B: 5 + 3 = 8   D: 2

The plurality winner is now B.

c. Yes, the irrelevant alternatives criterion has been violated because when Carol was removed the outcome was changed.

Level 2, page 837

21. Here is the vote count:

A: 25

B: 22

C: 20 + 11 = 31

D: 30

There is no majority winner, but C is the plurality winner. This violates the Condorcet criterion because the Condorcet candidate is A.

23. A majority is $\frac{108}{2} + 1 = 55$ votes.

A: 25

B: 22

C: 20 + 11 = 31

D: 30

The least number of first-place votes are for B, so we eliminate B. The results of the second round of voting are:

A: 25

C: 22 + 20 + 11 = 53

D: 30

Page 453

There still is not majority winner, so we eliminate A. The results of the third round of voting are:

C: $25 + 22 + 20 + 11 = 78$

D: 30

The winner is C. This violates the Condorcet criterion, since the Condorcet candidate is A.

25. We construct a table showing the one-on-one match-up winners.

A over B: $80 + 45 = 125$; B over A: $30 + 50 + 10 = 90$; A wins.

A over C: $80 + 45 = 125$; C over A: $30 + 10 + 50 = 90$; A wins.

A over D: $45 + 30 + 50 = 125$; D over A: $80 + 10 = 90$; A wins.

B over C: $80 + 30 = 110$; C over B: $45 + 10 + 50 = 105$; B wins.

B over D: $45 + 30 + 10 + 50 = 135$; D over B: 80; B wins

C over D: $45 + 30 + 10 + 50 = 135$; D over C: 80; C wins

We show this information in tabular form:

	A	B	C	D
A	—	A	A	A
B	A	—	B	B
C	A	B	—	C
D	A	B	C	—

We see that A wins over all the other candidates on a one-to-one match-up so A is the Condorcet candidate.

27.

				Total			
A	B	C	D	A	B	C	D
80(3	2	1	4) =	240	160	80	320
45(4	2	3	1) =	180	90	135	45
30(2	4	3	1) =	60	120	90	30
10(1	3	4	2) =	10	30	40	20
50(2	3	4	1) =	100	150	200	50
TOTAL:				590	550	545	465

We see that A wins the Borda count, which does not violates the Condorcet criterion.

29. See the solution for Problem 25 for details.

A wins over B, so A gets one point.   B wins over C, so B gets one point.

A wins over C, so A gets one point.   B wins over D, so B gets one point.

A wins over D, so A gets one point.   C wins over D, so C gets one point.

The score is A 3 points; B 2 points, C one point, and D zero points. A wins the pairwise comparison method. No, this does not violate the Condorcet criterion.

31. Here is the Borda count:

						Total				
E	M	H	I	T		E	M	H	I	T
15(5	1	3	4	2) =		75	15	45	60	30
6(3	5	2	4	1) =		18	30	12	24	6
6(3	4	5	1	2) =		18	24	30	6	12
6(3	4	1	2	5) =		18	24	6	12	30
TOTAL:						129	93	93	102	78

E is the Borda count winner.

33. Here is the Borda count:

			Total		
E	M	I		E	M	I
15(3	1	2) =		45	15	30
6(1	3	2) =		6	18	12
12(2	3	1) =		24	36	12
TOTAL:				75	69	54

E is the Borda count winner. No, the Borda count method does not violate the irrelevant alternatives criterion.

35. Here is the Borda count:

		Total	
E	M		E	M
15(2 | 1) =		30	15
18(1 | 2) =		18	36
TOTAL:			48	51

M is the Borda count winner. Yes, the Borda count method violates the irrelevant alternatives criterion because the pairwise winner is E.

37. A majority vote $\frac{93+1}{2} = 47$. Here is the result of the second (binding) day ballot:

A: 20

L: $1 + 35 + 4 + 3 = 43$

T: 30

S: 0

With the Hare method, S is eliminated for the second ballot:

A: 20

L: $1 + 20 + 4 + 3 = 43$

T: 30

A is eliminated from the second ballot. The third ballot results are:

T: 30

L: $1 + 35 + 20 + 4 + 3 = 63$

Page 455

Lillehammer (L) is the winner of the final vote. No, this does not violate the monotonicity criterion because the winning candidate from the first election Lillehammer is also the winner in this vote.

39. a.

					Total				
B	L	I	M	S	B	L	I	M	S
32(5	4	3	2	1) =	160	128	96	64	32
3(4	5	2	1	3) =	12	15	6	3	9
5(4	3	5	1	2) =	20	15	25	5	10
8(3	2	1	5	4) =	24	16	8	40	32
6(3	5	2	1	4) =	18	30	12	6	24
30(4	3	1	2	5) =	120	90	30	60	150
2(2	1	5	4	3) =	4	2	10	8	6
3(4	2	1	5	3) =	12	6	3	15	9
TOTAL:					370	302	190	201	272

Beijing (B) is the Borda count winner.

b. No, even though there was plenty of chance for the vote to change.

41. a. The runoff election was between Chirac and Le Pen (the top two percentages).

b. Answers vary.

c. Answers vary. Notice that the abstention rate went way down. One possibility is that 9% more of the voters who abstained in the first vote went to the polls and voted for Chirac. Also notice that 63% (16% + 47%) voted for other candidates who were not entered in second ballot. Almost all of these changed their vote to Chirac. Finally, notice that even with all the shifting of votes, Le Pen stayed fairly static (17% to 18%). This would indicate that all other factions formed a coalition to get Chirac elected.

43. Here is the Borda count:

			Total		
A	B	C	A	B	C
6(3	2	1) =	18	12	6
5(1	2	3) =	5	10	15
4(1	3	2) =	4	12	8
TOTAL:			27	34	29

The Borda count winner is Betty (B). This does not violate the Condorcet criterion because both methods picked Betty.

45. A obtains 0 points, B 2 points, and C 1 one point. Betty (B) wins the pairwise comparison method. (See the solution for Problem 42 for details.) None of the conditions in Arrow's impossibility theorem are violated.

47. Here is the Borda count:

				Total			
A	B	C	D	A	B	C	D
100(1	4	2	3) =	100	400	200	300
120(2	4	1	3) =	240	480	120	360
130(2	1	4	3) =	260	130	520	390
150(4	1	2	3) =	600	150	300	450
TOTAL:				1,200	1,160	1,140	1,500

The Borda count winner is Dave (D). This does not violate any of the fairness criteria.

49. We find the one-on-one pairings:

A over B: 130 + 150 = 280; B over A: 100 + 120 = 220; A wins; 1 point

A over C: 120 + 150 = 270; C over A: 100 + 130 = 230; A wins; 1 point

A over D: 150; D over A: 100 + 120 + 130 = 350; D wins; 1 point

B with C: B: 100 + 120 = 220; C: 130 + 150 = 280; C wins; 1 point

B with D: B: 100 + 120 = 220; D: 130 + 150 = 280; D wins; 1 point

C with D: C: 130; D: 100 + 120 + 150 = 370; D wins; 1 point

A has 2 points; B, 0 point; C, 1 point; and D, 3 points. Dave (D) is the pairwise winner. This does not agree with the plurality choice (Problem 46), but it does not violate any of the fairness criteria.

51. a. Pairwise comparison: A wins over B (33 to 7); A gets 1 point.

A wins over C (25 to 15); A gets 1 point.

C wins over B (21 to 19); C gets 1 point.

C wins over D (25 to 15); C gets 1 point.

D wins over A (21 to 19); D gets 1 point.

D wins over B (21 to 19); D gets 1 point.

There is no winner; A, C, and D each obtain 2 points.

   b. Tournaments:

A/B → A; A/C → A; A/D → D; D wins this tournament.

A/B → A; A/D → D; D/C → C; C wins this tournament.

A/C → A; A/B → A; A/D → D; D wins this tournament.

A/C → A; A/D → D; D/B → D; D wins this tournament.

A/D → D; D/B → D; D/C → C; C wins this tournament.

A/D → D; D/C → C; C/B → C; C wins this tournament.

B/C → C; C/A → A; A/D → D; D wins this tournament.

B/C → C; C/D → C; C/A → A; A wins this tournament.

B/D → D; D/A → D; D/C → C; C wins this tournament.

B/D → D; D/C → C; C/A → A; A wins this tournament.

Page 457

C/D $\rightarrow$ C; C/A $\rightarrow$ A; A/B $\rightarrow$ A; A wins this tournament.

C/D $\rightarrow$ C; C/B $\rightarrow$ C; C/A $\rightarrow$ A; A wins this tournament.

A, C, or D might win depending on the way they are paired.

c. Yes, both of these methods violate the condition of decisiveness. Some means for breaking ties should be provided.

53. a. Here is the Borda count:

				Total			
A	B	C	D	A	B	C	D
20(4	3	2	1) =	80	60	40	20
20(2	4	3	1) =	40	80	60	20
10(3	2	4	1) =	30	20	40	10
TOTAL:				150	160	140	50

The Borda count winner is B.

b. Suppose we eliminate candidate C (who is not a winning candidate). The votes become: (ABD)   (BAD)   (ABD)

$$\quad\quad 20 \quad\quad\quad 20 \quad\quad\quad 10$$

			Total		
A	B	D	A	B	D
20(3	2	1) =	60	40	20
20(2	3	1) =	40	60	20
10(3	2	1) =	30	20	10
TOTAL:			130	120	50

The winning candidate is A, which violates the irrelevant alternative criterion.

55. a. There is a total of 13 votes, so $\frac{13+1}{2} = 7$ is needed for a majority, so we see there is no majority winner. There is a plurality winner, namely C (with 6 votes).

b. Here is the Borda count:

			Total		
A	B	C	A	B	C
2(3	1	2) =	6	2	4
5(2	3	1) =	10	15	5
4(1	2	3) =	4	8	12
2(2	1	3) =	4	2	6
TOTAL:			24	27	27

There is no winner since B and C tie.

c. For the runoff election, eliminate the one with the most last-place votes: A has 4, B has 4 and C has 5, so we eliminate C. The tally now looks like:

Page 458

(AB) (BA) (BA) (AB)   We can combine these: (AB)   (BA)   We see that B
  2   5   4   2   4   9
wins the runoff election.

d. The insincere preferences are: (ACB) (BAC) (CBA) (CBA)
                                         2   5   4   2

A: 2 votes; B: 5 votes: C: 6 votes; no majority winner. The candidate with the most last place votes is: A: 6, B: 2; C: 5, so A is eliminated from the runoff. The tally now looks like: (CB) (BC) (CB) (CB)   We can combine these: (BC) (CB)
                     2   5   4   2   5   8

C wins this runoff election. Yes, they could change the outcome of the election by voting insincerely.

Level 3, page 840

57. Here are the voters preferences. "Motion: The meeting time will change to evenings."

	Yes	No
Jane	✓	
Linda	✓	
Ann	✓	
Melissa		✓

It looks like it will pass, but Melissa is ingenuous. After the motion is on the floor, she offers an amendment: "I move that we change the meeting time to Saturday mornings."

	Yes	No
Jane		✓
Linda	✓	
Ann	✓	
Melissa	✓	

The amendment is passed with a 3 to 1 vote. Now, the vote on the amended motion is:

	Yes	No
Jane		✓
Linda	✓	
Ann	✓	
Melissa		✓

It is a tie vote, so it does not pass and the old rule stands.

59. If the 90% vote is spread out evenly over the ten serious candidates, then a 10% vote for the radical candidate is theoretically enough to win the plurality vote.

17.3 Apportionment, page 841
New Terms Introduced in this Section

Adams' plan	Apportionment	Arithmetic mean	Geometric mean
Hamilton's plan	HH plan	Huntington-Hill's plan	Jefferson's plan
Lower quota	Modified divisor	Modified quota	Quota rule
Standard divisor	Standard quota	Upper quota	Webster's plan

Level 1, page 856

1. Answers vary; see Table 17.9.

3. Adams' plan favors the smaller states and Hamilton's and Jefferson's plans favor the larger states. You might also argue that Webster's and HH's plans also favor one or the other, but they are more equitable since they both round up sometimes and down other times.

5. If you round down, then lower the standard divisor to find the modified quotients.

7. **a.** Round 3.81 down for the lower quota: 3; round it up for the upper quota: 4
   b. $\overline{x} = \frac{3+4}{2} = 3.5$
   c. $\sqrt{3 \cdot 4} \approx 3.46$
   d. Compare 3.81 with $\overline{x} = 3.5$; it is larger, so we round up: 4; compare it with $\sqrt{3 \cdot 4} \approx 3.46$; it is larger, so we round up: 4

9. **a.** Round 1.46 down for the lower quota: 1; round it up for the upper quota: 2
   b. $\overline{x} = \frac{1+2}{2} = 1.5$
   c. $\sqrt{1 \cdot 2} \approx 1.41$
   d. Compare 1.46 with $\overline{x} = 1.5$; it is smaller, so we round down: 1; compare it with $\sqrt{1 \cdot 2} \approx 1.41$; it is larger, so we round up: 2

11. **a.** Round 2.49 down for the lower quota: 2; round it up for the upper quota: 3
   b. $\overline{x} = \frac{2+3}{2} = 2.5$
   c. $\sqrt{2 \cdot 3} \approx 2.45$
   d. Compare 2.49 with $\overline{x} = 2.5$; it is smaller, so we round down: 2; compare it with $\sqrt{2 \cdot 3} \approx 2.45$; it is larger, so we round up: 3

13. **a.** Round 1,695.4 down for the lower quota: 1,695; round it up for the upper quota: 1,696
   b. $\overline{x} = \frac{1,695 + 1,696}{2} = 1,695.5$
   c. $\sqrt{1,695 \cdot 1,696} \approx 1,695.50$
   d. Compare 1,695.4 with $\overline{x} = 1,695.5$; it is smaller, so we round down: 1,695; compare it with $\sqrt{1,695 \cdot 1,696} \approx 1,695.50$; it is smaller, so we round down: 1,695

Page 460

15. $\dfrac{52,000}{8} = 6,500$ **17.** $\dfrac{630}{5} = 126$ **19.** $\dfrac{1,450,000}{12} \approx 120,833.33$

21. $\dfrac{23,000,000}{125} = 184,000$

d	Manhattan	Bronx	Brooklyn	Queens	Staten Island
23. $\frac{81,000}{8} = 10,125$	$\frac{61,000}{d} \approx 6.02$	$\frac{2,000}{d} \approx 0.20$	$\frac{6,000}{d} \approx 0.59$	$\frac{7,000}{d} \approx 0.69$	$\frac{5,000}{d} \approx 0.49$
25. $\frac{3,438,000}{8} = 429,750$	$\frac{1,850,000}{d} \approx 4.30$	$\frac{201,000}{d} \approx 0.47$	$\frac{1,167,000}{d} \approx 2.72$	$\frac{153,000}{d} \approx 0.36$	$\frac{67,000}{d} \approx 0.16$
27. $\frac{7,324,000}{8} = 915,500$	$\frac{1,488,000}{d} \approx 1.63$	$\frac{1,204,000}{d} \approx 1.32$	$\frac{2,301,000}{d} \approx 2.51$	$\frac{1,952,000}{d} \approx 2.13$	$\frac{379,000}{d} \approx 0.41$

In Problems 29-32, the modified divisors, D, may vary.

29. a. $d = \dfrac{116,000}{10} = 11,600$ **b.** See q on table.

Precinct	q	Down	Q	Down
1st	$\dfrac{35,000}{d} \approx 3.02$	3	$\dfrac{35,000}{D} \approx 3.50$	3
2nd	$\dfrac{21,000}{d} \approx 1.81$	1	$\dfrac{21,000}{D} \approx 2.10$	2
3rd	$\dfrac{12,000}{d} \approx 1.03$	1	$\dfrac{12,000}{D} \approx 1.20$	1
4th	$\dfrac{48,000}{d} \approx 4.14$	4	$\dfrac{48,000}{D} \approx 4.80$	4
TOTAL		9		10

c. $3 + 1 + 1 + 4 = 9$

d. Let $D = 10,000$ and see Q from table; when rounded down we obtain the correct number of seats.

31. a. $d = \dfrac{800,000}{10} = 80,000$ **b.** See q on table.

Precinct	q	Down	Q	Down
1st	$\dfrac{135,000}{d} \approx 1.69$	1	$\dfrac{135,000}{D} \approx 2.08$	2
2nd	$\dfrac{231,000}{d} \approx 2.89$	2	$\dfrac{231,000}{D} \approx 3.55$	3
3rd	$\dfrac{118,000}{d} \approx 1.48$	1	$\dfrac{118,000}{D} \approx 1.82$	1
4th	$\dfrac{316,000}{d} \approx 3.95$	3	$\dfrac{316,000}{D} \approx 4.86$	4
TOTAL		7		10

c. $1 + 2 + 1 + 3 = 7$

d. Let $D = 65,000$ and see Q from table; when rounded down we obtain the correct number of seats.

In Problems 33-36, the modified divisors, D, may vary.

33. a. $d = \dfrac{116,000}{10} = 11,600$ **b.** See q on table.

Precinct	q	Up	Q	Up
1st	$\dfrac{35,000}{d} \approx 3.02$	4	$\dfrac{35,000}{D} \approx 2.69$	3
2nd	$\dfrac{21,000}{d} \approx 1.81$	2	$\dfrac{21,000}{D} \approx 1.62$	2
3rd	$\dfrac{12,000}{d} \approx 1.03$	2	$\dfrac{12,000}{D} \approx 0.92$	1
4th	$\dfrac{48,000}{d} \approx 4.14$	5	$\dfrac{48,000}{D} \approx 3.69$	4
TOTAL		13		10

c. $4 + 2 + 2 + 5 = 13$

d. Let $D = 13,000$ and see Q from table; when rounded up we obtain the correct number of seats.

35. a. $d = \dfrac{800,000}{10} = 80,000$ **b.** See q on table.

Precinct	q	Up	Q	Up
1st	$\dfrac{135,000}{d} \approx 1.69$	2	$\dfrac{135,000}{D} \approx 1.23$	2
2nd	$\dfrac{231,000}{d} \approx 2.89$	3	$\dfrac{231,000}{D} \approx 2.10$	3
3rd	$\dfrac{118,000}{d} \approx 1.48$	2	$\dfrac{118,000}{D} \approx 1.07$	2
4th	$\dfrac{316,000}{d} \approx 3.95$	4	$\dfrac{316,000}{D} \approx 2.87$	3
TOTAL		11		10

c. $2 + 3 + 2 + 4 = 11$

d. Let $D = 110,000$ and see Q from table; when rounded down we obtain the correct number of seats.

Level 2, page 857

37. $d \approx 37,084.51$; Maine (ME) is $\frac{96,643}{d} \approx 2.61$. The revised number for Massachusetts

(MA) is $475{,}199 - 96{,}643 = 378{,}556$, so that $\dfrac{378{,}556}{d} \approx 10.21$. This means that the standard quotas for the 16 states are:

	CT	DE	GA	KY	ME	MD	MA	NH	NJ	NY	NC	PA	RI	SC	VT	VA
	6.41	1.59	2.23	1.99	2.61	8.62	10.21	3.83	4.97	9.17	10.65	11.69	1.86	6.72	2.30	20.16
39.	6	1	2	1	2	8	10	3	4	9	10	11	1	6	2	20

The total is 96.

41. $n = 10;\ d = \dfrac{25{,}000}{10} = 2{,}500;\ D = 3{,}125$

Precinct	Population	q	lower quota	upper quota	Q	Adams' plan
North	8,700	3.48	3	4	2.78	3
South	5,600	2.24	2	3	1.79	2
East	7,200	2.88	2	3	2.30	3
West	3,500	1.40	1	2	1.12	2
TOTAL	25,000		8	12		10

43. $n = 10;\ d = \dfrac{25{,}000}{10} = 2{,}500$

Precinct	Population	q	lower quota	Hamilton's plan
North	8,700	3.48	3	4 (#2)
South	5,600	2.24	2	2
East	7,200	2.88	2	3 (#1)
West	3,500	1.40	1	1
TOTAL	25,000		8	10

45. $n = 10;\ d = \dfrac{25{,}000}{10} = 2{,}500,\ D = 2{,}500$ (no modification necessary)

Precinct	Population	q	lower quota, a	upper quota, b	$\sqrt{ab}$	HH's plan
North	8,700	3.48	3	4	3.46	4
South	5,600	2.24	2	3	2.45	2
East	7,200	2.88	2	3	2.45	3
West	3,500	1.40	1	2	1.41	1
TOTAL	25,000		8	12		10

47. $n = 26;\ d = \dfrac{62{,}000}{26} \approx 2{,}384.62,\ D = 2{,}210$

Precinct	Population	q	lower quota	upper quota	Q	Jefferson's plan
North	18,200	7.63	7	8	8.24	8
South	12,900	5.41	5	6	5.84	5
East	17,600	7.38	7	8	7.96	7
West	13,300	5.58	5	6	6.02	6
TOTAL	62,000		24	28		26

Page 463

49. $n = 26; d = \frac{62,000}{26} \approx 2,384.62, D = 2,384.62$ (no modification necessary)

Precinct	Population	q	round (nearest)	Webster's plan
North	18,200	7.63	8	8
South	12,900	5.41	5	5
East	17,600	7.38	7	7
West	13,300	5.58	6	6
TOTAL	62,000		26	26

51. $n = 16; d = \frac{62,000}{16} = 3,875, D = 4,425$

Precinct	Population	q	lower quota	upper quota	Q	Adams' plan
North	18,200	4.70	4	5	4.11	5
South	12,900	3.33	3	4	2.92	3
East	17,600	4.54	4	5	3.98	4
West	13,300	3.43	3	4	3.01	4
TOTAL	62,000		14	18		16

53. $n = 16; d = \frac{62,000}{16} = 3,875$

Precinct	Population	q	lower quota	Hamilton's plan
North	18,200	4.70	4	5 (#1)
South	12,900	3.33	3	3
East	17,600	4.54	4	5 (#2)
West	13,300	3.43	3	3
TOTAL	62,000		14	16

55. $n = 16; d = \frac{62,000}{16} = 3,875, D = 3,875$ (no modification necessary)

Precinct	Population	q	lower quota, a	upper quota, b	$\sqrt{ab}$	HH's plan
North	18,200	4.70	4	5	4.47	5
South	12,900	3.33	3	4	3.46	3
East	17,600	4.54	4	5	4.47	5
West	13,300	3.43	3	4	3.46	3
TOTAL	62,000		14	18		16

Level 3, page 858

57. Standard divisor: $d = \frac{16,630,000}{475} \approx 35,010.53$, modified divisor: $D = 34,720$

Region	Population	q	lower quota	upper quota	Q	Jefferson's plan
North	1,820,000	51.98	51	52	52.42	52
Northeast	2,950,000	84.26	84	85	84.97	84
East	1,760,000	50.27	50	51	50.69	50
Southeast	1,980,000	56.55	56	57	57.03	57
South	1,200,000	34.28	34	35	34.56	34
Southwest	2,480,000	70.84	70	71	71.43	71
West	3,300,000	94.26	94	95	95.05	95
Northwest	1,140,000	32.56	32	33	32.83	32
TOTAL	16,630,000		471	479		475

59. Standard divisor: $d = \frac{16,630,000}{475} \approx 35,010.53$ D = 35,010.53 (no modification necessary)

Region	Population	q	round (nearest)	Webster's Plan
North	1,820,000	51.98	52	52
Northeast	2,950,000	84.26	84	84
East	1,760,000	50.27	50	50
Southeast	1,980,000	56.55	57	57
South	1,200,000	34.28	34	34
Southwest	2,480,000	70.84	71	71
West	3,300,000	94.26	94	94
Northwest	1,140,000	32.56	33	33
TOTAL	16,630,000		475	475

17.4 Apportionment Paradoxes, page 858

New Terms Introduced in this Section

Alabama paradox   Balinski and Young's impossibility theorem   New states paradox
Population paradox

Level 1, page 863

1. An increase in the total number of to be apportioned resulting in a loss of items for a group is called the *Alabama paradox*.

3. When a reapportionment of an increased number of seats causes a shift in the apportionment of the existing states, it is known as the *new states paradox*.

SURVIVAL HINT: *Remember to use calculator values instead of rounded values in calculators.*

5. $n = 100$; $d = \frac{128,700}{100} = 1,287$, $D = 1,330$

		q	↓ quota	↑ quota	Q	Adams' plan	
A	68,500	53.22	53	54	51.50	523	← Violates the quota rule
B	34,700	26.96	26	27	26.09	27	
C	16,000	12.43	12	13	12.03	13	
D	9,500	7.38	7	8	7.14	8	
TOT.	128,700		98	102		100	

7. $n = 100$; $d = \frac{127,500}{100} = 1,275$, $D = 1,240$

		q	↓ quota	↑ quota	Q	Jefferson's plan	
A	68,500	53.73	53	54	55.24	55	← Violates the quota rule
B	34,700	27.22	27	28	27.98	27	
C	14,800	11.61	11	12	11.94	11	
D	9,500	7.45	7	8	7.66	7	
TOT.	127,500		98	102		100	

9. $n = 200$; $d = \frac{10,315}{200} = 51.575$, $D = 50.9$

		q	↓ quota	↑ quota	Q	Jefferson's plan	
A	1,100	21.3282	21	22	21.61	21	
B	1,100	21.3282	21	22	21.61	21	
C	1,515	29.3747	29	30	29.76	29	
D	4,590	88.9966	88	89	90.18	90	← Violates the quota rule
E	2,010	38.9724	38	39	39.49	39	
TOT.	10,315		197	202		200	

11. $n = 246$; $d = \frac{2,724}{246} \approx 11.07$; $\qquad\qquad$ $n = 247$; $d = \frac{2,724}{247} \approx 11.03$

		q	↓ quota	Hamilton's	q	Hamilton's	
A	181	16.35	16	17 (#1)	16.41	16	← Alabama paradox
B	246	22.22	22	22	22.31	22	
C	812	73.33	73	73	73.63	74 (#2)	
D	1,485	134.11	134	134	134.65	135 (#1)	
TOT.	2,724		245	246		247	

13. $n = 50$; $d = \frac{2,076}{50} = 41.52$; $\qquad\qquad$ $n = 51$; $d = \frac{2,076}{51} \approx 40.71$

Page 466

	q	↓ quota	Hamilton's	q	Hamilton's	
A	300	7.23	7	7	7.37	7	
B	301	7.25	7	8 (#1)	7.39	7	← Alabama paradox
C	340	8.19	8	8	8.35	8	
D	630	15.17	15	15	15.48	16 (#1)	
E	505	12.16	12	12	12.41	13 (#2)	
TOT.	2,076		49	50		51	

15. $n = 11$

	Original			Revised				
		q	H's plan		q	H's plan	Increase	% increase
A	55,200	1.64	1	61,100	1.584	2 (#2)	5,900	10.69%
B	124,900	3.71	4 (#1)	148,100	3.840	4 (#1)	23,200	18.57%
C	190,000	5.65	6 (#2)	215,000	5.575	5	25,000	13.16%
TOT.	370,100		11	424,200		11		
d:	33,645.45			38,563.64				

← Population paradox (greater than A)

17. $n = 13$

	Original			Revised				
		q	H's plan		q	H's plan	Increase	% increase
A	89,950	2.65	2	97,950	2.60	3 (#2)	8,000	8.89%
B	124,800	3.68	4 (#1)	144,900	3.84	4 (#1)	20,100	16.11%
C	226,000	6.67	7 (#2)	247,100	6.56	6	21,100	9.34%
TOT.	440,750		13	489,950		13		
d:	33,903.85			37,688.46				

← Population paradox (greater than A)

19. $n = 12$; $d \approx 16,999.58$ $n = 12 + 2 = 14$; $d = 17,302.5$

	Population	q	H's plan	Population	q	H's plan	
A	144,899	8.52	9 (#1)	144,899	8.37	8	← New states paradox
B	59,096	3.48	3	59,096	3.42	4 (#1)	
C				38,240	2.21	2	
TOT.	203,995		12	242,235		14	

21. $n = 50$; $d = 325,820$ $n = 50 + 4 = 54$; $d \approx 328,537.04$

	Population	q	H's plan	Population	q	HH's plan	
A	7,000,500	21.49	21	7,000,500	21.31	21	
B	9,290,500	28.51	29 (#1)	9,290,500	28.28	28	← New states paradox
C				1,450,000	4.41	5 (#1)	
TOT.	16,291,000		50	17,741,000		54	

Page 467

Level 2, page 864

23. a. $d = \dfrac{18,834}{300} = 62.78$

b.

City	No.	q	**c.** ↓ quota	↑ quota	**d.** $D = 62.4$	Jefferson's
Atlanta	12,520	$\dfrac{12,520}{d} \approx 199.43$	199	200		$\dfrac{12,520}{D} \approx 200$
Buffalo	4,555	$\dfrac{4,555}{d} \approx 72.55$	72	73		$\dfrac{4,555}{d} \approx 72$
Carson City	812	$\dfrac{812}{d} \approx 12.93$	12	13		$\dfrac{812}{d} \approx 13$
Denver	947	$\dfrac{947}{d} \approx 15.08$	15	16		$\dfrac{947}{d} \approx 15$
TOTAL	18,834		298	302		300

e. No, it does not violate the quota rule because all modified quotas are within the parameters specified in part **c.**

25. a. $d = \dfrac{20,330}{100} = 203.3$ $\qquad\qquad$ $d = \dfrac{22,830}{112} \approx 203.84$

b.

Districts	No.	q	**c.** Hamilton's	**d.**	q	Hamilton's
uptown	16,980	83.52	84 (#1)	16,980	83.30	83
downtown	3,350	16.48	16	3,350	16.43	17 (#1)
new				2,500	12.26	12
TOTAL	20,330		100	22,830		300

e. Yes, this illustrates the new states paradox because there was a shifting of the apportionments of the existing districts.

Level 3 Problem Solving, page 865

27. Yes; the best that child #1 can do is to slice off 1/3 of the cake. Anything less than that will result in a diminished slice for child #1.

The best that child #2 can do is to divide the larger piece into two parts; anything less than that will result in a diminished slice for child #2.

Child #3 will be faced with, at worst, three slices of equal size. If they are not equal size, then the other children did not make good cuts and child #3 will pick the largest piece.

29.

Age	Adams	Jeff	Web
Eldest	9	8	9
Middle	6	5	6
Youngest	2	1	2
TOTAL	17	14	17

We see that Adams' or Webster's plan gives the correct apportionment of the horses.

Chapter 17 Review Questions, page 866

Studying for a chapter examination is a personal process, one which nobody else can do for you. Simply take the time to review what you have done. Here are the new terms in Chapter 17.

Adams' plan [17.3]
Alabama paradox [17.4]
Apportionment [17.3]
Approval voting method [17.1]
Arithmetic mean [17.3]
Arrow's impossibility theorem [17.2]
Balinski and Young impossibility theorem [17.4]
Binary voting [17.1]
Borda count [17.1]
Condorcet candidate [17.2]
Condorcet criterion [17.2]
Condorcet's paradox [17.2]
Decisiveness [17.2]
Dictatorship [17.1]
Fairness criteria [17.2]
Fair voting principles [17.2]
Geometric mean [17.3]

Hamilton's plan [17.3]
Hare method [17.1]
HH plan [17.3]
Huntington-Hill's plan [17.3]
Independence of irrelevant alternatives [17.2]
Insincere voting [17.2]
Irrelevant alternatives criterion [17.2]
Jefferson's plan [17.3]
Lower quota [17.3]
Majority criterion [17.2]
Majority rule [17.1]
Modified divisor [17.3]
Modified quota [17.3]
Monotonicity criterion [17.2]
New states paradox [17.4]
Pairwise comparison method [17.1]

Pareto principle [17.1]
Plurality method [17.1]
Population paradox [17.4]
Quota rule [17.3]
Runoff election [17.1]
Sequential voting [17.1]
Social choice theory [17.1]
Standard divisor [17.3]
Standard quota [17.3]
Straw vote [17.2]
Symmetry [17.1; 17.2]
Tournament method [17.1]
Transitive law [17.2]
Unrestricted domain [17.1]
Upper quota [17.3]
Vote [17.1]
Webster's plan [17.3]

If you can describe the term, read on to the next one; if you cannot, then look it up in the text (the section number is shown in brackets). Next, study the types of problems listed at the end of Chapter 17.

TYPES OF PROBLEMS

Understand and be able to use voting notation. [17.1]
Decide how many different voting orders are possible for a set of candidates. [17.1]
Use the majority rule. [17.1]
Calculate the winner of an election using the plurality method. [17.1]
Calculate the winner of an election using the Borda count method. [17.1]
Calculate the winner of an election using the pairwise comparison method. [17.1]
Calculate the winner of an election using the tournament method. [17.1]
Calculate the winner of an election using the Hare method. [17.1]
Be able to discuss and recognize the majority criterion [17.2]
Be able to discuss and recognize the Condorcet criterion [17.2]
Be able to discuss and use the monotonicity criterion [17.2]
Be able to discuss and use the irrelevant alternatives criterion [17.2]
What are the fairness criteria? [17.2]
Be able to find the standard quota and divisor. [17.3]
Apportion using Adams' plan. [17.3]
Apportion using Jefferson's plan. [17.3]
Apportion using Hamilton's plan. [17.3]
Apportion using Webster's plan. [17.3]
Apportion using Huntington-Hill's plan. [17.3]
Be able to discuss and recognize the Alabama paradox. [17.4]
Be able to discuss and recognize the population paradox. [17.4]
Be able to discuss and recognize the new states paradox.[17.4]

Once again, see if you can verbalize (to yourself) how to do each of the listed types of problems.

　　Work all of Chapter 17 Review Questions (whether they are assigned or not). Work through all of the problems before looking at the answers, and *then* correct each of the problems. The entire solution is shown in the answer section at the back of the text. If you worked the problem correctly, move on to the next problem, but if you did not work it correctly (or you did not know what to do), look back in the chapter to study the procedure, or ask your instructor.

　　Finally, go back over the homework problems you have been assigned. If you worked a problem correctly, move on to the next problem, but if you missed it on your homework, then you should look back in the book or talk to your instructor about how to work the problem.

　　If you follow these steps, you should be successful with your review of this chapter.

We give all of the answers to the Chapter Review questions (not just the odd-numbered questions).

Chapter 17 Review Questions, page 866

1. There is a total of 45 votes, so a majority is $\dfrac{45+1}{2} = 23$; there is no majority.

   The plurality vote goes to C.

2. There is a total of $3! = 6$ arrangements, and 3 are shown here, so there are 3 possibilities that received no votes.

3. Here are the rankings: A: 15; B: 12; C: 18

   Eliminate B for the second round of votes:

   A: $15 + 12 = 27$

   C: 18

   A wins the majority in this round, so A is the winner.

4. A over B: 15; B over A: $18 + 12 = 30$; B wins 1 point.

   A over C: $15 + 12 = 27$; C over A: 18; A wins 1 point.

   B over C: $15 + 12 = 27$; C over B: 18; B wins 1 point.

   A has 1 point, B has 2 points, and C has 0 points, so B is the winner.

5. (1) The majority criterion is not violated because there is no majority winner.

   (2) We look for a Condorcet candidate by finding the one-on-one pairings:

   A with B: **B wins**

   A with C: **A wins**

   B with C: **B wins**

   In other words,

	A	B	C
A	—	B	A
B	B	—	B
C	A	B	—

   B is a Condorcet candidate. We see that the Hare method (Problem 3) violates the Condorcet criterion.

6. Here are the results (in percents):

   A: $22 + 23 = 45$

   B $15 + 29 = 44$

   C $7 + 4 = 11$

   Candidate A has a plurality.

7. We look for a Condorcet candidate by finding the one-on-one pairings:

   A over B: $22 + 23 + 4 = 49$; B over A: $15 + 29 + 7 = 51$; **B wins**

   A over C: $22 + 23 + 15 = 60$; C over A: $29 + 7 + 4 = 40$; **A wins**

   B over C: $22 + 15 + 29 = 66$; C over B: $23 + 7 + 4 = 34$; **B wins**

In tabular form,

	A	B	C
A	—	B	A
B	B	—	B
C	A	B	—

B is the Condorcet winner.

8. Three points for a first place vote, 2 points for a second place vote, and 1 point for a third place.

David Carr (Fresno State):	$34(3) + 60(2) + 58(1) = 280$
Eric Crouch (Nebraska):	$162(3) + 98(2) + 88(1) = 770$
Ken Dorsey (Miami):	$109(3) + 122(2) + 67(1) = 638$
Dwight Freeney (Syracuse):	$2(3) + 6(2) + 24(1) = 42$
Rex Grossman (Florida):	$137(3) + 105(2) + 87(1) = 708$
Joey Harrington (Oregon):	$54(3) + 68(2) + 66(1) = 364$
Bryant McKinnie (Miami):	$26(3) + 12(2) + 14(1) = 116$
Julius Peppers (North Carolina):	$2(3) + 10(2) + 15(1) = 41$
Antwaan Randle El (Indiana):	$46(3) + 39(2) + 51(1) = 267$
Roy Williams (Oklahoma):	$13(3) + 36(2) + 35(1) = 146$

The winner was Eric Crouch with 770 Borda points.

9. The votes are:

A: 7

B: 4

C: 5

D: 1

The least first-place votes were for D, so we delete D from the second round of votes, with these results:

A: 7

B: $4 + 1 = 5$

C: 5

There is a tie for last place. The Hare method fails to pick a winner unless we specify some way of breaking a tie. If we have a runoff between just B and C, we find:

B: $7 + 4 + 1 = 12$

C: 5

B is the winner, so for the next round we eliminate C, with these results:

A: $7 + 5 = 12$

B: $4 + 1 = 5$

We declare A the winner.

10. Here are the pairwise point counts.

A over B: $7 + 5 = 12$; B over A: $4 + 1 = 5$; A wins 1 point.

Page 472

A over C: 7 + 1 = 8; C over A: 5 + 4 = 9; C wins 1 point.
A over D: 7 + 5 = 12; D over A: 4 + 1 = 5; A wins 1 point.
B over C: 7 + 4 + 1 = 12; C over B: 5; B wins 1 point.
B over D: 5 + 4 = 9; D over B: 7 + 1 = 8; B wins 1 point.
C over D: 5 + 4 = 9; D over C: 7 + 1 = 8; C wins 1 point.
The final point count is:
A, 2 points; B, 2 points; C, 2 points; D, 0 points.
There is no pairwise winner without specifying a method of breaking a tie. If we have a runoff among A, B, and C, we find
A: 7
B: 4 + 1 = 5
C: 5
We declare A the winner.

11. If B pulls out of the race, we have the following preferences:

	(ADC)	(CAD)	(CDA)	(DAC)
	7	5	4	1

The first vote is:
A: 7
C: 5 + 4 = 9
D: 1

C has a majority, so C is the winner. If we compare this with the result of Problem 9 we see that this result does violate the irrelevant alternatives criterion.

12. The standard divisor is $\frac{790}{100} = 7.9$

13. The standard quotas are

School	q	Lower quota	Upper quota
Elsie Allen	$\frac{90}{d} \approx 11.39$	11	12
Maria Carrillo	$\frac{215}{d} \approx 27.22$	27	28
Montgomery	$\frac{268}{d} \approx 33.92$	33	34
Piner	$\frac{133}{d} \approx 16.84$	16	17
Santa Rosa	$\frac{84}{d} \approx 10.63$	10	11

14. Modified divisor is: $D = 8.15$:

School	Q	Adam's plan
Elsie Allen	$\frac{90}{D} \approx 11.39$	12
Maria Carrillo	$\frac{215}{D} \approx 26.38$	27
Montgomery	$\frac{268}{D} \approx 32.88$	33
Piner	$\frac{133}{D} \approx 16.31$	17
Santa Rosa	$\frac{84}{D} \approx 10.31$	11
TOTAL		100

15. Modified divisor is: $D = 7.67$:

School	Q	Jefferson's plan
Elsie Allen	$\frac{90}{D} \approx 11.73$	11
Maria Carrillo	$\frac{215}{D} \approx 28.03$	28
Montgomery	$\frac{268}{D} \approx 34.94$	34
Piner	$\frac{133}{D} \approx 17.34$	17
Santa Rosa	$\frac{84}{D} \approx 10.95$	10
TOTAL		100

16.

School	q	Lower quota	Hamilton's plan
Elsie Allen	$\frac{90}{d} \approx 11.39$	11	11
Maria Carrillo	$\frac{215}{d} \approx 27.22$	27	27
Montgomery	$\frac{268}{d} \approx 33.92$	33	34 (#1)
Piner	$\frac{133}{d} \approx 16.84$	16	17 (#2)
Santa Rosa	$\frac{84}{d} \approx 10.63$	10	11 (#3)
TOTAL		97	100

17. Modified divisor is: $D = 8$:

School	Q	Webster's plan
Elsie Allen	$\frac{90}{D} = 11.25$	11
Maria Carrillo	$\frac{215}{D} = 26.875$	27
Montgomery	$\frac{268}{D} = 33.5$	34
Piner	$\frac{133}{D} = 16.625$	17
Santa Rosa	$\frac{84}{D} = 10.5$	11
TOTAL		100

Page 474

18. Modified divisor is $d = 7.9$

School	q	$\downarrow$ quota, a	$\uparrow$ quota, b	$\sqrt{ab}$	HH's plan
Elsie Allen	11.39	11	12	$\sqrt{11 \cdot 12} \approx 11.489$	11
Maria Carrillo	27.22	27	28	$\sqrt{27 \cdot 28} \approx 27.495$	27
Montgomery	33.92	33	34	$\sqrt{33 \cdot 34} \approx 33.496$	34
Piner	16.84	16	17	$\sqrt{16 \cdot 17} \approx 16.492$	17
Santa Rosa	10.63	10	11	$\sqrt{10 \cdot 11} \approx 10.488$	11
TOTAL					100

19. No, not from the given information. We can see that the data does not violate the quota rule, and without more information, we cannot check to see whether any of the other paradoxes might be violated.

20. The sum of the number of violent crimes per 100 residents for the five areas is 62.68. Thus, $d = \dfrac{62.68}{180} = 0.348\overline{2}$. If we round to the nearest unit, we have the appropriate apportionment.

Precinct	q	$\downarrow$ quota	$\uparrow$ quota	Webster's plan
Downtown	$\frac{24.45}{d} \approx 70.2$	70	71	70
Fairground	$\frac{10.04}{d} \approx 28.8$	28	29	29
Columbus Square	$\frac{9.75}{d} \approx 27.999$	27	28	28
Downtown West	$\frac{9.43}{d} \approx 27.1$	27	28	27
Peabody	$\frac{9.01}{d} \approx 25.9$	25	29	26
TOTAL				180

CHAPTER 18 The Nature of Calculus

Chapter Overview
You might look at the title of the chapter and ask, "Why is such an advanced topic included in a liberal arts textbook?" Don't stop there! This chapter is NOT like a calculus course, and I have been successfully teaching this material to liberal arts students for years. If you look again at the Preface, you will see that I've said, "I have attempted to give insight into what mathematics is, what it accomplishes, and how it is pursued as a human enterprise."

Calculus is one the great ideas not only in the history of mathematics, but also in the history of the world. We could not have the world in which we live without calculus, and I believe that every educated person should have some idea about the *nature of calculus*.

18.1 What Is Calculus?, page 871

New Terms Introduced in this Section

Calculus	Derivative	Differential calculus	Integral
Integral calculus	Limit	Mathematical modeling	Secant line
Tangent line			

Level 1, page 879

1. You should write at least a paragraph describing each of the topics, limits, derivatives, and integrals.
3. Theoretically, she never reaches the wall, but in reality she must hit the wall because the woman has physical dimensions.
5. It looks like the limit of the sequence is an infinite number of repeated threes; we guess that it the limit is $0.\overline{3} = \frac{1}{3}$.
7. It looks like the limit of the sequence is an infinite number of repeated nines; we guess that it the limit is $0.\overline{9} = \frac{3}{3} = 1$. You many not believe that $0.9999\cdots = 1$, but notice that if $0.\overline{3} = \frac{1}{3}$ then simply multiply both sides by 3.
9. It looks like this sequence of decimals is approaching the irrational number $\pi \approx 3.14169265359$.

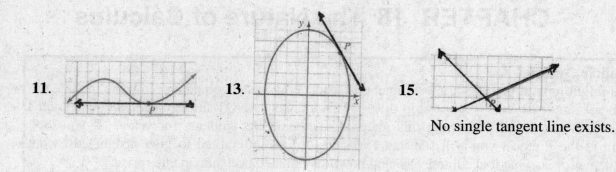

11.   **13.**   **15.**

No single tangent line exists.

Level 2, page 879

Answers to Problems 16-27 may vary.

17. $\lim\limits_{n\to\infty} \dfrac{2n}{3n+1}$ Consider larger and larger values of n:

n	1	10	100	1,000
$\dfrac{2n}{3n+1}$	$\dfrac{2(1)}{3(1)+1} = \dfrac{1}{2}$	$\dfrac{2(10)}{3(10)+1} = \dfrac{20}{31}$	$\dfrac{2(100)}{3(100)+1} = \dfrac{200}{301}$	$\dfrac{2(1,000)}{3(1,000)+1} = \dfrac{2,000}{3,001}$

It looks like the limit is $\frac{2}{3}$.

19. $\lim\limits_{n\to\infty} \dfrac{n+1}{2n}$ Consider larger and larger values of n:

n	1	10	100	1,000
$\dfrac{n+1}{2n}$	$\dfrac{1+1}{2(1)} = 1$	$\dfrac{10+1}{2(10)} = 0.55$	$\dfrac{100+1}{2(100)} = 0.505$	$\dfrac{1,000+1}{2(1,000)} = 0.5005$

It looks like the limit is $0.5 = \frac{1}{2}$.

21. $\lim\limits_{n\to\infty} \dfrac{3n^2+1}{2n^2-1}$ Consider larger and larger values of n:

n	1	10	100
$\dfrac{3n^2+1}{2n^2-1}$	$\dfrac{3(1)^2+1}{2(1)^2-1} = 4$	$\dfrac{3(10)^2+1}{2(10)^2-1} = 1.51$	$\dfrac{3(100)^2+1}{2(100)^2-1} = 1.50$

It looks like the limit is $1.5 = \frac{3}{2}$.

23. Look at the underlying grid. It looks like the shared area is about 10 or 11 "squares" (actually they are rectangles with area $4 \times \frac{1}{2} = 2$). Thus, we estimate that the area of is about 20 to 22 units.

25. Look at the underlying grid. It looks like the shared area is about 8 or 9 "squares" (actually they are rectangles with area $1 \times \frac{1}{2} = \frac{1}{2}$). Thus, we estimate that the area of is about 4 to 5 units.

27. Look at the underlying grid. It looks like the shared area is about 5 or 6 "squares" (actually they are rectangles with area $1 \times \frac{1}{2} = \frac{1}{2}$). Thus, we estimate that the area of is about $2\frac{1}{2}$ to 3 units.

Level 3, page 880

29. Answers vary, but the income and expense amounts need to be $100,000.

18.2 Limits, page 880

New Terms Introduced in this Section

Converge	Diverge	General term	Infinite limit
Limit of a sequence	Predecessor	Sequence	Successor

Level 1, page 885

1. The limit of a sequence means that successive terms of a sequence are getting closer and closer to the number called the limit.

3. Given $\{1 + (-1)^n\}$;
   if $n = 1$: $1 + (-1)^1 = 0$
   if $n = 2$: $1 + (-1)^2 = 2$
   if $n = 3$: $1 + (-1)^3 = 0$
   if $n = 4$: $1 + (-1)^4 = 2$
   if $n = 5$: $1 + (-1)^5 = 0$. The first five terms of the sequence are 0, 2, 0, 2, 0.

5. Given $\left\{ \dfrac{3n + 1}{n + 2} \right\}$;

   if $n = 1$: $\dfrac{3(1) + 1}{1 + 2} = \dfrac{4}{3}$

   if $n = 2$: $\dfrac{3(2) + 1}{2 + 2} = \dfrac{7}{4}$

   if $n = 3$: $\dfrac{3(3) + 1}{3 + 2} = \dfrac{10}{5} = 2$

   if $n = 4$: $\dfrac{3(4) + 1}{4 + 2} = \dfrac{13}{6}$

   if $n = 5$: $\dfrac{3(5) + 1}{5 + 2} = \dfrac{16}{7}$. The first five terms of the sequence are $\frac{4}{3}, \frac{7}{4}, 2, \frac{13}{6}, \frac{16}{7}$.

Page 479

7. $\lim\limits_{n\to\infty} \dfrac{8{,}000n}{n+1} = \lim\limits_{n\to\infty} \dfrac{8{,}000}{1+\frac{1}{n}}$

$= \dfrac{8{,}000}{1+0}$

$= 8{,}000$

9. $\lim\limits_{n\to\infty} \dfrac{2n+1}{3n-4} = \lim\limits_{n\to\infty} \dfrac{2+\frac{1}{n}}{3-\frac{4}{n}}$

$= \dfrac{2+0}{3-0}$

$= \dfrac{2}{3}$

11. $\lim\limits_{n\to\infty} \dfrac{5n+8}{n} = \lim\limits_{n\to\infty} \dfrac{5+\frac{8}{n}}{1}$

$= \dfrac{5+0}{1}$

$= 5$

13. $\lim\limits_{n\to\infty} \dfrac{8n^2+800n+5{,}000}{2n^2-1{,}000n+2} = \lim\limits_{n\to\infty} \dfrac{8+\frac{800}{n}+\frac{5{,}000}{n^2}}{2-\frac{1{,}000}{n}+\frac{2}{n^2}}$

$= \dfrac{8+0+0}{2-0+0}$

$= 4$

15. $\lim\limits_{n\to\infty} \dfrac{8n^2+6n+4{,}000}{n^3+1} = \lim\limits_{n\to\infty} \dfrac{\frac{8}{n}+\frac{6}{n^2}+\frac{4{,}000}{n^3}}{1+\frac{1}{n^3}}$

$= \dfrac{0+0+0}{1+0}$

$= 0$

Level 2, page 885

17. $\lim\limits_{n\to\infty} \frac{1}{3n} = 0$

19. Since we know $0.9999\cdots = 1$, we guess that $0.69999\cdots = 0.7$

Page 480

21. $\displaystyle\lim_{n\to\infty}\frac{3n^2-7n+2}{5n^4+9n^2}=\lim_{n\to\infty}\frac{\frac{3}{n^2}-\frac{7}{n^3}+\frac{2}{n^4}}{5+\frac{9}{n^2}}$

$$=\frac{0+0+0}{5+0}$$
$$=0$$

23. $\displaystyle\lim_{n\to\infty}\frac{2n^4+5n^2-6}{3n+8}=\lim_{n\to\infty}\frac{2+\frac{5}{n^2}-\frac{6}{n^4}}{\frac{3}{n^3}+\frac{8}{n^4}}$

As $n\to\infty$ the numerator approaches 2 and the denominator approaches 0. Thus, the quotient becomes an ever increasing positive number, so we conclude that the limit does not exist, and since this quotient grows toward positive infinity we write "∞."

25. $\displaystyle\lim_{n\to\infty}\frac{15+9n-6n^2}{2n^2-4n-1}=\lim_{n\to\infty}\frac{\frac{15}{n^2}+\frac{9}{n}-6}{2-\frac{4}{n}-\frac{1}{n^2}}$

$$=\frac{0+0-6}{2-0-0}$$
$$=-3$$

27. $\displaystyle\lim_{n\to\infty}\frac{-21n^3+52}{-7n^3+n^2+20n-9}=\lim_{n\to\infty}\frac{-21+\frac{52}{n^3}}{-7+\frac{1}{n}+\frac{20}{n^2}-\frac{9}{n^3}}$

$$=\frac{-21+0}{-7+0+0-0}$$
$$=3$$

Level 3, page 885

29. The sequence is $12, 6, 3, \frac{3}{2}, \frac{3}{4}$; there is 0.75 mg of the drug present. At the end of n hours, there is $24(\frac{1}{2})^n$ mg present.

18.3 Derivatives, page 886
New Terms Introduced in this Section

Average rate of change	Derivative	Derivative of e^x	Differentiable function
Instantaneous rate of change	Secant line	Tangent line	Velocity

Level 1, page 895

1. a. $\text{AVERAGE}=\dfrac{(80-40)\text{ ft}}{(8-1)\text{ sec}}\approx 5.71$ ft/s

b. $\text{AVERAGE}=\dfrac{(100-40)\text{ ft}}{(5-1)\text{ sec}}=15$ ft/s

Page 481

c. AVERAGE $= \dfrac{(80 - 40) \text{ ft}}{(2 - 1) \text{ sec}} = 40$ ft/s

d. It looks like the slope of the tangent line has a rise of 4 squares and a run of 1 square for a slope of 40 ft/s.

3. **a.** AVERAGE $= \dfrac{(26.7 - 0) \text{ mi}}{(6{:}36 - 6{:}09) \text{ hr}}$

   $= \dfrac{26.7 \text{ mi}}{\left(\frac{36}{60} - \frac{9}{60}\right) \text{ hr}}$

   $= \dfrac{26.7 \text{ mi}}{\frac{27}{60} \text{ hr}}$

   ≈ 59 mph

   b. AVERAGE $= \dfrac{(35 - 26.7) \text{ mi}}{(7{:}03 - 6{:}36) \text{ hr}}$

   $= \dfrac{8.3 \text{ mi}}{\frac{27}{60} \text{ hr}}$

   ≈ 18 mph

   c. AVERAGE $= \dfrac{(50 - 35) \text{ mi}}{(7{:}28 - 7{:}03) \text{ hr}}$

   $= \dfrac{15 \text{ mi}}{\frac{25}{60} \text{ hr}}$

   $= 36$ mph

   d. AVERAGE $= \dfrac{(50 - 0) \text{ mi}}{(7{:}28 - 6{:}09) \text{ hr}}$

   $= \dfrac{50 \text{ mi}}{\frac{79}{60} \text{ hr}}$

   ≈ 38 mph

5. Let $ be trillions of dollars.

   a. AVERAGE $= \dfrac{\$7.6360 - \$0.51530}{(1996 - 1960) \text{ yr}}$

   ≈ 0.198 \$/yr

   b. AVERAGE $= \dfrac{\$7.6360 - \$1.0155}{(1996 - 1970) \text{ yr}}$

   ≈ 0.255 \$/yr

Page 482

c. AVERAGE $= \dfrac{\$7.6360 - 2.7320\$}{(1996 - 1980)\ \text{yr}}$

$\approx 0.307\ \$/\text{yr}$

d. AVERAGE $= \dfrac{\$7.6360 - \$5.5461}{(1996 - 1990)\ \text{yr}}$

$\approx 0.348\ \$/\text{yr}$

e. AVERAGE $= \dfrac{\$7.6360 - \$7.2654}{(1996 - 1995)\ \text{yr}}$

$\approx 0.371\ \$/\text{yr}$

f. If we look at the sequence, we conclude that it is changing at the rate greater than of $371 billion per year.

7. 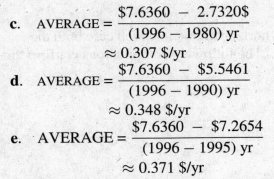 From the graph, it looks like the slope is 3/2.

9. 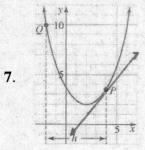 From the graph, it looks like the slope is slope is −2.

11. 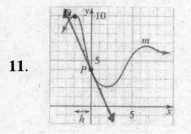 From the graph, it looks like the slope is slope is −2.

Level 2, page 896

13. You can notice that $f(x) = 5$ is the graph of a horizontal line in which case both the average and instantaneous rates of change is 0. The following computations confirm this expectation.

a. $\dfrac{f(x+h) - f(x)}{h} = \dfrac{f(-3+6) - f(-3)}{3 - (-3)}$

$= \dfrac{f(3) - f(-3)}{6}$

$= \dfrac{5 - 5}{6}$

$= 0$

b. $\displaystyle\lim_{h \to 0} \dfrac{f(x+h) - f(x)}{h} = \lim_{h \to 0} \dfrac{f(-3+h) - f(-3)}{h}$

$= \displaystyle\lim_{h \to 0} \dfrac{5 - 5}{h}$

$= \displaystyle\lim_{h \to 0} 0$

$= 0$

15. Given $f(x) = -2x^2 + x + 4$

a. $\dfrac{f(x+h) - f(x)}{h} = \dfrac{f(4+5) - f(4)}{9 - 4}$

$= \dfrac{f(9) - f(4)}{5}$

$= \dfrac{[-2(9)^2 + 9 + 4] - [-2(4)^2 + 4 + 4]}{5}$

$= \dfrac{-149 - (-24)}{5}$

$= \dfrac{-125}{5}$

$= -25$

b. $\displaystyle\lim_{h \to 0} \dfrac{f(x+h) - f(x)}{h} = \lim_{h \to 0} \dfrac{f(4+h) - f(4)}{h}$

$= \displaystyle\lim_{h \to 0} \dfrac{[-2(4+h)^2 + (4+h) + 4] - [-2(4)^2 + 4 + 4]}{h}$

$= \displaystyle\lim_{h \to 0} \dfrac{-2(16 + 8h + h^2) + 4 + h + 4 - (-32 + 4 + 4)}{h}$

Page 484

$$= \lim_{h \to 0} \frac{-32 - 16h - 2h^2 + 8 + h + 24}{h}$$

$$= \lim_{h \to 0} \frac{-15h - 2h^2}{h}$$

$$= \lim_{h \to 0} \frac{h(-15 - 2h)}{h}$$

$$= \lim_{h \to 0} (-15 - 2h)$$

$$= -15$$

17. We carry out the five-step process.

Step 1: $f(x) = \frac{1}{3}x^3$

Step 2: $f(x + h) = \frac{1}{3}(x + h)^3$

$$= \frac{1}{3}(x^3 + 3x^2h + 3xh^2 + h^3)$$

$$= \frac{1}{3}x^3 + x^2h + xh^2 + \frac{1}{3}h^3$$

Step 3: $f(x + h) - f(x) = \frac{1}{3}x^3 + x^2h + xh^2 + \frac{1}{3}h^3 - \frac{1}{3}x^3$

$$= x^2h + xh^2 + \frac{1}{3}h^3$$

Step 4: $\dfrac{f(x + h) - f(x)}{h} = \dfrac{h(x^2 + xh + \frac{1}{3}h^2)}{h}$

$$= x^2 + xh + \frac{1}{3}h^2$$

Step 5: $\lim_{h \to 0} \dfrac{f(x + h) - f(x)}{h} = \lim_{h \to 0}\left(x^2 + xh + \frac{1}{3}h^2 \right)$

$$= x^2$$

19. Use the derivative of the exponential function property.

$f(x) = e^{-6x}$, so $f'(x) = -6e^{-6x}$

21. We carry out the five-step process.

Step 1: $f(x) = 3 + 2x - 3x^2$

Step 2: $f(x + h) = 3 + 2(x + h) - 3(x + h)^2$

$$= 3 + 2x + 2h - 3(x^2 + 2xh + h^2)$$

$$= 3 + 2x + 2h - 3x^2 - 6xh - 3h^2$$

Step 3: $f(x + h) - f(x) = (3 + 2x + 2h - 3x^2 - 6xh - 3h^2) - (3 + 2x - 3x^2)$

$$= 2h - 6xh - 3h^2$$

Page 485

Step 4: $\dfrac{f(x+h)-f(x)}{h} = \dfrac{h(2-6x-3h)}{h}$

$$= 2 - 6x - 3h$$

Step 5: $\displaystyle\lim_{h \to 0} \dfrac{f(x+h)-f(x)}{h} = \lim_{h \to 0}(2-6x-3h)$

$$= 2 - 6x$$

23. $y' = 4x$ from Example 5; at $x = 4$, $m = 4(4) = 16$.

When $x = 4$, $y = 2(4)^2 = 32$, so at (4, 32) the equation of the tangent line is

$$y - 32 = 16(x - 4)$$
$$y - 32 = 16x - 64$$
$$16x - y - 32 = 0$$

25. First we find the derivative of $y = f(x) = 3x^2 + 4x$.

$$\lim_{h \to 0} \dfrac{f(x+h)-f(x)}{h} = \lim_{h \to 0} \dfrac{[3(x+h)^2 + 4(x+h)] - [3x^2 + 4x]}{h}$$

$$= \lim_{h \to 0} \dfrac{3(x^2 + 2xh + h^2) + 4x + 4h - 3x^2 - 4x}{h}$$

$$= \lim_{h \to 0} \dfrac{3x^2 + 6xh + 3h^2 + 4x + 4h - 3x^2 - 4x}{h}$$

$$= \lim_{h \to 0} \dfrac{h(6x + 4 + 3h)}{h}$$

$$= \lim_{h \to 0} (6x + 4 + 3h)$$

$$= 6x + 4$$

At $x = 0$, $m = 4$. When $x = 0$, $y = 3(0)^2 + 4(0) = 0$, so at (0, 0) the equation of the tangent line is

$$y - 0 = 4(x - 0)$$
$$4x - y = 0$$

27. a. The height of the tower is found when $t = 0$: $h(0) = -16(0)^2 + 96(0) + 176 = 176$.

  b. The velocity is the derivative of the h. We don't want to confuse the h for the height at time t, so we find the limit as $k \to 0$.

Page 486

$$\lim_{k \to 0} \frac{h(t+k) - h(t)}{k} = \lim_{k \to 0} \frac{[-16(t+k)^2 + 96(t+k) + 176] - [-16t^2 + 96t + 176]}{k}$$

$$= \lim_{k \to 0} \frac{-16(t^2 + 2tk + k^2) + 96t + 96k + 176 + 16t^2 - 96t - 176}{k}$$

$$= \lim_{k \to 0} \frac{-16t^2 - 32tk - 16k^2 + 96t + 96k + 176 + 16t^2 - 96t - 176}{k}$$

$$= \lim_{k \to 0} \frac{k(-32t - 16k + 96)}{k}$$

$$= \lim_{k \to 0} (-32t - 16k + 96)$$

$$= -32t + 96$$

29. a. AVERAGE $= \dfrac{C(200) - C(100)}{200 - 100}$ The average is \$8,900/item.

$$= \frac{1{,}180{,}000 - 290{,}000}{100}$$

$$= 8{,}900$$

b. AVERAGE $= \dfrac{C(110) - C(100)}{110 - 100}$ The average is \$6,200/item.

$$= \frac{352{,}000 - 290{,}000}{10}$$

$$= 6{,}200$$

c. AVERAGE $= \dfrac{C(101) - C(100)}{101 - 100}$ The average is \$5,930/item.

$$= \frac{295{,}930 - 290{,}000}{1}$$

$$= 5{,}930$$

d. AVERAGE $= \dfrac{C(x+h) - C(h)}{h}$

$$= \frac{30(x+h)^2 - 100(x+h) - [30x^2 - 100x]}{h}$$

$$= \frac{30x^2 + 60xh + 30h^2 - 100x - 100h - 30x^2 + 100x}{h}$$

$$= \frac{60xh + 30h^2 - 100h}{h}$$

$$= 60x - 100 + 30h$$

Page 487

This is in dollars/item. In particular, if $x = 100$, then this is $5{,}900 + 30h$.

18.4 Integrals, page 897

New Terms Introduced in this Section

Antiderivative	Antiderivative of a sum	Antiderivative of e^x	Area function
Area under a curve	Constant multiple	Definite integral	Dummy variable
Indefinite integral	Integral of a sum	Integrand	Limits of integration

Level 1, page 904

1. The figure is a rectangle, so we use the formula $A = \ell w$ where $w = 5$ and $\ell = x$. Thus, $A(x) = 5x$, so $A(8) = 5(8) = 40$.

3. The figure is a triangle, so we use the formula $A = \frac{1}{2}bh$, where $b = x$ and $h = 3x$. Thus, $A(x) = \frac{1}{2}x(3x) = \frac{3}{2}x^2$, so $A(4) = \frac{3}{2}(4)^2 = 24$.

5. This is a trapezoid; the area of a trapezoid is $A = \frac{1}{2}h(b_1 + b_2)$:
   $A(x) = \frac{1}{2}(3)(2 + 11) = \frac{39}{2}$.

7. Let $f(x) = 6$ and consider the region bounded by the x-axis, the y-axis, the horizontal line $y = 6$, and the vertical line x units to the right of the y-axis.

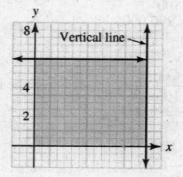

The area of the rectangle is $6x$, so $\int 6\,dx = 6x + C$.

9. Let $f(x) = x + 5$ and consider the region bounded by the x-axis, the y-axis, the line $y = x + 5$, and the vertical line x units to the right of the y-axis.

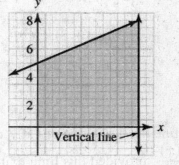

The area of the trapezoid is $A = \frac{1}{2}h(b + B)$. For this example $h = x$, $b = 5$, and $B = x + 5$, so $\displaystyle\int (x + 5)\,dx = \frac{1}{2}x[5 + (x + 5)] + C$

$$= \frac{1}{2}x(x + 10) + C$$

$$= \frac{1}{2}x^2 + 5x + C$$

11. Let $f(x) = 3x + 4$ and consider the region bounded by the x-axis, the y-axis, the line $y = 3x + 4$, and the vertical line x units to the right of the y-axis.

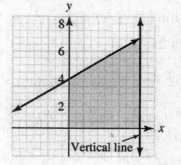

Page 489

The area of the trapezoid is $A = \frac{1}{2}h(b + B)$. For this example $h = x$, $b = 4$, and

$$B = 3x + 4, \text{ so } \int (3x + 4)\, dx = \frac{1}{2}x[4 + (3x + 4)] + C$$
$$= \frac{1}{2}x(3x + 8) + C$$
$$= \frac{3}{2}x^2 + 4x + C$$

Level 2, page 905

13. For a given function f, we define the derivative of f at x, denoted by $f'(x)$ to be

$$f'(x) = \lim_{h \to 0} \frac{f(x + h) - f(x)}{h} \text{ provided this limit exits.}$$

15. The area function, $A(t)$, is the area bounded below by the x-axis, above by a function $y = f(x)$, on the left by the y-axis, and on the right by the vertical line $y = t$.

17. If $f(x) = x^2$, then $f'(x) = 2x$ (use the five-step rule).
$$f(x) = x^2$$
$$f(x + h) = (x + h)^2 = x^2 + 2xh + h^2$$
$$f(x + h) - f(x) = 2xh + h^2$$
$$\frac{f(x + h) - f(x)}{h} = 2x + h$$
$$\lim_{h \to 0} \frac{f(x + h) - f(x)}{h} = \lim_{h \to 0} (2x + h) = 2x$$
Thus, x^2 is an antiderivative of $2x$.

19. If $f(x) = 2x^3$, then $f'(x) = 6x^2$ (use the five-step rule).
$$f(x) = 2x^3$$
$$f(x + h) = 2(x + h)^3 = 2x^3 + 6x^2h + 6xh^2 + 2h^3$$
$$f(x + h) - f(x) = 6x^2h + 6xh^2 + 2h^3$$
$$\frac{f(x + h) - f(x)}{h} = 6x^2 + 6xh + 2h^2$$
$$\lim_{h \to 0} \frac{f(x + h) - f(x)}{h} = \lim_{h \to 0} (6x^2 + 6xh + 2h^2) = 6x^2$$
Thus, $2x^3$ is an antiderivative of $6x^2$.

21. We use Example 6 for the antiderivative.
$$\int_{1}^{9} x^2 \, dx = \frac{1}{3} x^3 \Big|_{1}^{9}$$
$$= 242 \frac{2}{3}$$

23. We use the area function from Problem 3.
$$\int_{3}^{8} 3x \, dx = A(8) - A(3)$$
$$= \frac{3}{2}(8)^2 - \frac{3}{2}(3)^2$$
$$= \frac{192}{2} - \frac{27}{2}$$
$$= \frac{165}{2}$$
$$= 82.5$$

25. $\int_{0}^{2} e^{0.5x} \, dx = 2e^{0.5x} \big|_{0}^{2} = 2e - 2$

27. $\int_{1}^{5} (1 + 6x) = \int_{1}^{5} dx + 6 \int_{1}^{5} x \, dx$
$$= x \big|_{1}^{5} + 6 \left(\frac{x^2}{2} \right) \Big|_{1}^{5}$$
$$= (5 - 1) + 3(5)^2 - 3(1)^2$$
$$= 4 + 75 - 3$$
$$= 76$$

Level 3, page 905

29. $\int_{0}^{10} 200 e^{0.5t} \, dt = 200(2) e^{0.5t} \big|_{0}^{10}$
$$= 400 e^5 - 400$$
$$\approx 58{,}965.3$$

This represents the increase in the number of bacteria after 10 hours (or net change).

Page 491

Chapter 18 Review Questions, page 906

Studying for a chapter examination is a personal process, one which nobody else can do for you. Simply take the time to review what you have done. Here are the new terms in Chapter 18.

Antiderivative [18.4]	Differential calculus [18.1]	Limit [18.1]
Antiderivative of a sum [18.4]	Diverge [18.2]	Limit of a sequence [18.2]
Antiderivative of e^x [18.4]	Dummy variable [18.4]	Limits of integration [18.4]
Area function [18.4]	General term [18.2]	Mathematical modeling [18.1]
Area under a curve [18.4]	Indefinite integral [18.4]	Predecessor [18.2]
Average rate of change [18.3]	Infinite limit [18.2]	Secant line [18.1]
Calculus [18.1]	Instantaneous rate of	Sequence [18.2]
Constant multiple [18.4]	change [18.3]	Successor [18.2]
Converge [18.2]	Integral [18.1]	Tangent line [18.1, 18.3]
Definite integral [18.4]	Integral calculus [18.1]	Velocity [18.3]
Derivative [18.1; 18.3]	Integral of a sum [18.4]	
Differentiable function [18.3]	Integrand [18.4]	

If you can describe the term, read on to the next one; if you cannot, then look it up in the text (the section number is shown in brackets). Next, study the types of problems listed at the end of Chapter 18.

TYPES OF PROBLEMS

Describe the limit process, including Zeno's paradox. [18.1]
Describe the derivative, including the tangent problem. [18.1]
Describe the integral, including the area problem. [18.1]
Be able to guess the limit of a sequence. [18.1]
Be able to draw the line tangent to a curve at a specified point. [18.1]
Approximate an area by using rectangles. [18.1]
Explain the process of mathematical modeling. [18.1]
Write out the first 5 terms of a sequence when given a general term. [18.2]
Find the limit of a sequence. [18.2]
Solve applied problems involving limits of sequences. [18.2]
Estimate a rate of change by looking at a graph. [18.3]
Estimate the slope of a tangent line by looking at a graph. [18.3]
Find an average rate of change for a given function over an interval. [18.3]
Find an instantaneous rate of change for a given function at a particular point. [18.3]

Page 492

Find the derivative by using the definition of derivative. [18.3]
Find the equation of a tangent line. [18.3]
Evaluate an area function. [18.4]
Find an antiderivative. [18.4]
Find the area under a curve. [18.4]
Approximate the value of a definite integral by using areas. [18.4]

Once again, see if you can verbalize (to yourself) how to do each of the listed types of problems.

    Work all of Chapter 18 Review Questions (whether they are assigned or not). Work through all of the problems before looking at the answers, and *then* correct each of the problems. The entire solution is shown in the answer section at the back of the text. If you worked the problem correctly, move on to the next problem, but if you did not work it correctly (or you did not know what to do), look back in the chapter to study the procedure, or ask your instructor.

    Finally, go back over the homework problems you have been assigned. If you worked a problem correctly, move on to the next problem, but if you missed it on your homework, then you should look back in the book or talk to your instructor about how to work the problem.

    We give all of the answers to the Chapter Review questions (not just the odd-numbered questions).

Chapter 18 Review Questions, page 906

1. Answers vary. The main ideas of calculus are limits, derivatives, and integrals.

    (1) $\lim\limits_{n \to \infty} a_n = L$ means that the sequence a_n becomes closer and closer to the number L as n becomes larger and larger.

    (2) The derivative illustrates the idea of a tangent line.

    (3) The integral is used to find the area under a curve.

2. The series illustrates the idea of a tangent line ("instantaneous growth rate"). The first shows 5 measurements, the second 20 measurements, and the third 40 measurements. The closer the measurements are together, the better the approximation at a particular point.

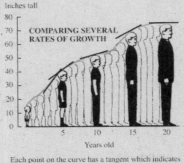

Page 493

3. $\lim\limits_{n \to \infty} \frac{1}{n} = 0$

4. As $n \to \infty$, the terms of the sequence $(-1)^n$ oscillate between -1 and 1, so the limit does not exist.

5. $\lim\limits_{n \to \infty} \dfrac{3n^4 + 20}{7n^4} = \lim\limits_{n \to \infty} \dfrac{\frac{3n^4}{n^4} + \frac{20}{n^4}}{\frac{7n^4}{n^4}}$

$$= \lim\limits_{n \to \infty} \dfrac{3 + \frac{20}{n^4}}{7}$$

$$= \dfrac{3}{7}$$

6. $\lim\limits_{n \to \infty} (2n + 3) = \infty$

7. $\lim\limits_{n \to \infty} \left(1 + \frac{1}{n}\right)^n$

Construct a table of values:
$n = 1;$ 2
$n = 10;$ 2.5937424601
$n = 10^3;$ 2.71692393224
$n = 10^{10}$ 2.71828182832

It looks like the limit is e. In fact, this limit is often given as the *definition* of e.

8. We use the five-step process:
Step 1: $f(x) = 9$
Step 2: $f(x + h) = 9$
Step 3: $f(x + h) - f(x) = 0$
Step 4: $\dfrac{f(x + h) - f(x)}{h} = 0$
Step 5: $\lim\limits_{h \to 0} \dfrac{f(x + h) - f(x)}{h} = \lim\limits_{h \to 0} 0$
$$= 0$$

Thus $f'(x) = 0$.

9. We use the five-step process:
Step 1: $f(x) = 10 - 2x$
Step 2: $f(x + h) = 10 - 2(x + h)$
$$= 10 - 2x - 2h$$
Step 3: $f(x + h) - f(x) = 10 - 2x - 2h - (10 - 2x)$
$$= -2h$$
Step 4: $\dfrac{f(x + h) - f(x)}{h} = \dfrac{-2h}{h}$
$$= -2$$

Page 494

Step 5: $\displaystyle\lim_{h \to 0} \frac{f(x+h) - f(x)}{h} = \lim_{h \to 0} (-2)$

$\qquad\qquad\qquad\qquad\qquad\qquad = -2$

Thus $f'(x) = -2$.

10. We use the five-step process:

Step 1: $f(x) = x - 10$

Step 2: $\qquad\qquad f(x+h) = (x+h) - 10$

Step 3: $\qquad f(x+h) - f(x) = (x+h-10) - (x-10)$

$\qquad\qquad\qquad\qquad\qquad = h$

Step 4: $\dfrac{f(x+h) - f(x)}{h} = \dfrac{h}{h}$

$\qquad\qquad\qquad\qquad\qquad = 1$

Step 5: $\displaystyle\lim_{h \to 0} \frac{f(x+h) - f(x)}{h} = \lim_{h \to 0} 1$

$\qquad\qquad\qquad\qquad\qquad = 1$

Thus $f'(x) = 1$.

11. We use the five-step process:

Step 1: $f(x) = 6 - 4x^2$

Step 2: $\qquad\qquad f(x+h) = 6 - 4(x+h)^2$

$\qquad\qquad\qquad\qquad\qquad = 6 - 4(x^2 + 2xh + h^2)$

$\qquad\qquad\qquad\qquad\qquad = 6 - 4x^2 - 8xh - 4h^2$

Step 3: $\qquad f(x+h) - f(x) = (6 - 4x^2 - 8xh - 4h^2) - (6 - 4x^2)$

$\qquad\qquad\qquad\qquad\qquad = -8xh - 4h^2$

Step 4: $\dfrac{f(x+h) - f(x)}{h} = \dfrac{-8xh - 4h^2}{h}$

$\qquad\qquad\qquad\qquad\qquad = -8x - 4h^2$

Step 5: $\displaystyle\lim_{h \to 0} \frac{f(x+h) - f(x)}{h} = \lim_{h \to 0} (-8x - 4h^2)$

$\qquad\qquad\qquad\qquad\qquad = -8x$

Thus $f'(x) = -8x$.

12. From the derivative of e^x property, if $f(x) = 44e^{0.5x}$, $f'(x) = 44(0.5)e^{0.5x} = 22e^{0.5x}$.

13. $\displaystyle\int 16x \, dx = 16\left(\frac{x^2}{2}\right) + C$

$\qquad\qquad = 8x^2 + C$

14. $\displaystyle\int \pi \, dx = \pi x + C$

Page 495

15. $\displaystyle\int_2^3 (6x^2 - 2x)\,dx = 6\int_2^3 x^2 dx - 2\int_2^3 x\,dx$

$$= 6\left(\frac{x^3}{3}\right)\Big|_2^3 - 2\left(\frac{x^2}{2}\right)\Big|_2^3$$
$$= 2(3^3 - 2^3) - (3^2 - 2^2)$$
$$= 2(27 - 8) - (9 - 4)$$
$$= 38 - 5$$
$$= 33$$

16. $\displaystyle\int_0^2 (x - 3x^2)\,dx = \int_0^2 x\,dx - 3\int_0^2 x^2\,dx$

$$= \left(\frac{x^2}{2}\right)\Big|_0^2 - 3\left(\frac{x^3}{3}\right)\Big|_0^2$$
$$= \left(\frac{2^2}{2} - \frac{0^2}{2}\right) - (2^3 - 0^3)$$
$$= 2 - 8$$
$$= -6$$

17. $\displaystyle\int_1^2 e^x\,dx = e^2 - e \approx 4.67$

18. Since $f(x) = x^2$, $f'(x) = 2x$; at $x = 1$, $m = f'(1) = 2(1) = 2$.
Also, when $x = 1$, $y = (1)^2 = 1$. The equation of the line passing through $(1, 1)$ with slope 2. Use the point-slope form of the equation of a line:

$$y - k = m(x - h)$$
$$y - 1 = 2(x - 1)$$
$$y - 1 = 2x - 2$$
$$2x - y - 1 = 0$$

19. If $d(t) = \frac{1}{5}t^2 + 5t$. The velocity is the derivative, which we find using the five-step process.

Step 1: $\quad d(t) = \dfrac{1}{5}t^2 + 5t$

Step 2: $\quad d(t + h) = \dfrac{1}{5}(t + h)^2 + 5(t + h)$

$$= \frac{1}{5}(t^2 + 2th + h^2) + 5t + 5h$$
$$= \frac{1}{5}t^2 + \frac{2}{5}th + \frac{1}{5}h^2 + 5t + 5h$$

Page 496

Step 3:
$$d(x+h) - d(x) = (\frac{1}{5}t^2 + \frac{2}{5}th + \frac{1}{5}h^2 + 5t + 5h) - (\frac{1}{5}t^2 + 5t)$$
$$= \frac{2}{5}th + \frac{1}{5}h^2 + 5h$$

Step 4:
$$\frac{d(x+h) - d(x)}{h} = \frac{\frac{2}{5}th + \frac{1}{5}h^2 + 5h}{h}$$
$$= \frac{2}{5}t + \frac{1}{5}h + 5$$

Step 5:
$$\lim_{h \to 0} \frac{d(x+h) - d(x)}{h} = \lim_{h \to 0} \left(\frac{2}{5}t + \frac{1}{5}h + 5 \right)$$
$$= \frac{2}{5}t + 5$$

The velocity at time $t = 5$ hours, we find $d'(5) = \frac{2}{5}(5) + 5$
$$= 7$$

The velocity is 7 mph.

20. Total amount used is $\int_0^t 32.4e^{6t/125} \, dx = 675(e^{6t/125} - 1)$

Solve
$$670 = 675(e^{6t/125} - 1)$$
$$\frac{670}{675} = e^{6t/125} - 1$$
$$\frac{1,345}{675} = e^{6t/125}$$
$$\frac{6t}{125} = \ln\left(\frac{1,345}{675}\right)$$
$$t = \frac{125}{6}\ln\left(\frac{1,345}{675}\right)$$
$$\approx 14.363$$

This says that the oil reserves will be depleted in 2014.

Epilogue

Overview

This epilogue asks the question "Why Not Math?" whereas the prologue (at the beginning of the book) asks the question "Why Math?" These two "bookends" frame the material covered in a typical liberal arts mathematics class.

   A theme of this book is problem solving, and this problem set was designed to give an overview of variety of ideas and concepts from the natural sciences, social sciences, humanities, business, and economics. If you did not work the problems in the Prologue, now is a good time to look as some of those problems as well.

Epilogue Problem Set

1. Answers vary. You should write at least a half a page and be sure to support your opinions.

3. **a.** social science **b.** natural science **c.** natural science **d.** humanities **e.** social science

5. **a.** social science **b.** natural science **c.** social science **d.** social science **e.** natural science
   f. natural science

7. all natural science

9. The given number is 8.58×10^{20}. From Example 2, one light year is 5.87×10^{12} miles, so
   distance of the star is $\dfrac{8.58 \times 10^{20}}{5.86 \times 10^{12}} = \dfrac{8.58}{5.86} \times 10^{8}$
   $$= 1.46 \times 10^{8}$$

11. Venus and Neptune have the smallest eccentricity and therefore have the most circular orbits.

13. If we assume the densities of the two types of eggs are approximately the same, then the weight, which should be related to the volume ($V = s^3$), the ostrich eggs should be $3^3 = 27$ times as great as the chicken egg, or $27(2) = 54$. We expect the ostrich   egg to weigh 54 oz.

15. Consider $(p + q)^2$ where $p = 0.48$, $q = 0.55$: $(p + q)^2 = p^2 + 2pq + q^2$
    genotype $SS = (0.55)^2 = 0.3025$,
    genotype wS or $Sw = 2(0.55)(0.45) = 0.4950$,
    genotype $ww = (0.45)^2 = 0.2025$

17. Since this is measuring area, the scaling factor is $s^2 = 2,000$ or $s \approx 44.721$, so the area of the
    unmagnified cell is $\dfrac{20}{44.721} \approx 0.45 \text{ cm}^2$.

19.

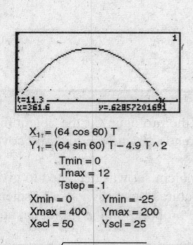

$X_{1T} = (64 \cos 60)\, T$

$Y_{1T} = (64 \sin 60)\, T - 4.9\, T\wedge 2$

     Tmin = 0

     Tmax = 12

     Tstep = .1

Xmin = 0   Ymin = -25

Xmax = 400   Ymax = 200

Xscl = 50   Yscl = 25

21. a. $a = \sqrt{2.015 \times 10^{16}} \approx 141{,}950{,}696;\ b = \sqrt{1.995 \times 10^{16}} \approx 141{,}244{,}469$

     Since a and b are almost the same, we note the ellipse is almost circular.

$$\epsilon = \sqrt{1 - \frac{b^2}{a^2}} \approx 0.0996$$

     b. Let each unit be $\sqrt{10^{16}} = 10^8$.

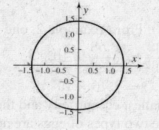

23. $4! = 24$

25. a.

$$
\begin{array}{c}
\ \\ a \\ b \\ c \\ d
\end{array}
\begin{array}{cccc}
A & B & C & D \\
\left[\begin{array}{c}(1,2)\\(2,3)\\(2,4)\\(3,1)\end{array}\right. & \begin{array}{c}(3,1)\\(4,4)\\(3,3)\\(2,2)\end{array} & \begin{array}{c}(4,1)\\(3,2)\\(4,4)\\(4,3)\end{array} & \left.\begin{array}{c}(2,2)\\(1,1)\\(1,3)\\(1,4)\end{array}\right]
\end{array}
$$

Page 500

b. Answers vary. Here is one of each:

$$
\begin{array}{c}
 & A & B & C & D \\
a & \begin{bmatrix} (1,2) & (3,1) & (4,1) & (2,2) \\ (2,3) & (4,4) & (3,2) & (1,1) \\ (2,4) & (3,3) & (4,4) & (1,3) \\ (3,1) & (2,2) & (4,3) & (1,4) \end{bmatrix} \end{array}
\quad \text{unstable}
$$

$$
\begin{array}{c}
 & A & B & C & D \\
a & \begin{bmatrix} (1,2) & (3,1) & (4,1) & (2,2) \\ (2,3) & (4,4) & (3,2) & (1,1) \\ (2,4) & (3,3) & (4,4) & (1,3) \\ (3,1) & (2,2) & (4,3) & (1,4) \end{bmatrix} \end{array}
\quad \text{stable}
$$

27.

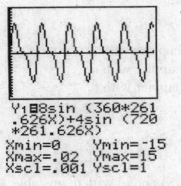

```
Y1◼8sin (360*261
.626X)+4sin (720
*261.626X)
Xmin=0    Ymin=-15
Xmax=.02  Ymax=15
Xscl=.001 Yscl=1
```

29. a. The matrix of their stated choices is:

$$
\begin{array}{c}
 & A & B & C \\
a & \begin{bmatrix} (3,1) & (1,3) & (2,2) \\ (2,2) & (3,1) & (1,3) \\ (1,3) & (2,2) & (3,1) \end{bmatrix} \end{array}
$$

There are $3! = 6$ possibilities.

$$
\begin{array}{c}
 & A & B & C \\
a & \begin{bmatrix} (3,1) & (1,3) & (2,2) \\ (2,2) & (3,1) & (1,3) \\ (1,3) & (2,2) & (3,1) \end{bmatrix} \end{array}
\quad \text{stable; each school has its first choice.}
$$

Page 501

$$\begin{array}{c} \quad\;\; A \qquad\; B \qquad\; C \\ \begin{array}{c} a \\ b \\ c \end{array} \left[\begin{array}{ccc} (3,1) & (1,3) & (2,2) \\ (2,2) & (3,1) & (1,3) \\ (1,3) & (2,2) & (3,1) \end{array} \right] \end{array}$$ a would rather be with C and C with a; unstable.

$$\begin{array}{c} \quad\;\; A \qquad\; B \qquad\; C \\ \begin{array}{c} a \\ b \\ c \end{array} \left[\begin{array}{ccc} (3,1) & (1,3) & (2,2) \\ (2,2) & (3,1) & (1,3) \\ (1,3) & (2,2) & (3,1) \end{array} \right] \end{array}$$ c would rather be with B and B with c; unstable.

$$\begin{array}{c} \quad\;\; A \qquad\; B \qquad\; C \\ \begin{array}{c} a \\ b \\ c \end{array} \left[\begin{array}{ccc} (3,1) & (1,3) & (2,2) \\ (2,2) & (3,1) & (1,3) \\ (1,3) & (2,2) & (3,1) \end{array} \right] \end{array}$$ stable; each person has his or her first choice

$$\begin{array}{c} \quad\;\; A \qquad\; B \qquad\; C \\ \begin{array}{c} a \\ b \\ c \end{array} \left[\begin{array}{ccc} (3,1) & (1,3) & (2,2) \\ (2,2) & (3,1) & (1,3) \\ (1,3) & (2,2) & (3,1) \end{array} \right] \end{array}$$

a prefers B, but B is happier with c.
b prefers C, but C is happier with a.
c prefers A, but A is happier with b.
stable

$$\begin{array}{c} \quad\;\; A \qquad\; B \qquad\; C \\ \begin{array}{c} a \\ b \\ c \end{array} \left[\begin{array}{ccc} (3,1) & (1,3) & (2,2) \\ (2,2) & (3,1) & (1,3) \\ (1,3) & (2,2) & (3,1) \end{array} \right] \end{array}$$ b would rather be with A and A with b; unstable.

Page 502